ESTIMATING FOR RESIDENTIAL CONSTRUCTION

Second Edition

ESTIMATING FOR RESIDENTIAL CONSTRUCTION

SECOND EDITION

David J. Pratt

DELMAR
CENGAGE Learning

Australia • Brazil • Japan • Korea • Mexico • Singapore • Spain • United Kingdom • United States

Estimating for Residential Construction, 2E
David J. Pratt

Vice President, Career and Professional Editorial: Dave Garza

Director of Learning Solutions: Sandy Clark

Acquisitions Editor: Stacy Masucci

Managing Editor: Larry Main

Senior Product Manager: John Fisher

Senior Editorial Assistant: Andrea Timpano

Vice President, Career and Professional Marketing: Jennifer Baker

Marketing Director: Deborah Yarnell

Marketing Manager: Katie Hall

Marketing Coordinator: Mark Pierro

Production Director: Wendy Troeger

Production Manager: Mark Bernard

Senior Content Project Manager: David Plagenza

Senior Art Director: Casey Kirchmayer

Technology Project Manager: Joe Pliss

Compositor: MPS Limited, a Macmillan Company

For product information and technology assistance, contact us at
Cengage Learning Customer & Sales Support, 1-800-354-9706

For permission to use material from this text or product, submit all requests online at **www.cengage.com/permissions**.
Further permissions questions can be e-mailed to
permissionrequest@cengage.com

Library of Congress Control Number: 2011920495

ISBN-13: 978-1-1113-0887-2

ISBN-10: 1-1113-0887-X

Delmar
5 Maxwell Drive
Clifton Park, NY 12065-2919
USA

Cengage Learning is a leading provider of customized learning solutions with office locations around the globe, including Singapore, the United Kingdom, Australia, Mexico, Brazil, and Japan. Locate your local office at: **international.cengage.com/region**

Cengage Learning products are represented in Canada by Nelson Education, Ltd.

To learn more about Delmar, visit **www.cengage.com/delmar**

Purchase any of our products at your local college store or at our preferred online store **www.CengageBrain.com**

Notice to the Reader

Printed in the United States of America
1 2 3 4 5 6 7 15 14 13 12 11

CONTENTS

Chapter 4   **MEASURING CONCRETE WORK   77**

Chapter 5   **MEASURING CARPENTRY WORK   91**

Chapter 6   **MEASURING MASONRY AND FINISHES   119**

Chapter 13   ESTIMATES FOR REMODELING WORK   263

CHAPTER 14   COMPUTER ESTIMATING   291

PREFACE

Intended Use & Level

Estimating for Residential Construction, 2E is intended for 2-year and 4-year post-secondary construction programs. It is written to provide the reader with the resources necessary to learn how to estimate the construction costs of residential buildings and remodelling projects using modern technology and following the methods employed by estimators in the residential sector. The book is primarily intended for the person who is beginning to learn the practice of cost estimating whether they are a student taking a course in estimating at college, or someone who has recently assumed estimating responsibilities for a builder or consultant. The text will also be of interest to construction managers, supervisors, and practicing estimators who, from time to time, may wish to refer to a source of estimating data or simply investigate how other estimators approach this subject.

Approach

The goal of the text is to describe an easy to follow step-by-step method of estimating that can be used to produce an accurate construction cost estimate in the minimum of time; a cost estimate that is easy to review and one that is recognized by experienced estimators as the product of a professional. While the text concentrates on the kind of small projects typically undertaken by homebuilders, the techniques developed here can be applied to the many different types of projects ranging in scope from a single detached home to multi-unit high-rise buildings. The examples offered relate to a single-family house, a 4-unit apartment building, and a small renovation project. These projects were selected for their relative simplicity so that the reader can concentrate on the estimating technique rather than having to spend time unravelling complicated detail.

The estimating process presented is not intended to be some radical, new method of estimating, instead, it is estimating as it is currently pursued by professional estimators doing essentially what estimators have always done but with far more speed, accuracy, and thoroughness using all the modern innovations that advance the efficiency and effectiveness of the process.

Note that all prices used in the text are for illustrative purposes only, actual prices of construction work vary considerably from place to place and from time to time and should be carefully considered before using in actual estimates.

Text Layout

- *Chapter 1* introduces readers to the estimating process.
- *Chapter 2* explains the types of calculations used in estimating and describes the general principles of measuring work and preparing quantity takeoffs, which is one of the main components of the estimating process. In chapter 3 the quantity takeoff of excavation work is discussed in detail and an example of a full estimate begins in this chapter and continues in the chapters that follow. This example takes the reader through all of the stages of an estimate of a house from initial takeoff to completion of the cost summary. The house drawings used for this example can be found in appendix D.
- *Chapters 4 through 7* consider the takeoff process for concrete, carpentry, masonry, plumbing, HVAC, and electrical work.
- *Chapters 8 through 10* concern the pricing of the builder's work including the pricing of equipment involved in the work.
- In *Chapter 11* the pricing of subcontracted work is studied and *Chapter 12* deals with estimate summaries and bids.
- *Chapter 13* examines the estimating process for remodeling work and works through an example of a complete estimate for a renovation project.
- The final chapter, *Chapter 14,* presents a full estimate with step-by-step directions for a 4-unit apartment building using ICE software from MC^2.
- *Appendix A* contains a glossary of terms.
- *Appendix B* describes the metric units used in estimating and provides unit conversion factors for English to metric and metric to English conversions.
- *Appendix C* provides an overview of the equations used in estimating.
- *Appendix D* consists of a set of house drawings used for takeoff examples.
- *Appendix E* consists of a set of drawings for a 4-unit apartment building used for the ICE software example.

Key Features

There are also many features to help enhance learning for the reader. These features can be found integrated throughout the textbook:

- *Key Terms* are highlighted and defined in each chapter and provide the reader with the necessary terminology for effective communication in the construction field.
- *Examples* illustrate important estimations in a step-by-step approach that allows the reader to see how costs are worked out in the preparation of bids.
- *Review Questions* and *Practice Problems* appear at the end of each chapter and allow students to apply what they have learned in the chapter.
- *Recommended Resources* provide helpful links and references that contain additional information on the content presented in the chapter.

New to This Edition

- Coverage of Green Building and Sustainable Construction and the NAHB National Green Building Program has been added to Chapter 1.
- Chapter 14 is new to this edition and covers the inclusion of extensive computer applications. There is also the inclusion of examples of the use of MC^2 and Timberline software in residential estimating. This new edition also has an online student companion site containing Excel spreadsheets for use in estimating.

Accessing the Student Companion Site from CengageBrain

You can obtain Excel spreadsheets for estimating from the Student Companion site at CengageBrain by following these steps:

1. With your computer's web browser, go to www.cengagebrain.com.
2. Enter the name of the author, title of the book, or the book's ISBN (international standard book number) in the Search window.
3. Locate the desired book and then click on Access Now.
4. Under Book Resources on the left side of the screen will be a list of the assets available for download.

Supplement

An *Instructor's Resource CD* providing solutions to all end-of-chapter review questions, chapter presentations in PowerPoint, and a testbank is also available. *Order #: 1-1113-0888-8.*

About the Author

David J. Pratt is a Professor of Civil Engineering Technology at Southern Alberta Institute of Technology. Professor Pratt has had a long career as a construction consultant and before that as estimator for a number of building and construction companies. He is a member of the Canadian Institute of Quantity Surveyors and holds a degree in Quantity Surveying from Liverpool College of Building together with a degree in Economics from the University of Calgary.

Acknowledgements

I would like to thank the large number of people who contributed to this text. Particular thanks goes to the following:

- Ken Head of Head Construction who has been building houses for as long as I have been estimating them.
- Trevor Canuel for preparing house drawings.
- Sue Zegarelli of Partners Composition for copyediting.
- The large group of professional estimators, many of whom are former students of mine, who continue to give me insight into the estimator's task.

In addition, many thanks go to the reviewers, who provided guidance throughout the development of the text:

Joe Dusek
Triton Community College
River Grove, IL

Robert Irion
AAA Construction School
Jacksonville, FL

Richard Harrington
SUNY–Delhi
Delhi, NY

Chester Melton
Isothermal Community College
Spindale, NC

Darwin Olson
Anoka-Hennepin Technical College
Anoka, MN

Timothy Ray
Midlands Technical College
Columbia, SC

Sean Quinn
Lansing Community College
Lansing, MI

Mark Walsh
Indian Hills Community College
Centerville, IN

Kevin Ward
McEachern High School
Powder Springs, GA

Brenda Yamin
Erie Community College
Williamsville, NY

1

INTRODUCTION TO BUILDING COST ESTIMATES

OBJECTIVES

After reading this chapter and completing the review questions, you should be able to:

- Explain what a building estimate is and what is needed to compile a building estimate.
- Identify the different types of builder and contracts found in the residential construction industry.
- Identify the different uses of estimates in the residential construction.
- Explain how value analysis may be used to try to maximize the value obtained from a project.
- Explain how bills of material are prepared.
- Outline the role of architects and designers in the residential sector.
- Explain the purpose of soils reports and site visits in the estimating process.

KEY TERMS

apartment	feasibility estimate	spec builders and specs home
architects	firm price contract	
bill of materials	green building	specifications
condominium	lump sum contracts	takeoff and taking off
cost plus contracts	prime consultants	unit price contracts
custom homes	production builders	value analysis
designers	row housing	
duplexes	semi-custom homes	

Introduction

In general terms, an estimate is an evaluation of a future cost; a building cost estimate is an attempt to determine the likely cost of some building work before the work is done. In order to compile such a cost estimate, the estimator needs to answer two basic questions:

1. How much work is required to be done?
2. What will it cost to do this work?

In the construction industry, the process of measuring the amount of work to be done is called taking off and the product obtained from this process is referred to as a **takeoff**. After the work is measured, the takeoff may then be processed and priced in a number of different ways depending upon what the estimate is to be used for.

Estimates in the Residential Construction Industry

There are a number of different types of estimates in the residential construction industry, each serving a different purpose. The type of estimate required in any given situation depends upon the residential market served by the builder preparing the estimate, the nature of the contract with the home purchaser or owner, and the purpose of the estimate (see Figure 1.1).

Builders and Residential Markets

There are basically three distinct groups of builders serving the residential construction market: those that build new homes for sale to home buyers; those that work under contract to the property owner to build a custom home or renovate an existing property for the owner; and those that construct multi-unit residential buildings for owners/developers.

Spec Builders

This first group of homebuilders buys land for building homes, often in the form of housing lots in a new subdivision. Spec builders decide on the type of houses to build and make many in-house decisions about the design of the homes they will build. Because this business involves a certain amount of speculation, this group is often referred to as speculative builders (**spec builders**), and a home built before it

Types of Estimate	Residential Market	Types of Contract
• Feasibility Estimate	• Speculative Homes	• Home Purchase
• Preliminary Estimate	• Production Homes	• Firm Price Lump Sum
• Bill of Material	• Semi-Custom Homes	• Variable Price Lump Sum
• Detailed Estimate	• Custom Homes	• Cost Plus Percentage
• Budget Estimate	• Townhouses	• Cost Plus Fixed Fee
• Cost Control Estimate	• Condominiums	• Cost Plus Variable Fee

Figure 1.1   Estimates, Residential Markets, and Contracts

has been sold is called a **spec home.** Some spec builders construct a series of homes of basically the same design, almost like on a production line. These builders are often called **production builders.**

The main marketing tool of the spec builder is the show home. The builder constructs a home of the design they have decided upon and use it to show prospective buyers what they will receive if they buy from this builder. When a sale is made, the builder enters into a contract with the buyer to construct a similar home on one of the builder's lots for an agreed sale price. The buyer will generally have some choice about the construction of their new home. The buyer will at least be allowed to select kitchen cabinets, floor finishes, and paint colors. Sometimes spec builders offer much more flexibility with regard to design; they may start with a standard plan, then allow the buyer to make changes in size, add or delete features, and so on to meet the buyer's particular needs. Because this homebuyer has so much influence in the final design, these houses are referred to as **semi-custom homes.**

Custom Home Builders

The builders of **custom homes** do not take the same risks as spec builders. They do not have to buy building lots, nor do they need to construct show homes, and the owner that hires them usually makes all the design decisions. Some owners employ architectural consultants to prepare plans and specifications for their home. House plans are also available from plan service organizations that offer off-the-shelf designs, or the owner may simply take their ideas directly to a builder, and together compile a design. Some of the larger builders have in-house designers to provide this service on custom home projects.

If the owner did not go directly to a homebuilder, the next step, once a design is established, is to find a builder to construct their home. Here owners and their consultants may organize a competitive bid if they wish to find the best price to complete the project. Up to six contractors that the owner and/or consultant consider to be qualified and financially sound are asked to submit prices to construct the home as detailed in the plans and specifications. Then, once all bids are received, the owner usually awards the contract to the lowest bidder.

On the other hand, the owner may choose a particular builder to construct the home because they believe this builder has certain qualities that distinguish it from the competition. In which case, the project price will probably be higher than if a bid competition was used, but the owner may be willing to pay the extra to have this builder do the work. The builder will still usually be required to quote a firm price for the job, and a savvy owner will request a price breakdown and details of the builder's estimate. The owner and/or their consultant can then go over the price breakdown and may call upon the builder to explain some of the prices. A solid, well-presented estimate can help the builder obtain the job in this situation.

In all of these situations, the builder who is selected to do the work will be required to enter into a contract with the owner. This will usually be a **firm price contract,** but there are possible alternative arrangements.

Multi-Unit Residential Contractors

Multi-unit residential projects include:

- **Duplexes**
- **Row housing** or townhouses comprising two, three, four, or more attached units
- **Apartment** buildings
- **Condominium** developments

Builders of multi-unit projects usually work for owner/developers who hire them on a contract basis. Under terms of these agreements, the builder is required to complete the project as detailed in the plans and specifications prepared by a consultant for a stipulated lump-sum amount. Builders will usually be selected following competitive bid and, on the larger jobs, they will often be required to be bonded just as contractors are on commercial projects. While homebuilders may construct some of the small two, three, or four unit jobs, the larger projects are usually out of their league and are built by specialized or commercial construction companies.

Types of Contract

Contracts used in the residential construction industry can be grouped into three main categories:

- Lump sum contracts
- Cost plus contracts
- Unit price contracts

Stipulated Lump Sum Contracts

Under the terms of a stipulated **lump sum contract,** often referred to as a **firm price contract,** the builder agrees to complete the project as described in the plans and specifications for a fixed sum. As long as the scope of work is not changed, the builder will be paid the agreed amount, no more, no less. For this arrangement to succeed, the scope of work has to be precisely defined in the plans and specifications. Any uncertainties will usually lead to disputes between the owner and the builder. Because changes are almost inevitable on building contracts, the terms of the contract will include a mechanism for making scope changes and adjusting the contract price to reflect these changes.

Under the terms of these contracts, the risk of cost overruns lies with the builder. If the cost of materials turns out to be higher than estimated, the builder will pay the extra cost. When bidders anticipate the possibility of price increases, they often add contingency sums to their bid price to compensate for the risk they are taking. If prices do not rise, these contingencies will simply add to the builder's profit on the job.

In order to shift contract risks to the owner, some builders have modified the terms of lump sum contracts (by means of an escalation clause) to allow price increases to be added to the original contract sum. This results in what might be called a variable price lump sum contract since the final price will be adjusted for price changes for materials, equipment, and possible subtrades.

Stipulated lump sum contracts are commonly used on custom home projects and on multi-unit residential projects. When a competitive bid has been used to select a builder, the successful bidder will usually be awarded a firm price contract. Also, the agreement made by spec builders with homebuyers can be considered a kind of stipulated price contract since the builder agrees to complete the home for the sale price. However, there may be provisions in the sales agreement, similar to the escalation clause previously mentioned, that allow the agreed price to rise when there are unanticipated price increases.

Cost Plus Contracts

Cost plus contracts allow the builder to be compensated for all the costs of constructing the project plus an agreed upon fee. Under these terms, the owner bares the risk of cost overruns, but the owner will not be paying for any contingency sums previously mentioned when price increases fail to materialize. If for some reason the project scope of work is difficult to pin down, as is the case with some renovation projects, the owner may feel that there is an advantage in choosing a cost plus contract. This type of contract is also used in situations where the owner has hired a builder directly without going through a bid competition. In this case, either the owner places a great deal of trust in the builder or is willing to spend time closely monitoring costs as the work progresses.

The amount of fee included in a cost plus contract may be a percentage of the cost of the work, a fixed sum, or a variable sum. Owners do not generally favor percentage fees because this type of agreement provides no incentive for cost control. In fact, the higher the costs, the more profit the builder will make. There is also no incentive to control costs when fees are fixed, but at least it will not result in rewards for inefficiency. In order to provide a cost control incentive, provisions have to be included that allow the fee to increase when savings are made and to decrease when project costs exceed expectations. These provisions result in what is known as a variable fee cost plus contract.

Unit Price Contracts

Unit price contracts are appropriate when there is some uncertainty about the amount of work to be done. These contracts are not often used on housing projects since on most jobs the amount of work involved is established before a builder is hired. There may, however, be situations, typically related to earthwork, where some task is required but the quantity of this work is not known. For example, a builder may have to remove an existing underground sewer pipe as part of a project. The location of the pipe may be known but depth below grade is not. Under the terms of a unit price contract, the builder would not bid a sum for the whole job; instead he would quote unit prices for the work involved. In this example, the builder may quote $15.00 per cubic yard for the trench excavation and $25.00 per cubic yard for common backfill.

A unit price contract, therefore, has two parts:

1. Prices per unit of measurement for the different types of work involved in the project, and
2. Measurement of the actual work completed.

At the end of the job, the amount of work completed is measured and recorded, and then the builder is paid his unit prices multiplied by the quantity of work done. So, continuing the example, if our builder excavates and backfills 100 cubic yards on the project, he will be paid:

$$100 \text{ cu. yds.} \times \$15.00 = \$1,500.00$$
$$100 \text{ cu. yds.} \times \$25.00 = \underline{\$2,500.00}$$
$$\text{Total} \quad \underline{\$4,000.00}$$

Uses of Estimates

Estimates are used for a number of different purposes in residential construction:

- Determining the feasibility of a project
- Calculating an approximate price of a project
- Providing value analysis

- Setting sales prices
- Calculating bid prices
- Determining project budgets in cost control

Project Feasibility

A question that almost always arises when someone is considering the construction of a new home is: Do we have sufficient money to build this house? So, when this person consults a builder with their ideas about the new home, they inevitably raise the question of cost. At this early stage, the builder will have very little information about the proposed house. Perhaps the location, a rough idea of size and style of design may be known, but usually not much more. Based in this scant data, the builder now has to put together an estimate to determine if the concept is financially feasible. Hence, this first estimate is often called the **feasibility estimate.**

For a developer who is considering a sizable investment in a new residential project not only will the projected cost be a prime consideration, but they will also be concerned about the value of the development. The value of a proposed building is basically what it is worth once it has been constructed. Value can be appraised from the profits that are expected to flow after project completion, or, if the concept of profits is not applicable to the venture, it can be based on an assessment of the benefits that are expected to materialize from the new facility. In either case, an attempt will be made to place a monetary value on the finished project. As costs and benefits are usually extended over a number of years, monetary value will normally be determined by means of "present worth" analysis or other "time value of money" concepts.

Using this analysis, a feasible project can be defined as one where the anticipated value of all benefits exceeds the estimated total cost of putting the project in place. The cost profile of a commercial project includes many constituent parts including the cost of the land required, the cost of financing the project, the legal and general administrative costs, the cost of designing and administrating the work and, of course, the construction cost of the work. Further costs may also need to be considered, such as commissioning costs, operating costs and, possibly, marketing costs. Some costs are relatively easy to establish.

The cost of land and financing, for instance, is not difficult to determine since current market prices and interest rates are normally easy to establish, but the amount of what is most often the major cost component—the construction cost—is far more difficult to ascertain with any certainty. We will see that the most accurate way to predetermine the cost of construction work is by means of a detailed estimate using the methods employed by builders. However, a detailed estimate requires complete design details, and as we have said, these are generally not available at this early stage of a project.

So a feasibility or conceptual estimate is normally produced from merely the notion the owner has of what he or she would like to see constructed. If the owners' analysis has begun with the assessed value of the project, they should be in a position to say that the project is viable if it can be built for a certain price, where this price is the maximum amount the owner is willing to pay to obtain the benefits that are anticipated from the project. Alternatively, analysis may lead an owner to conclude that a certain structure of definite size and scope is necessary to generate the specific benefits that are sought. Typically the owner's financial situation reveals that there is only a certain amount of funds available to spend on this structure. In this case, the obvious question will be "can I build the structure for this amount?"

In either case, the owner needs an estimate of the most likely cost of the work. This estimate, because of the lack of design details, must be prepared using one of the approximate estimating techniques that follow, but this is not to say a crude approximation of costs will suffice. The feasibility decision, which may involve many thousands if not millions of dollars, is of major importance to most owners and developers, so the accuracy of the predicted costs used in the calculations is crucial if the following decisions are to be sound. Builders should note that the price they come up with at this time is the one the client always remembers.

Value Analysis

In the process of **value analysis,** the estimated cost of a component or element of a building is compared with the perceived value of that element to consider if the sum allocated to that part of the building is justified by the value provided by the component. At the time of the conceptual estimate, the estimator will have made numerous assumptions about these elements based on his discussions with the owner and his perception of the owner's needs. For instance, the exterior cladding of the building may have been assumed to be vinyl siding.

During the design stage, it might be suggested that the cladding be changed to brick masonry. Cost estimates of the alternatives and the relative benefits of the two systems will be evaluated to determine if the extra cost of the more expensive brick cladding is justified by the increased value of brick masonry over vinyl siding exterior. Then, if a decision to spend the extra amount on exterior cladding is made *and* the overall budget is still to be maintained, a saving in another element must be found to balance the additional cost of the cladding. His process is repeated so that the final design represents the greatest value for the money invested in the new project.

Estimates for Setting Sales Prices

For the spec builder, estimates are used when setting the basic sales price of a new home. If cost-based pricing is used, the builder begins with the price of the lot, and then adds the estimated cost of construction together with markup to arrive at the sales price. Adjustments for the homebuyer's choices will be made at the time of the sale.

Alternatively, some builders use a value-based approach to pricing. With this method of pricing, the builder first investigates the current market prices of the types of home he intends to construct. This determines the sales price of the home, but, in order to make a profit, the builder has to construct the house for something less that this sales price. Therefore, an estimate is also required when value-based pricing is used. In this case, the full cost of a certain design of home is estimated to determine if it can be built for the market price including the required profit. If the estimated cost including markup exceeds the market price, design modifications may be have to be made to make the deal profitable. Alternatively, the builder may simply choose not to build that particular design of house.

For semi-custom homes, the builder will have a cost estimate and sales price for the standard design; the sales price will then be adjusted for all of the owner's modifications to customize the home. Each design change will call for an estimate of the cost of the new feature, which will be compared with the cost of the standard feature. The estimates of these design modifications may be presented to homebuyers in a format similar to a menu. For instance, buyers may be offered a list of alternative kitchen layouts each one priced separately, then a list of alternative floor finishes, and so on. The sales person merely adjusts the base price of the home to account for each of the owner's choices to arrive at the final price.

Estimates for Calculating Bid Prices

Custom home builders use estimates to determine their bid prices for the work. Owners usually invite a number of builders to quote a price to construct their home in accordance with the drawings and specifications provided by the owner. In this situation, each builder will estimate the cost of the project and submit a bid to the owner before the bid closing date. The owner will generally proceed to award a contract to the builder who quotes the lowest amount for the work.

Builders in the multi-unit residential sector also obtain most of their work by means of competitive bids. On larger projects, owners and developers often hire architects to prepare the plans and specifications and go on to administer the project to completion. The type of estimating used in this part of the residential industry is very similar to the commercial construction estimating process that is outlined in some detail in Fundamentals of Construction Estimating.[1]

Cost Control Estimates

In a recent publication of the National Association of Home Builders,[2] "sloppy cost control" was listed as one of the top-ten ways builders mess up their businesses. According to this book, "inaccurate and inefficient estimating procedure" is the number one financial control problem in the home building business. Obviously the homebuilder who obtains his work in bid competitions relies on accurate estimating in order to make a profit; if the costs are underestimated, profits will quickly disappear. All builders need to establish budgets for their projects, and then obtain feedback (preferably on a monthly basis) to determine if budgets are being met on each job. This process begins with an accurate estimate from which budgets may be established. Further estimating is then needed each month to determine how much work has been done and how much it has cost to do that work.

A budget document is just another way of displaying an estimate. It usually follows a simpler format than the estimate recap used for pricing because fewer details are required and there is no need to divide it into general expenses and trades. See Figure 1.2 for an example of how a budget may be presented.

Methods of Estimating

Methods of estimating can be divided into two main categories: preliminary estimating methods and detailed estimating methods. The *price per unit* and the *price per unit area* are the two most frequently used methods of preparing preliminary estimates. These are used in the early stages of a project to provide at least a rough estimate of the construction cost at a time when there is no detailed information available about the nature of the proposed project. The detailed estimating method, because it offers so much greater accuracy, is the method of choice for most builders calculating prices when drawings and specifications are available. Certainly it should be the only method of estimating used to prepare a bid price.

Price Per Unit

The price per unit method of estimating is used on multi-unit projects. It begins with an analysis of the cost of projects that have been completed. After

---

1. *Fundamentals of Construction Estimating* by David Pratt, Delmar Thomson Learning, New York, 2nd Edition, 2004

2. *Building With Attitude* by Al Trellis and Paul Sharp, Home Builder Press, Washington DC, 1999

PROJECT BUDGET

JOB . . . . DATE . . . .

No.	DESCRIPTION	LABOR $	MATERIALS $	EQUIPMENT $	SUBTRADE $	TOTAL $
1	Schedule					
2	Permits					
3	Project Supervision					
4	Project Signs					
5	Survey and Plot Plan					
6	Stake Out					
7	Site Security					
8	Equipment Rentals					
9	Municipal Charges					
10	Excavation and Backfill					
11	Soils Testing					
12	Plumbing—Rough In					
13	Temporary Site Services					
14	Temporary Hoarding					
15	Temporary Heating					
16	Concrete					
17	Formwork					
18	Safety and First Aid					
19	Rentals					
20	Framing					
21	Heating—Rough In					
22	Windows					
23	Doors					
24	Hardware					
25	Roofing					
26	Exterior Finish					
27	Electrical—Rough In					
28	Cabinets					
29	Bathroom Accessories					
30	Appliances					
31	Heating Finish					
32	Plumbing Finish					
33	Permanent Connections					
34	Flooring					
35	Drywall					
36	Painting					
37	Electrical Finish					
38	Cleanup					
39	Snow Removal					
40	New Home Warranty					

Figure 1.2 Example of Budget Format

completing an apartment building, for example, the analyst would divide the construction cost of the new building by number of suites it contains to determine the cost per suite. This unit price can then be used as a guide to price future apartment projects by multiplying the number of suites in the proposed facility by this unit price per suite.

An accurate forecast of cost will be obtained only if the future project is very similar to those previously analyzed with regard to such items as the design of the building, the price of resources used, the total project size, the quality of finish, the geographical location, and the time of year when the work was undertaken. Clearly, differences between the previous projects and the new one can be expected in most of these areas, but making the proper price adjustments for these project differences is very difficult which seriously reduces the accuracy of price per unit estimates.

Using this estimating method can generate a rough estimate quickly, so is often used to determine the very first notion of a price in early discussions of a multi-unit project and as a crude means of comparing the known costs of different buildings.

Examples—Price Per Unit

1. A builder recently completed a 16-unit apartment building for a construction cost of $3,113,100, so the *cost per unit* was $3,113,100/16 = $194,569. Using this analysis, the cost of a proposed 12-unit apartment can be quickly calculated as: 12 × $194,569 = $2,334,828.
2. In the second example, a developer is considering the construction of a 40-unit senior citizens home. The construction cost of a similar project containing 30 units was $7,552,000, so the *cost per unit* was $7,552,000/30 = $251,733. The developer believes that prices have risen by 10% since the construction of the first building, so the cost per unit is now expected to be $251,733 × 1.10 = $276,907; therefore, the anticipated cost of the new 40-unit home would be 40 × $276,907 = $11,076,280.

Obviously, further price adjustments will be required if the proposed facility is going to differ in any respect from the previously completed project, but this method does provide a quick way to establish at least an approximation of the construction cost, which can be useful in the early stages of the new project.

Price Per Unit Area

With the price per unit area estimating method, the unit of analysis is the square foot or square meter of gross floor area of the project. The gross floor area is defined as the area of all floors measured to the outside of containing walls. As with the cost per unit method, first the construction cost of completed projects have to be analyzed. This time, the known project construction cost is divided by the gross floor area of the project to obtain a cost per unit area. This unit rate can then be applied to future projects to estimate their construction cost.

This is the most common method of preliminary estimating because it is easy to understand and a large amount of published data on prices per square foot is available that can supplement data obtained from a builder's analysis of its own projects. Because one of the first details of design to be defined is always the floor area of a proposed building, this method of estimating can be applied very early in the development of a project. However, virtually all the same variable factors that make accuracy difficult to attain with the price per unit method of estimating also apply to this price per square foot analysis.

Examples—Price Per Unit Area

1. Most custom home builders keep a record of the cost of constructing their projects. So, when a prospective client asks what it will cost to build their house, the builder can refer back to the record of a similar house constructed in the past and do a simple calculation to obtain at least a rough idea of what the construction cost will be. For example, say a house of 2,000 square feet was previously built for $243,000; the cost per square foot would be $243,000/2,000 = $121.50. If the client wants to build a similar residence but of 2,400 square feet floor area, the cost can quickly be estimated to be (2,400 × $121.50) $291,600.

2. A sophisticated cost record of a project would include a breakdown of the total cost so that the cost per square foot for each major component could be determined. This information could then be used to adjust the total price to reflect the particular requirements of the client. If we return to the example above, the client may indicate that she would like a specific type of carpet that the builder knows is more expensive than the one used in the previous project. The cost record of the previous job indicates that the amount spent on carpets was $9,000, which works out to be $4.50 per square foot. Then, if the carpet required for the new home is calculated to be approximately $14.00 per square foot, the price adjustment would be as follows:

Estimated basic price (as above)	= $291,600
Less carpet allowance 2,400 × $4.50	= <$10,800>
Add new carpet allowance 2,400 × $14.00	= $33,600
Revised estimated price	= $314,400

   With good cost records and clear details of what the client is looking for, any number of these adjustments could be made to arrive at a better estimate of the cost of the client's particular project requirements.

Detailed Estimates

The process of detailed estimating is the subject matter of a majority of this book, but before we begin to examine all the particular aspects of this topic, it may be useful to consider what the essence of this subject is. Whether prepared by hand, by computer spreadsheet, or by means of a totally computerized system, a detailed estimate can be analyzed in terms of six distinct procedures:

1. Quantity Takeoff—The work to be performed by the builder is measured in accordance with standard rules of measurement.
2. Recap Quantities—The quantities of work taken off are sorted and listed to comply with the trade breakdown to facilitate the process of pricing.
3. Pricing the Recap—Prices for the required labor, equipment, and materials are entered against the quantities to determine the estimated cost of the builder's own work.
4. Pricing Subcontractor's Work—Prices are obtained from competing subtrades who quote to perform the work of their trade then, usually, the lowest bid from each trade is entered into the estimate.
5. Pricing General Expenses—The costs of the anticipated project overheads are calculated and added to the estimate.
6. Summary—All the estimated prices are summarized and the contractor's markup is added.

The detailed estimating method, rather than any of the other estimating methods considered, is far more likely to produce a price that is an accurate forecast of the actual costs of building a project. Because the very survival of builders and subtrades who obtain work by offering firm price bids often rests on the accuracy of their estimating, it will come as no surprise that detailed estimating is the method of choice for bid preparation in the highly competitive building industry.

The basis of a detailed estimate is the accurate assessment of the work in the form of a quantity takeoff that can only be obtained from the full design of the project. Because of this requirement and also because detailed estimating is such a time-consuming process, feasibility estimates are usually prepared by the other quicker, but less accurate methods previously considered.

Materials Estimates

A detailed construction estimate can also be used to generate a **bill of materials.** This document lists the quantities of all materials required to construct a project. It is compiled directly from the estimator's takeoff of the work. Bills of material are usually divided into categories that correspond to the type of materials obtained from specialized suppliers, such as:

- Gravel materials
- Concrete materials
- Carpentry materials

Copies of the appropriate categories can then be sent to individual suppliers for pricing. Categories can further be divided into packages, for example, carpentry materials may comprise:

- Cribbing package
- Main floor framing package
- Roof framing package

When computer estimating is used, producing a bill of materials report from a takeoff can be as easy as pressing a button, and, if up-to-date database prices have been maintained, the bill of materials can be fully priced. Alternatively, unpriced bills of material can be sent to suppliers who simply have to apply unit prices to the quantities of materials listed to produce their quotes.

Architects and Designers

The principal role of **architects** and **designers** in the residential sector is to create designs for new building and renovation projects. Consultant architects are seldom appointed on custom homes these days, except perhaps for some larger and expensive undertakings since the expense would be far beyond the means of most homeowners. Architects are, however, employed as **prime consultants** on most large multi-unit developments where the size of such an undertaking justifies the expense. Prime consultants will not only prepare plans and specifications on these projects, but they will usually also provide administration services throughout the course of the work. Indeed, on major developments, architects or architectural companies often play the lead role in moving a project from the conceptual phase through to final completion.

In the case of single custom homes, some owners obtain designs from plan service companies for a fraction of the fee an architect would charge; other owners rely on builders who often have in-house design staff to prepare plans and specifications for new homes. Many of the larger spec builders also have in-house design capabilities so that they can ensure that their designs continue to be fashionable. Spec builders with smaller organizations tend to rely more on plan services when home designs need to be updated.

Green Building and Sustainable Construction

"Green," "sustainable," "eco-friendly," "energy-efficient," "organic" are all terms we have come to be familiar with over recent years and the homebuilding industry has certainly not escaped this phenomenon in our society. The **green building** movement has grown and is still evolving; homebuilders are aware of this and most are responding to their customers' needs in this area. Homebuyers have become concerned about such matters as energy consumption, renewable materials, water usage, and indoor air quality. Consequently, many builders are now addressing these issues in the design of their projects, the products and materials used, and even the construction methods employed to build their homes. Government agencies have also responded to the green movement by publishing guidelines and, in some cases, offering incentives to builders who adopt green construction.

Knowledgeable homebuyers realize that building green comes with a price tag, but they also appreciate that spending extra for an energy-efficient heating system soon pays off with reduced energy consumption costs; that water is becoming a scarce resource and needs to be conserved; that many products in new homes emit unhealthy gases; and that consuming nonreplaceable resources is not such a good idea. So these prospective homeowners are looking for assurances that the homes they buy meet at least minimum standards that deal with these issues.

There are many green building standards that define how a "green" house should be built, what materials should be used, and what performance standards should be met by the completed house. Perhaps the most well-known set of standards and rating system for green building come from the US Green Building Council's (USGBC's) Leadership in Energy and Environmental Design (LEED) Green Building Rating System.

LEED Certification

The USGBC, a construction industry organization whose membership includes owners, designers, and contractors, developed the LEED rating system as an objective standard for certifying that a building is environmentally friendly. Because of the wide variety of construction projects, LEED has developed several different versions of LEED including:

- New commercial construction
- Commercial interiors
- Existing buildings: operation and maintenance
- Core and shell
- Homes
- Neighborhood development
- Portfolio program
- LEED for schools
- LEED for retail
- LEED for health care

In each of these categories, LEED programs award certificates when a project achieves the required points for achievement in green construction. Points are awarded in all aspects of the construction process, for example, in homebuilding points can be obtained for such things as:

- Use of efficient insulation
- Adoption of high-performance windows
- Tight construction of the building envelope
- Use of efficient heating and cooling equipment

Various levels of certification are attainable depending on how many points have been awarded out of a total of 100:

- Certification: 26–32 points
- Silver: 33–38 points
- Gold: 39–51 points
- Platinum: 52–69 points

Homebuilders generally, however, view LEED certification as expensive; there are a series of fees for registration, certification review, etc. Many feel that LEED is appropriate only for certain types of projects such as high-end custom homes. An alternative green building rating system is offered by the National Association of Home Builders; this seeks to be more in line with the needs of its membership.

NAHB National Green Building Program

National Green Building Certification is an application of the ICC 700-2008 National Green Building Standard that has been officially approved by the American National Standards Institute (ANSI). Like LEED, the NAHBGreen program uses a points system to rate projects and, on the basis of number of points achieved, offers four green certification levels: Bronze, Silver, Gold, and Emerald.

Residential buildings, remodeling projects, and developments can be Green Certified by addressing key green construction areas including:

- Lot design, preparation, and site development
- Resource efficiency
- Energy efficiency
- Water efficiency
- Indoor environmental quality
- Operation, maintenance, and homeowner education

After all points have been confirmed in the field by an impartial third party, the NAHB Research Center is able to certify a home under the NAHBGreen program.

Green Building and Estimating

The question is: How will adoption of green building techniques affect your cost estimate of the work? Most green building requirements will automatically be dealt with by following the estimating procedures described in this text. For example, if a green design calls for energy-efficient wall and roof systems, the cost of building to meet these specifications will be reflected in the labor, material, and equipment prices used in the estimate; the cost of providing such items as ventilation energy recovery equipment will be reflected in the price of the mechanical subcontractor supplying these units; and so on.

There may, however, be some additional costs for building green such as the cost of sorting recyclable items at the site and the cost of fees if LEED or NAHBGreen certification is pursued. These types of costs are added to the project's general expenses in the estimate; general expenses are discussed in Chapter 8.

As noted earlier, some municipalities and government agencies are offering incentives for builders to implement green construction. These may be in the form of tax rebates or reduced building permit fees, so the estimator needs to be aware of these schemes in order to make an adjustment to the price of the project.

Drawings and Specifications

The role of drawings and **specifications** is to communicate to the builder exactly what is to be built. On single-home residential jobs, all the information necessary to build the project is usually contained in a set of working drawings together with supplementary notes regarding items such as cabinets, floor coverings, paint colors, light fixtures, and any other selections made by the owner. Specifications are usually noted on the drawings of these smaller projects.

On larger projects, drawings will be supplemented by a separately bound set of specifications that go into much detail about the materials and workmanship required on the project. Where the owner is seeking bids from builders for the project, information about contracts and bid procedures will also be found in the specifications. If a soils report has been compiled, a copy of this document may also be provided for information to builders.

A set of drawings for a new home will usually include the following:

- A plot plan that provides information about the site of the home and shows property lines together with the location of the building on the lot
- Plan views of each level of the proposed building from which the reader can obtain information about the size and layout of the house
- Building elevations that show what the building will look like as viewed from the front, sides, and back
- Cross-sections that show the construction of foundations, floors, walls, and roofs
- Additional sketches and information about building components

For example, full set of home plans can be found in Appendix D of this text.

Soils Reports

The soils report is a resource that is available on most large projects and also on some smaller jobs. This document provides information about the subsurface conditions at the site obtained from bore holes and/or other investigations made by a soils engineer. The purpose of the report is to furnish data about the site for use in the design of the foundation system. Indeed, the report usually includes advice about the type of foundation systems suitable for the conditions encountered. However, the information contained in the report can also be very helpful to the excavator since it not only discloses the type of soil that will probably be found at the site, but also provides an indication of the moisture content of this soil. Both of these factors can have a profound effect on the cost of excavation work.

A soils report usually begins with an introduction identifying the name and location of the project, the name of the owner and the name of the prime consultant, and it states the objective of the study. Details of the investigation then follow including the date of the site was examined, the method of investigation undertaken, and general details of the borehole procedures. Next, the report addresses the nature of the subsurface conditions encountered and goes on to provide comments and recommendations about the foundation system together with detailed information about the test hole results.

An estimator is cautioned to read the contract documents carefully to determine the true status of the soils report because, while some contracts contemplate extra compensation to a contractor who is mislead by information contained in the report, many contracts state that the contractor should not rely upon information provided by the soils report but should make its own investigation of subsurface conditions.

Site Visits

Whether it is a custom home to be built or a multi-unit residential project, the bid price for the work will always depend, at least to some extent, on the characteristics of the particular site where the building is to be constructed. A site visit is always an important part of a bid estimate but, if there is no soils report on the project, what is uncovered on the visit may be crucial to the accuracy of an estimate. While an experienced estimator will certainly know what to look for and assess when investigating the site, the presence of a more senior estimator and/or a project superintendent on the visit will always be valuable if only to offer alternative viewpoints in the evaluation of the data obtained.

Figure 1.3 lists the items that are normally considered on the site visit. The condition of the site and any soil information that can be observed will impact the price of the excavation work estimated. Many of the other items examined are included in the general expenses of the project; pricing these items is dealt with in Chapter 8. It is recommended that photographs or videos be taken on the site visit since the condition of the site may change in the time between the bid and the start of work. Also, the condition of structures adjacent to the site should be carefully documented and photographed at this time, especially if these structures are in poor condition. This information could be important in determining if damage to these occurred before or after the project work began.

Environmental concerns are another important area for the estimator to consider on the site visit. Complying with exacting environmental requirements can be a very costly proposition for a contractor, so the estimator must review the contract documents thoroughly to assess the extent of liability for possible environmental problems the contractor is assuming under the terms of the contract. Every effort should then be made to investigate the history of the site from local information and by reviewing soils reports and any environmental assessments that might be available. Special attention should be paid to the possibility of soil contamination in the location of the site. Only by assembling accurate data can the contractor quantify the financial risk imposed by the contract regarding environmental issues, then an appropriate sum can be added to the bid to cover this risk.

SITE VISIT CHECKLIST

Estimator: _____   Date: _____

Job Number: _____   Location: _____

Project Name: _____   _____

Distance from Office: _____

Weather Conditions: _____

Access and Roads: _____

Sidewalk Crossings: _____

Site Conditions: _____

_____

_____

_____

Adjacent Structures: _____

_____

_____

_____

Obstructions: _____

Shoring or Underpinning: _____

Depth of Topsoil: _____

Soil Data: _____

Ground Water: _____

Soil Disposal Location: _____

Distance to Borrow Pit: _____

Local Sand & Gravel: _____

Electrical Service: _____

Telephone: _____

Sewer & Water Services: _____

Parking and Storage: _____

Security Needs: _____

Temporary Fences Required: _____

Garbage Disposal: _____

Toilets: _____

Site History: _____

Possible Contamination: _____

Other Comments: _____

_____

_____

_____

_____

Figure 1.3   Site Visit-Checklist

SUMMARY

- A building cost estimate is an attempt to determine the likely cost of some building work before the work is done.
- In residential construction, the type of estimate used depends upon the residential market served by the builder preparing the estimate, the nature of the contract with the home purchaser or owner, and the purpose of the estimate.
- The three main groups of builders serving the residential construction market are:
  - Spec builders
  - Custom home builders
  - Multi-unit residential contractors
- Contracts used by homebuilders and residential contractors include:
  - Stipulated lump sum
  —Firm price
  —Variable price
  - Cost plus
  —Percentage fee
  —Fixed fee
  —Variable fee
  - Unit Price
- Residential estimates are prepared for many different purposes, including:
  - Determining the feasibility of a project
  - Calculating an approximate price of a project
  - Setting sales prices
  - Calculating bid prices
  - Determining project budgets in cost control
- When projects are no more than ideas, feasibility estimates are prepared in response to questions such as: Do we have sufficient money to build this house?
- In addition to cost, developers of commercial projects are also interested in value when considering construction of a new project.
- The estimated cost of a component or element of a building is compared with the perceived value of that element in the value analysis process.
- The spec builder uses estimates when setting the sales price of a new home.
- Sales prices may be set by means of cost-based pricing or by value-based pricing.
- Custom home builders and multi-unit contractors prepare their bid prices by means of detailed estimates.
- Cost control should be an important consideration of all builders.
- Estimates are used to set project budgets and monitor budgets during construction of a project.
- Preliminary estimates can be prepared by several different methods:
  - Price per unit
  - Price per unit area
- Detailed estimates include six distinct procedures:
  - Quantity takeoff
  - Recap quantities
  - Price recaps
  - Price subcontractor's work
  - Price general expenses
  - Summarize and add markup

- Detailed construction estimates can also be used to generate bills of material that list the quantities of all materials required to construct a project.
- The principal role of architects and designers in the residential sector is to create designs for new building and renovation projects.
- Green building has become an established feature of the homebuilding industry and its scope is still expanding. Many homebuilders identify benefits in having their projects certified as meeting the standards of the LEED or the NAHBGreen rating systems. The cost of accomplishing these green building standards is generally automatically accounted for in the estimating process, but additional costs may need to be considered in the project's general expenses.
- Drawings and specifications are used in residential projects to communicate to the builder exactly what is to be built. They usually comprise:
  - A plot plan
  - Plan views of each level of the proposed building
  - Building elevations
  - Cross-sections
  - Additional sketches and information
- Soils reports, prepared for large and some small projects, provide information about the subsurface conditions at the site obtained from bore holes and/or other investigations made by a soils engineer.
- A site visit is always an important part of a bid estimate but, if there is no soils report on the project, what is uncovered on the visit may be crucial to the accuracy of an estimate.

RECOMMENDED RESOURCES

Information	Web Page Address
National Association of Home Builders—Provides information about the homebuilding industry together with education, books, and publications for homebuilders. You will also find much information about the NAHB National Green Building Program on this web site.	http://www.nahb.org/default.aspx
Builder Central—Provides information for and about homebuilders.	http://www.buildercentral.com/
B4UBUILD.COM—A web site for use by anyone involved with residential construction. It contains information about the building process, house plans, construction contracts, and much more.	http://www.b4ubuild.com/
US Green Building Council—LEED rating system.	http://www.usgbc.org/DisplayPage.aspx?CategoryID=19
Much information can also be obtained by using key words such as **Building Estimates, Home Plans, Green Building,** and **Value Analysis** in **web search engines**.	

REVIEW QUESTIONS

1. Explain what a building cost estimate is.
2. Explain the following terms:
   a. Spec builder
   b. Production builder
   c. Custom home builder
3. Compare the risks taken by a spec builder with those taken by a custom home builder.
4. Describe the process by which builders are selected for large multi-unit residential projects.
5. What is a firm price contract?
6. How can you provide incentive for a builder to exercise cost control with a cost plus contract?
7. How would a builder determine the feasibility of an owner's custom home proposal?
8. How can value analysis be used with the design of a house?
9. Describe the value-based approach to setting the sales price of a home.
10. How is estimating used in the cost control process?
11. Describe the following methods of preliminary estimating:
    a. Cost per unit
    b. Cost per unit area
12. What is the most accurate method of estimating the cost of a building project, and why is it more accurate than other methods?
13. What is listed on a bill of materials, and what format is usually adopted for a bill of materials?
14. On what kind of project would you find architects in the role of prime consultant?
15. What is usually contained in a set of drawings for a new house?
16. On a site visit, why is it important to investigate the condition of structures adjacent to the site of the new project?

PRACTICE PROBLEMS

1. Based on the following information, how much would the amount paid to this subcontractor be?
   a. The subtrade's bid indicated that they would excavate trenches for $15.00 per cubic yard, they would lay pipes for $2.75 per foot, and they would backfill trenches for $12.00 per cubic yard.
   b. The quantity of work completed on the job was 120 cubic yards of excavation and backfill and 92 feet of pipe.
2. What would the range of prices be for a 2,200 square foot home when the builder's cost ranges from $65.00 to $120.00 per square foot?
3. A 3,000 square feet custom home was recently constructed at a cost of $331,500, which includes $23,000 for heating and ventilating the house. What would be the estimated cost of a new home of 3,500 square feet, if it was similar to the previous home in all respects except that a more expensive heating and ventilating system is required that is expected to cost $9.00 per square foot?

2

ARITHMETIC AND PRINCIPLES OF MEASUREMENT

OBJECTIVES

After reading this chapter and completing the review questions, you should be able to:

- Describe the different ways of measuring the work of a construction project.
- Calculate the centerline length of perimeter footings and walls.
- Calculate the areas and perimeters of rectangles, triangles, and other shapes encountered in buildings.
- Calculate the volume footings, walls, and trenches.
- Describe the quantity takeoff process.
- Describe what is measured in the takeoff and the composition of takeoff items.
- Explain the level of accuracy desired in estimating.
- Describe how an estimate is organized in order to minimize errors.

KEY TERMS

assembly	method of measurement
computer estimating	pricing sheet
Excel spreadsheets	quantity sheet
extensions	specifications
mensuration	takeoff item

Introduction

There are a number of different ways that the work of a construction project is measured:

1. By counting the number of pieces; for example, the number of doors
2. By measuring the length of an item, such as a pipe
3. By measuring an area of a material; for instance, the area of floor sheathing
4. By measuring the volume of a material, such as concrete
5. By measuring the weight of materials, such as structural steel beams

In order to accurately perform the many calculations required for these measurements, an estimator needs a thorough knowledge and skill in **mensuration** (application of the principles of measurement required to calculate lengths, areas, and volumes). Even when items are enumerated, the estimator frequently has to perform calculations to arrive at the quantity of items required; anchor bolts and joists are typical examples. While the number of anchor bolts for columns can usually by counted directly from the plans, the anchor bolts along the length of a wall plate may not all be shown on the drawings. Instead, there is often a note that bolts are required at 24-inches on center along the length of the plate. Items like joists are also often not fully detailed on the drawings. In such cases, the number of joists required has to be determined from the specified spacing of these components.

Calculating Number of Items

When items like anchor bolts, joists, and studs are spaced out at certain intervals along a straight line, we calculate the number required by first dividing the length of the line by the spacing, then rounding up to the next whole number and adding one.

EXAMPLE 2.1

Figure 2.1 shows the top of a wall that is 7' long. Suppose that we want to lay out joists on this wall spaced 16" (1.33') from the center of one joist to the next.

7' / 1.33' = 5.26

Rounding up gives six joists.
Adding the required end joist gives seven joists in total.

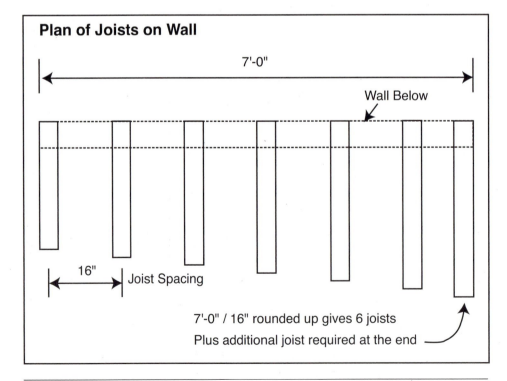

Figure 2.1   Calculating Number of Joists

EXAMPLE 2.2

In the next example, anchor bolts are required to be installed at a spacing of 24" from bolt to bolt along the top of a wall 13'-0" long. The end bolts are also required to be 2" from the end of the wall as shown on Figure 2.2.

To calculate the number of bolts, we first deduct the 2" from each end of the wall: 13'-0" less 4" leaves 12'-8". Next, we divide the 12'-8" by 24" (12'-8" / 2'-0"). This gives 6.3, which we round up to 7. We then add one to account for the end bolt, so the required number of bolts is 8.

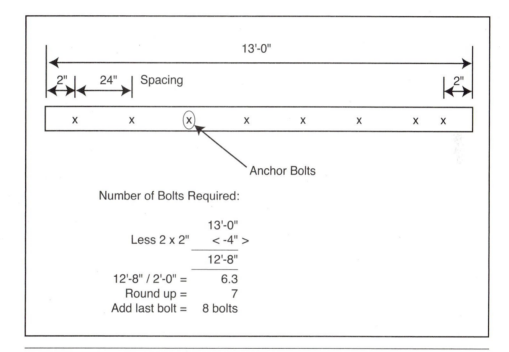

Figure 2.2   Calculating Number of Anchor Bolts

Length Calculations

Whenever possible, the length of an item of work is obtained directly from the dimensions given on the drawings. Sometimes dimensions are missing, in which case the estimator may be able to determine the length by reference to dimensions shown on other plans. You may have to add and/or subtract a number of dimensions from a drawing to arrive at the required length. Also, the estimator has to remember to read a set of drawings as a whole. This means you may have to search through a number of plans to find the dimensions that enable you to calculate the length of something shown on a particular drawing.

As a last resort, it is occasionally necessary to scale off a length from the drawings. Scaling is not recommended because, even when plans indicate that they are drawn to a certain scale, they may not be accurate or, as sometimes happens in the design process, changes are made to the plans that cause some lengths to be out of scale.

Perimeter Calculations

Many buildings have perimeter walls. In these situations, the volume of material in the wall is obtained by multiplying together the centerline length of the wall, the width of the wall, and the height of the wall. Once it is determined, the centerline length of the wall is often used for a number of calculations, so it is useful for an estimator to be able to quickly and accurately calculate the length of a perimeter centerline.

EXAMPLE 2.3

A perimeter length is calculated from the figured dimensions that are provided on the plan of the structure. Figure 2.3 shows the plan of a basement wall with the dimensions to the exterior of the walls. The length of the outside perimeter can be determined in this fashion:

$$
\begin{array}{rcl}
2 \times 42\text{'-}0" & = & 84\text{'-}0" \\
2 \times 26\text{'-}0" & = & 52\text{'-}0" \\
\text{Thus, the exterior perimeter} & = & 136\text{'-}0"
\end{array}
$$

To arrive at the centerline length, it is necessary to deduct the thickness of the wall at each corner, that is $4 \times 8" = 2\text{'-}8"$, which makes the total centerline $133\text{'-}4"$. The reason for making this adjustment at the corners can be seen from the enlarged detail of a corner shown in Figure 2.3.

EXAMPLE 2.4

When the first perimeter length was calculated, it was measured to the exterior corner. However, to obtain the centerline perimeter, one half the width of the wall has to be deducted from each dimension. So, at this corner, $2 \times 8"/2$ ($8"$) has to be deducted. Because altogether there are four corners, the total adjustment is $4 \times 8"$, which is $32"$ or $2\text{'-}8"$.

The full calculation of the centerline perimeter would appear like this:

$$
\begin{array}{rcl}
2 \times 42\text{'-}0" & = & 84\text{'-}0" \\
2 \times 26\text{'-}0" & = & \underline{52\text{'-}0"} \\
& = & 136\text{'-}0" \\
\text{Less } 4 \times 8" & = & \underline{<2\text{'-}8">} \\
& = & \underline{133\text{'-}4"}
\end{array}
$$

In Figure 2.4, the plan of the building is not a full rectangle, but a portion of the building is cut away. This cutout will certainly reduce the floor area within the walls, but it will not affect the length of the perimeter wall. In fact, we do not need to know the size of the cutout in order to calculate the length of the wall centerline.

Adding two times the full building length to two times the full building width and making an adjustment for four exterior corners will still give the centerline length. Although there are five exterior corners on this plan, the effect on the centerline length of an interior corner balances the effect of an exterior corner, so they cancel each other leaving just four exterior corners. This hypothesis that interior corners cancel the effect of exterior corners is demonstrated on Figure 2.5.

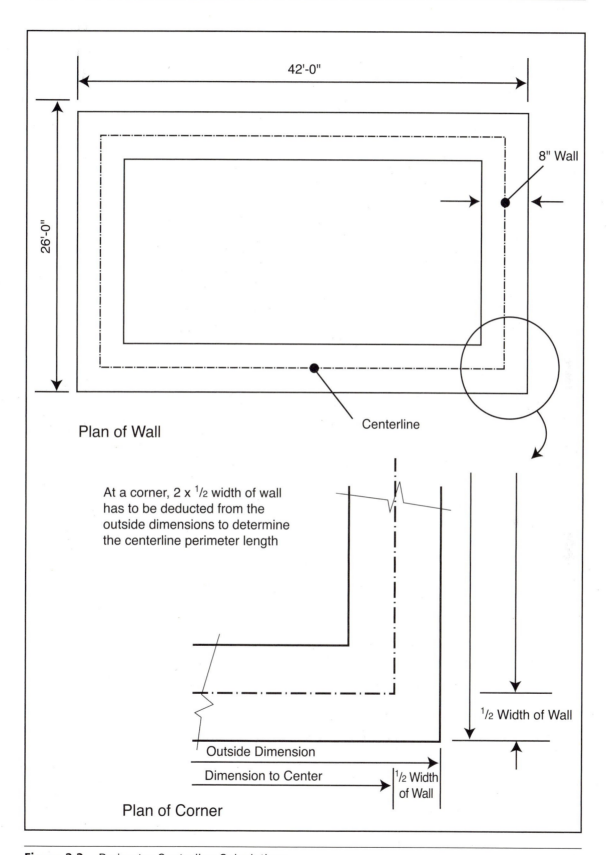

Figure 2.3   Perimeter Centerline Calculations

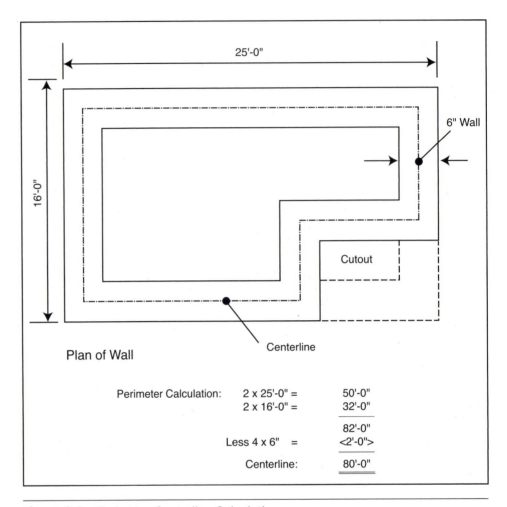

Figure 2.4   Perimeter Centerline Calculations

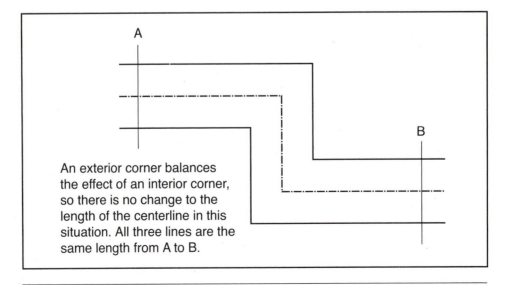

Figure 2.5   Interior and Exterior Corners

Figure 2.6 and Figure 2.7 show further examples of perimeter calculations for a number of different building shapes.

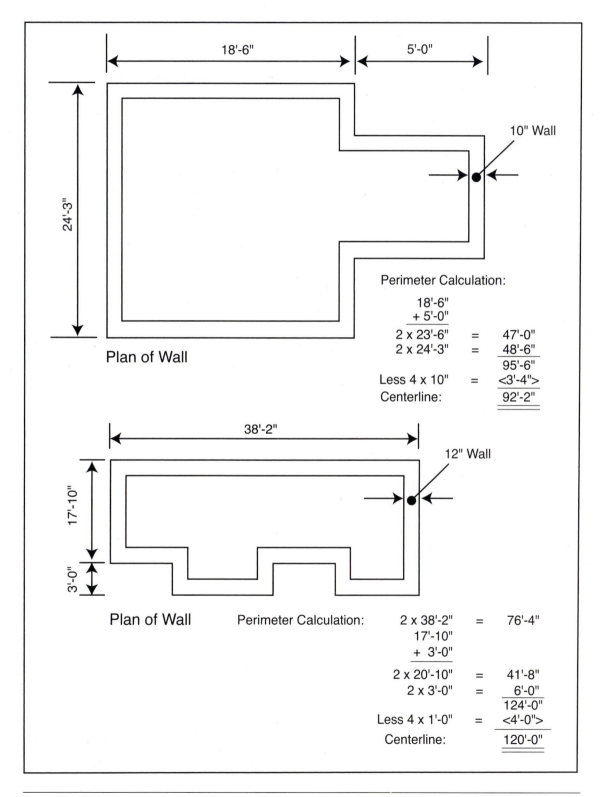

Figure 2.6   Perimeter Centerline Calculations

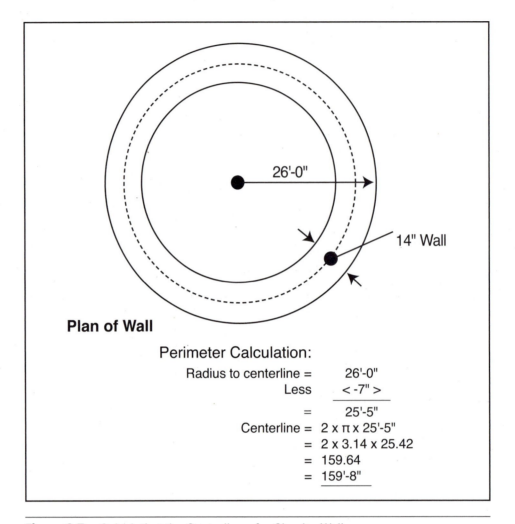

Plan of Wall

Perimeter Calculation:

Radius to centerline =	26'-0"
Less	< -7" >
=	25'-5"
Centerline =	2 x π x 25'-5"
=	2 x 3.14 x 25.42
=	159.64
=	159'-8"

Figure 2.7   Calculating the Centerline of a Circular Wall

Perimeters—Quick Method

To speed up the takeoff process, many residential estimators use the external perimeter rather than centerline perimeter for calculations. In Example 2.3, the external perimeter was quickly calculated as 136 feet; using this method, this length would be used to calculate trench, footing, and wall volumes rather than taking time to make the adjustment to determine the centerline perimeter. This approach saves time and, at least for small buildings, the error in the resulting quantities is not significant.

Calculating Areas

Most of the areas we are required to calculate in residential estimating will be rectangular in shape, but occasionally the estimator may encounter triangles, parallelograms, trapezoids, and other quadrilateral shapes. Even part or whole circles are occasionally encountered in buildings. Figure 2.8 provides the formulae for calculating the perimeter and the area of a number of different shapes.[1]

---

1. See Appendix C for a list of formulas to calculate lengths, areas, and volumes

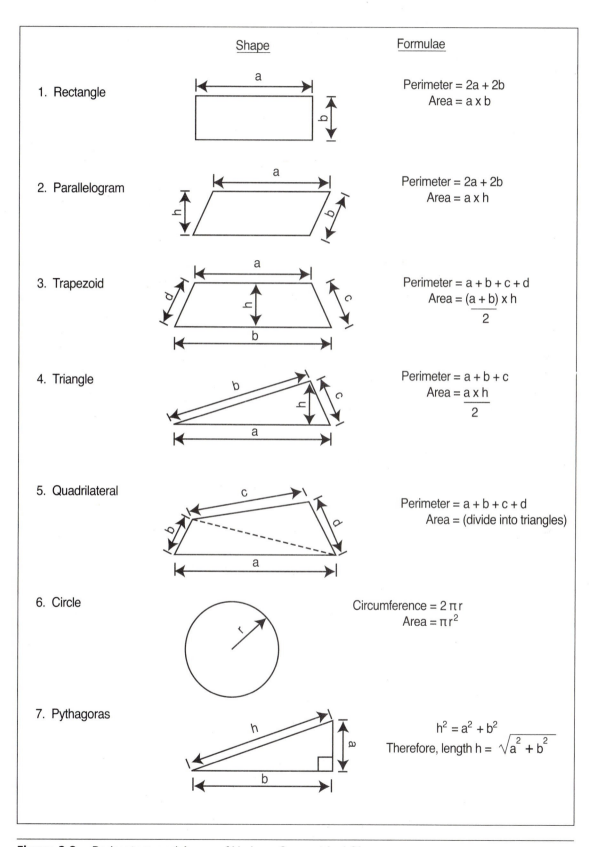

Figure 2.8   Perimeters and Areas of Various Geometrical Shapes

Inches		Feet	Inches		Feet
1"	=	0.08'	7"	=	0.58'
2"	=	0.17'	8"	=	0.67'
3"	=	0.25'	9"	=	0.75'
4"	=	0.33'	10"	=	0.83'
5"	=	0.42	11"	=	0.92'
6"	=	0.50'	12"	=	1.00'

Figure 2.9   Feet–Inches Conversions

We have also included the Pythagorean formula in Figure 2.8. This formula is often required to calculate the length of sloping lines, such as those found on roof systems.

Note that inches are converted to decimals of a foot for area calculations so that the results are stated in square feet. Working to two decimal places generally provides sufficient accuracy for estimating purposes; Figure 2.9 shows the decimal equivalents of inches.

Figure 2.10, Figure 2.11, and Figure 2.12 show calculations of the area and perimeter of some of the different shapes found on buildings.

Calculating Volumes

Pad footings are usually shaped like a rectangular box, so the volume of concrete in the footing is calculated by multiplying the footing length by the width by the depth. This formula can also be used to calculate the amount of concrete in a rectangular column.

Circular columns are cylindrical, and their volume is determined using the following formula:

Volume $= \pi r^2 \times h$, where r is the radius and h is the height of the column.

See Figure 2.13 for examples of calculating volumes of pad footings and columns.

To calculate the volume of concrete in a continuous footing to the perimeter of a building, use this formula:

Volume = Centerline Length × Width × Depth

The same formula can be used to calculate the volume of a perimeter trench or the volume of a perimeter wall. In fact, when a wall sits on top of a perimeter footing that is in a trench, the wall, the footing, and the trench all have the same centerline length. Consequently, the estimator has to be particularly careful when calculating a perimeter centerline length for a building since this single dimension can be used for a number of different calculations in an estimate.

Volumes of material are usually required to be stated in cubic yards, but dimensions are normally given in feet and inches. So, to determine cubic yards, first calculate volumes by multiplying together length, width, and depth each in feet to give cubic feet, and then divide by 27 (since there are 27 cubic feet in a cubic yard) to establish the number of cubic yards.

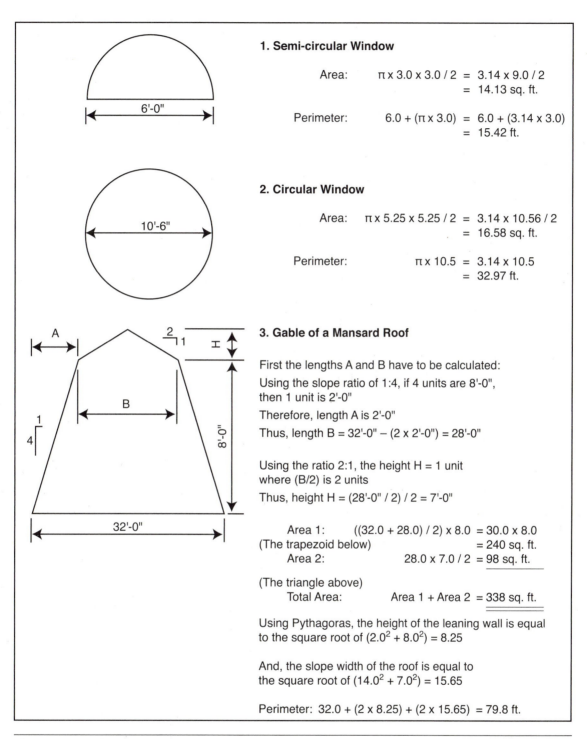

1. Semi-circular Window

Area: $\pi \times 3.0 \times 3.0 / 2$ = $3.14 \times 9.0 / 2$
                                        = 14.13 sq. ft.

Perimeter: $6.0 + (\pi \times 3.0)$ = $6.0 + (3.14 \times 3.0)$
                                         = 15.42 ft.

2. Circular Window

Area: $\pi \times 5.25 \times 5.25 / 2$ = $3.14 \times 10.56 / 2$
                                            = 16.58 sq. ft.

Perimeter: $\pi \times 10.5$ = 3.14×10.5
                                  = 32.97 ft.

3. Gable of a Mansard Roof

First the lengths A and B have to be calculated:

Using the slope ratio of 1:4, if 4 units are 8'-0",
then 1 unit is 2'-0"

Therefore, length A is 2'-0"

Thus, length B = 32'-0" − (2 x 2'-0") = 28'-0"

Using the ratio 2:1, the height H = 1 unit
where (B/2) is 2 units

Thus, height H = (28'-0" / 2) / 2 = 7'-0"

Area 1: $((32.0 + 28.0) / 2) \times 8.0$ = 30.0×8.0
(The trapezoid below)   = 240 sq. ft.
Area 2: $28.0 \times 7.0 / 2$ = 98 sq. ft.

(The triangle above)
Total Area: Area 1 + Area 2   = 338 sq. ft.

Using Pythagoras, the height of the leaning wall is equal
to the square root of $(2.0^2 + 8.0^2)$ = 8.25

And, the slope width of the roof is equal to
the square root of $(14.0^2 + 7.0^2)$ = 15.65

Perimeter: $32.0 + (2 \times 8.25) + (2 \times 15.65)$ = 79.8 ft.

Figure 2.10   Area and Perimeter Calculations

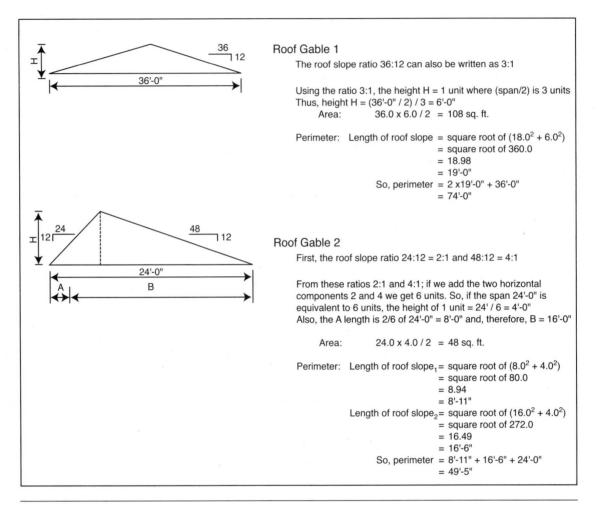

Roof Gable 1

The roof slope ratio 36:12 can also be written as 3:1

Using the ratio 3:1, the height H = 1 unit where (span/2) is 3 units
Thus, height H = (36'-0" / 2) / 3 = 6'-0"
Area: 36.0 x 6.0 / 2 = 108 sq. ft.

Perimeter: Length of roof slope = square root of $(18.0^2 + 6.0^2)$
                              = square root of 360.0
                              = 18.98
                              = 19'-0"
So, perimeter = 2 x19'-0" + 36'-0"
                              = 74'-0"

Roof Gable 2

First, the roof slope ratio 24:12 = 2:1 and 48:12 = 4:1

From these ratios 2:1 and 4:1; if we add the two horizontal
components 2 and 4 we get 6 units. So, if the span 24'-0" is
equivalent to 6 units, the height of 1 unit = 24' / 6 = 4'-0"
Also, the A length is 2/6 of 24'-0" = 8'-0" and, therefore, B = 16'-0"

Area: 24.0 x 4.0 / 2 = 48 sq. ft.

Perimeter: Length of roof slope$_1$ = square root of $(8.0^2 + 4.0^2)$
                              = square root of 80.0
                              = 8.94
                              = 8'-11"
Length of roof slope$_2$ = square root of $(16.0^2 + 4.0^2)$
                              = square root of 272.0
                              = 16.49
                              = 16'-6"
So, perimeter = 8'-11" + 16'-6" + 24'-0"
                              = 49'-5"

Figure 2.11 Area and Perimeter Calculations

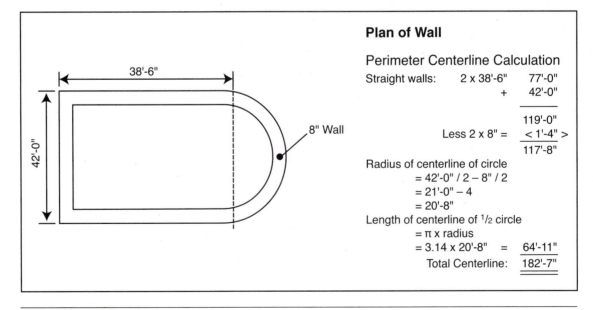

Plan of Wall

Perimeter Centerline Calculation
Straight walls: 2 x 38'-6"   77'-0"
                                +   42'-0"

                                          119'-0"
                Less 2 x 8" =   < 1'-4" >
                                          117'-8"
Radius of centerline of circle
        = 42'-0" / 2 – 8" / 2
        = 21'-0" – 4
        = 20'-8"
Length of centerline of $1/2$ circle
        = π x radius
        = 3.14 x 20'-8"   =   64'-11"
                Total Centerline: 182'-7"

Figure 2.12 Perimeter Calculations

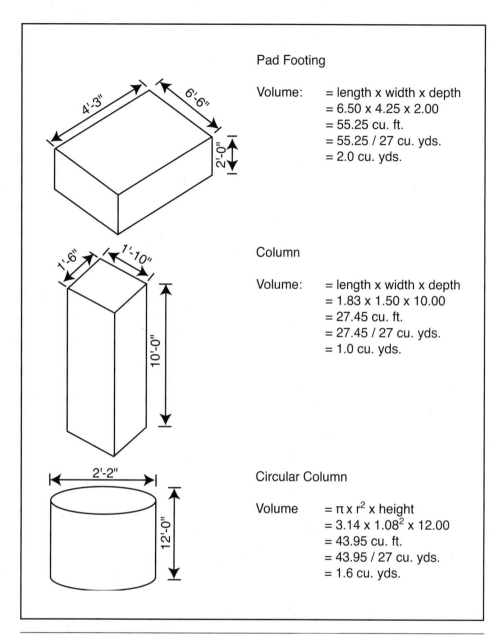

Pad Footing

Volume: = length x width x depth
             = 6.50 x 4.25 x 2.00
             = 55.25 cu. ft.
             = 55.25 / 27 cu. yds.
             = 2.0 cu. yds.

Column

Volume: = length x width x depth
             = 1.83 x 1.50 x 10.00
             = 27.45 cu. ft.
             = 27.45 / 27 cu. yds.
             = 1.0 cu. yds.

Circular Column

Volume   = π x r^2 x height
             = 3.14 x 1.08^2 x 12.00
             = 43.95 cu. ft.
             = 43.95 / 27 cu. yds.
             = 1.6 cu. yds.

Figure 2.13   Volume Calculations

When the sides of a trench are cut back to a slope (as they often are for safety reasons), the volume of the trench is calculated using the average width of the trench:

Volume = Centerline Length × Average Width × Depth

Similarly, when the depth of trench varies, the volume is found by using the average depth:

Volume = Centerline Length × Average Width × Average Depth

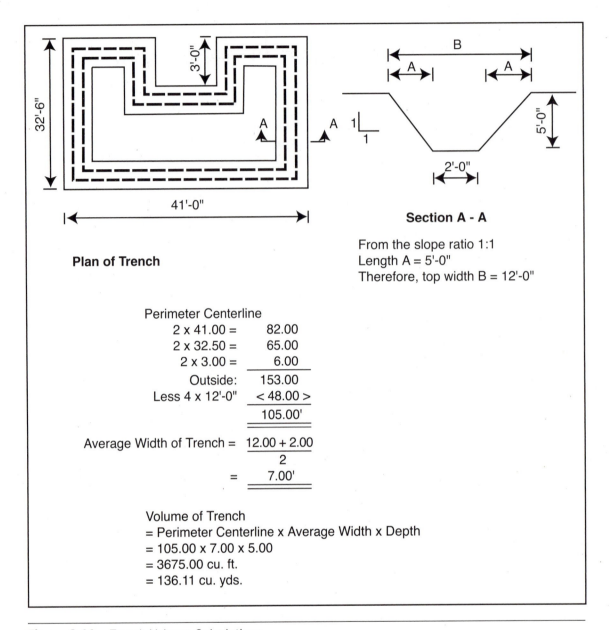

Figure 2.14   Trench Volume Calculations

Figure 2.14 shows an example of calculating the volume of perimeter trenches with cutback sides.

The Quantity Takeoff

An estimate begins with a quantity takeoff. A quantity takeoff is a process of measuring the work of the project in the form of a series of quantified work items. To prepare the takeoff, an estimator has to break down the design that is shown on the drawings and described in the **specifications** into pre-defined tasks (work items). These items correspond to the operations the builder will perform to complete the work of the project. Many estimators maintain a catalog of standard items that represent work tasks encountered on the kind of projects they estimate. This catalog can be in a book form or in the form of a database in a software program. Junior estimators especially often use such a catalog of standard work items as a checklist to help ensure that all the categories of work in the project are accounted for in the takeoff process.

Each item considered in the takeoff is measured according to a uniform set of rules with the object of producing a list of work items and their associated quantities in a format familiar to estimators. Because a standard format is used, estimators are able to more easily evaluate the work of each item and proceed to put a price on this work. In each subsequent chapter, where the measurement of the work of a particular trade is studied, we have included "measuring notes" which outline detailed rules regarding how the items of work of that trade are to be measured. These rules of measurement are based on the **Method of Measurement** followed by professional quantity surveyors and estimators. It is strongly recommended that estimators, at least those in the same company, apply a standard method of measuring work such as this, because price information that is gathered can only be shared if it relates to work that has been measured in the same fashion.

What Is Measured

The estimator should note that the work measured in the takeoff process for a residential project is generally the quantity of materials required for the job. The objective of a materials takeoff is to compute the amount of material that needs to be purchased in order to construct each particular item of the project. However, a quantity takeoff for a cost estimate has to go a little further than a purely materials takeoff because additional information is often required for pricing.

For example, "160 square feet of ½" G1S ply" may be adequate for a material takeoff, but information about what the plywood is to be used for is required in a takeoff for a cost estimate. This is because the labor price of installation varies according to whether the material is to be applied to floors, walls, or soffits.

Also, there are a number of work items measured in a cost estimate takeoff that do not involve materials. For example, the item "Hand Troweling" has only a labor price associated with it; what is measured for this work item is the plan area of concrete to be troweled, there is no material to consider.

Takeoff Items and General Rules of Measurement

1. **Takeoff items** comprise two components:
   a. Dimensions that define the size or quantity of the item in accordance with the required units of measurement for that item; and
   b. A description that classifies the item for pricing.
2. Dimensions are entered onto the takeoff in the order length, width, and depth (or height).
3. Dimensions are written in feet to two decimal places, thus, 5'-10" would be written into a dimension column as 5.83.
4. If a dimension does not come directly from a drawing, the estimator shall provide evidence in the form of side calculations showing how the dimension was determined. This is done for even the simplest calculation.
5. Dimensions are rounded to the nearest whole inch. However, to avoid compounding rounding errors, fractions of an inch are not rounded in side calculations until the end result of the calculation is obtained.
6. Dimensions figured on the drawings, or those calculated from figured dimensions are used in preference to scaled dimensions. The estimator should scale drawings to obtain dimensions only as a last resort because drawings are not

always accurately drawn to scale. (But see discussion of the use of digitizers in Chapter 3.)

7. Deductions listed with the dimensions are written in red or are enclosed in brackets and are noted as deductions.

8. **Extensions** (the result of multiplying dimensions together) are calculated to the nearest whole number whether it is linear feet, square feet, or cubic feet.

9. Totals in the extensions column are rounded off to the nearest whole number.

10. Throughout the takeoff, headings are inserted to indicate such factors as the trade being taken-off, the location of the work under consideration, or the phase in which the work is classified. Side notes are also recommended to explain what is being measured, especially when the work is complex or unusual. All of these headings and notations help provide an audit trail so that the estimator or any other interested party can more easily review the takeoff.

11. Headings and side notes into the takeoff should not be overlooked when using a **computer estimating** system, especially if dimensions are entered into the computer directly from the drawings without making hand-written notes. The computer may be able to print a report that shows the takeoff items in the order in which they were measured, but on a large job without headings and notes in the takeoff, it will be almost impossible to review this data to determine which part of the project the items relate to and whether or not all the work of the project has been accounted for.

12. It is recommended that the estimator use highlighter pens to check off items of work as the takeoff progresses. This allows her/him to identify what has been measured and distinguishes what remains to be considered.

13. Takeoff descriptions contain sufficient information for the estimator to later price the work. Estimators use abbreviations extensively in takeoffs to increase the speed of the process. Detail that can easily be added at the pricing stage may also be omitted from descriptions at the time of the takeoff.

    To illustrate this point, consider a project that requires three mixes of concrete: mix A for foundations, mix B for columns, and mix C for all other concrete, and mix A consists of 3,500 psi concrete with type V cement and 4 percent air entrainment. The estimator will describe the use of the concrete in the takeoff in terms of footings, foundation walls, columns up to the second floor, and slabs-on-grade, but it wastes time to include a full mix description with each concrete takeoff item. The type of mix and its price has to be considered at the pricing stage, so why not leave the mix categorization until that time? See the examples of recaps in Chapter 11 for a description of how this mix classifying can easily be performed.

14. Generally, when describing a takeoff item, the estimator does not mention in the description or measure separately any of the following items because they will be dealt with when the takeoff items are priced:
    a. Transportation or any other costs associated with the delivery of the materials involved
    b. Unloading materials
    c. Hoisting requirements
    d. Labor setting, fitting, or fixing in position
    e. Lapping, cutting, or waste of materials
    f. Stripping formwork
    g. Form oil
    h. Rough hardware
    i. Scaffold and **falsework**

15. On special projects that require extensive or complicated scaffold systems, it may be necessary to take off the quantities of scaffold required. Otherwise normal scaffold requirements, and all the other items listed previously, are accounted for later in the pricing process.

16. The estimator should be aware that while all the above rules are in place to obtain the consistency and objectivity required, there is always an occasional situation where the work is not routine and a better result may be obtained if the rules are relaxed for once. To continue to be effective, an estimator has to preserve a certain flexibility of approach keeping in mind their major goal is to price the work of the project as efficiently as possible.

Accuracy of Measurement

The quantity takeoff should accurately reflect the amount of work involved in a project, but how accurately should the work be measured? There is no clear answer to this question since the level of accuracy pursued by the estimator depends on the costs and benefits of attaining high accuracy. Devoting extra time to improving the accuracy of the measurement of certain items of work may not be justified. All we can say is that the takeoff has to be as accurate as possible given the nature of the work being measured and the cost of attaining high accuracy.

To demonstrate this, consider the measurement of concrete work items. It is not difficult to calculate the quantities of concrete with a high degree of accuracy since concrete items are usually well detailed on contract drawings. Also, the time spent carefully measuring concrete work can be justified because items of concrete are relatively expensive. In contrast, excavation items are not detailed well (if at all) on drawings, and the unit prices of these items are usually quite low. With the excavation trade, the estimator usually has to ascertain the dimensions of the work by applying judgment based on experience; this results in an assessment that may be quite different from the volumes of work actually excavated. Clearly, there is little benefit in spending much time developing and carefully measuring what is only a theoretical impression of the excavation requirements of a project. However, the estimator still needs to make a reasonable evaluation of excavations because, even though the price per cubic yard may be as low as $10.00 when there are 1,000 cubic yards of material to excavate and haul away for a row of new houses, the total price of the work is significant.

Because of this, an estimator constantly has to balance the cost of achieving high accuracy against the value obtained from the increase in accuracy. This is not an easy task, particularly for those new to estimating. Our advice to the estimator who is in some doubt about whether the time spent improving the accuracy of a measurement is justified is to err on the side of caution and spend the extra time because a more accurate takeoff is always a better takeoff.

Takeoff Order

The order of the takeoff will generally follow the sequence of the work tasks of the project. However, some estimators find it preferable to measure concrete work, before excavation work even though excavation activities usually have to precede concrete work on the job. The reason for considering concrete work first is that the sizes and details of concrete items are clearly defined on the drawings, whereas excavation requirements have to be assessed. Therefore, measuring concrete first allows the estimator to become familiar with the project and, consequently, more efficient later at assessing the excavation requirements. As an estimator, you must decide which of these alternative approaches to adopt.

Takeoff Strategy and the Use of Assemblies

An estimator needs a strategy to deal with large projects. Without a systematic approach, the estimator can easily become lost in the takeoff process and reach a point where it is not clear exactly what has and what has not been measured. This can cause extreme frustration to the estimator and can lead to high stress and, in some cases, absolute panic. A simple yet effective strategy that can be used with any estimate consists of first dividing the project into manageable parts, proceeding through the takeoff one part at a time, and, within each of these parts, measuring the work as a sequence of assemblies. How to divide up the project is entirely dependent on its nature; high-rise jobs, for instance, are most easily divided floor by floor or by groups of floors. Where the project comprises a number of buildings, it is usually best to do the takeoff building by building. The estimator will often find that the contract drawings for large projects are conveniently divided into parts, so it makes sense to complete the takeoff of one part and then move on to the next.

Once the project is divided into manageably sized parts, assemblies are distinguished, within each of these parts. An **assembly** is a component of the work that can be considered in isolation from the other components of the work. The notion of an assembly will develop as the reader progresses through the takeoffs that follow. For now, though, consider the example of a perimeter foundation wall as an assembly in the estimate of a house. The basic idea is that the estimator measures all the work involved in an identified assembly, then moves on to consider the next assembly. So, in this example, the estimator may takeoff all of the following items associated with this one assembly before passing on to the next assembly:

1. The concrete in the wall
2. The forms to the sides of the wall
3. The forms to openings and blockouts in the wall
4. The rubbed finish on the exposed concrete of the wall
5. The reinforcing steel in the wall

The major advantages of taking off by assemblies are that the assembly is evaluated only once and that a number of items within the assembly can share the same dimensions. If this procedure was not adopted and the estimator was to measure all the concrete on the project and then all the forms followed by all the cement finishes and embedded miscellaneous items, there is a great deal of wasted time reassessing each assembly each time it is encountered. The same assembly dimensions would be referred to over and over again in this series of takeoffs and, possibly, recalculated each time they are needed.

Another advantage of this approach is that takeoff by assemblies, or "work packages," is used in many of the better computer estimating programs that produce a very powerful and speedy estimating tool. A student who becomes familiar with the concept and use of assemblies when learning estimating will be more adept where these features are encountered in computer programs and will be better able to develop and customize them to obtain maximum benefits from their use.

Also, the process of scheduling the project is made easier when assemblies have been used in the takeoff because the activity breakdown used by schedulers corresponds quite closely to the assemblies measured by the estimator. Some computer estimating packages can integrate with scheduling packages to allow the scheduler to directly transfer the estimate assemblies to the schedule.

Estimating Stationery (Standard Forms)

Specially printed forms designed for each of the various estimating procedures are used to increase productivity and contribute to the accuracy of an estimate. There are a number of different formats of these forms available to the estimator, the decision on which to adopt depends upon the method of estimating that is used. Figure 2.15 shows the form of **quantity sheet** used for quantity takeoffs in the following chapters, and Figure 2.16 shows the form of **pricing sheets** used.

In this book, the quantity takeoff is recapped onto separate sheets for pricing so one form of stationery is used for the takeoff, and a second for the recap. The process of recapping quantities onto a separate pricing sheet is adopted because it makes the task of pricing an estimate far more efficient since all related items are gathered together in a concise manner. To price the takeoff directly would entail a long and tedious process with a great deal of repetition, especially with estimates of larger projects. On the other hand, there may be some advantage to directly pricing the takeoff on small projects. Figure 2.17 shows an example of a combined takeoff and pricing sheet that could be used for this purpose.

Computer Estimating

Although the stationery described above was set up for estimating by hand, it can also be used on a computer spreadsheet; in fact, the takeoff and pricing examples that follow in the text are all prepared using **Excel spreadsheets**. This practice not only reduces the incidence of calculating errors but also greatly speeds up the estimating process, especially when the *copying* facility can be exploited. Once you have an Excel takeoff, recap, and summary completed, that is, linked together as demonstrated later in the text, you can use this template over and over for future estimates by merely adjusting the dimensions and prices to reflect specific requirements for each job. Note that the takeoff, pricing, and summary templates used in the text are all available online in a student companion to this text.

The use of specialized computer software can also greatly increase the efficiency of the estimating process. When using a program such as ICE from MC^2 Software, once the takeoff data is entered, the computer performs all the arithmetical functions needed to arrive at the required total net quantities; items are then automatically priced from the database, sorted, and summarized into the required format to arrive at the estimated cost of the project. Adding waste factors, converting takeoff quantities to order units, and all the other adjustments that the estimator would make to arrive at the final price of the work are handled by the program.

Especially in the case of larger projects, all this automation can leave estimators wondering exactly what they have done; the software does offer a number of ways of auditing a takeoff to facilitate checking of input to verify that all aspects of the job have been accounted for in the measurement process. Reviewing the estimate can be further enhanced by inserting headings and explanatory notes in the takeoff; this makes it much easier for a second estimator to quickly check for errors and/or

QUANTITY SHEET   SHEET No.

JOB . . . .   DATE . . . .

ESTIMATOR . . . .   EXTENDED . . . .   EXT. CHKD . . . .

DESCRIPTION	DIMENSIONS					
	TIMES					

Figure 2.15   Quantity Sheet Stationery

PRICING SHEET   SHEET No. [   ]

JOB . . . .   DATE . . . .

ESTIMATED . . . .

No.	DESCRIPTION	QUANTITY	UNIT	UNIT PRICE	LABOR	UNIT PRICE	MATERIALS	UNIT PRICE	EQUIP.	SUBS.	TOTAL

Figure 2.16   Pricing Sheet Stationery

ESTIMATE SHEET							Page No.							

JOB . . . .   DATE . . . .

ESTIMATED . . . .   EXTENDED . . . .   CHECKED . . . .

No.	DESCRIPTION	No. Pieces	DIMENSIONS			Extension	Total Quantity	UNIT PRICE	LABOR	UNIT PRICE	MATERIALS	UNIT PRICE	TOTAL

Figure 2.17 Combined Takeoff and Pricing Sheets

omissions in a takeoff. However, some estimators still like to prepare takeoff notes on paper before entering items and dimensions into the computer. These notes provide a record of how the estimator proceeded and give details of how the numbers entered into the computer were obtained. For example, the computer audit trail may confirm that a certain number was inputted as the length of a wall, but backup notes would be useful if a question arises about where the number came from for the length of this wall.

A fully worked example of an estimate prepared using MC² ICE software for a 4-unit bi-level building is included in Chapter 14.

SUMMARY

- The work of a project is measured in a number of different ways in an estimate:
  - By counting the number of pieces; for example, the number of doors
  - By measuring the length of an item, such as a pipe
  - By measuring an area of a material, such as sheathing
  - By measuring the volume of a material, such as concrete
  - By measuring the weight of materials, such as structural steel beams
- The estimator requires knowledge and skill in mensuration (measurement of lengths, areas, and volumes) to perform the many calculations involved when measuring work.
- As an estimator, you need to be able to calculate the perimeter and area of a variety of shapes and compute the volume of solid figures of various shapes.
- From the dimensions of a building, it is also useful for the estimator to accurately calculate the building perimeter that is used to determine the quantities of trench, footing, and wall around a building.
- A quantity takeoff is defined as a process of measuring the work of a project in the form of a list of quantified work items measured in accordance with a uniform set of rules.
- Note that quantities of materials are measured in the takeoff process, but the takeoff for a cost estimate is more than just a list of materials.
- There are two systems of measurement used in the North American construction industry: the traditional English units, and metric units; however, residential projects are for the most part designed using English units.
- Takeoff items comprise dimensions and a description; dimensions are always recorded in the order: length, width, and height.
- Descriptions contain sufficient information for the estimator to price the work.
- The quantity takeoff should accurately reflect the amount of work in the project. While high accuracy is possible with well-detailed concrete items, a margin of error has to be allowed with excavation work since it is never well detailed on project drawings.
- The order of the takeoff will generally follow the sequence of construction, although some estimators prefer to measure concrete work before an assessment of earthwork requirements is undertaken.
- A strategy is needed for completing a takeoff of larger projects:
  - Divide the project into manageable parts
  - Consider one part of the project at a time
  - Measure the work as a series of assemblies
  - Complete the takeoff for one assembly before passing on to the next
- Specially designed forms designed for each of the various estimating procedures is used to increase productivity and contribute to the accuracy of an estimate.
- The computer is an important estimating tool; it can greatly increase the efficiency of the estimating process and also lead to improved accuracy.
- Takeoffs prepared using computer software can be audited to check for errors and/or omissions, but "takeoff notes" are a useful supplement to record exactly how an estimate was prepared.

RECOMMENDED RESOURCES

Information	Web Page Address
■ Use a Web Search Engine such as "Ask" to get information about measuring areas and volumes.	http://www.ask.com
■ Much information about ICE estimating software can be obtained from the MC² web page.	http://www.mc2-ice.com/

REVIEW QUESTIONS

1. What units of measurement would you use to measure the following items of work?
   a. Floor sheathing
   b. Trench excavation
   c. Footing drains
   d. Duplex electrical receptacles
2. Define *quantity takeoff*.
3. What is the difference between a materials takeoff and a takeoff used for a cost estimate?
4. In what order are dimensions entered into a takeoff?
5. Why is an estimator advised to insert headings and side notes into a takeoff?
6. How accurate should a takeoff be?
7. What is the usual order followed in a takeoff? Also, what is a common exception to this?
8. What is the advantage to doing a takeoff by assembly?

PRACTICE PROBLEMS

1. If anchor bolts were required to be inserted 6" from the end of a wall and spaced at 36" on center along the length of the wall, how many bolts would be required for a 33'-0" long wall?
2. Calculate the area and the length of perimeter of each shape on Figure 2.18.
3. How many cubic yards of concrete is required for ten circular columns each 12'-0" high and 20" diameter?
4. Calculate the volume, in cubic yards, of a trench that is 240'-0" long and 6'-0" deep with cut-back sides so that the top width of the trench is 12'-0" and the bottom width is 3'-0".
5. Figure 2.19 shows the plan of a concrete wall that is 10'-0" high. Calculate the volume of concrete in this wall.

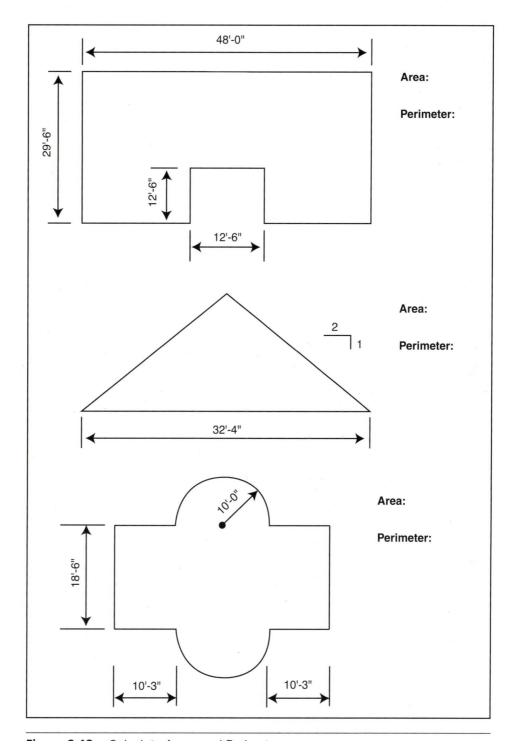

Figure 2.18   Calculate Areas and Perimeters

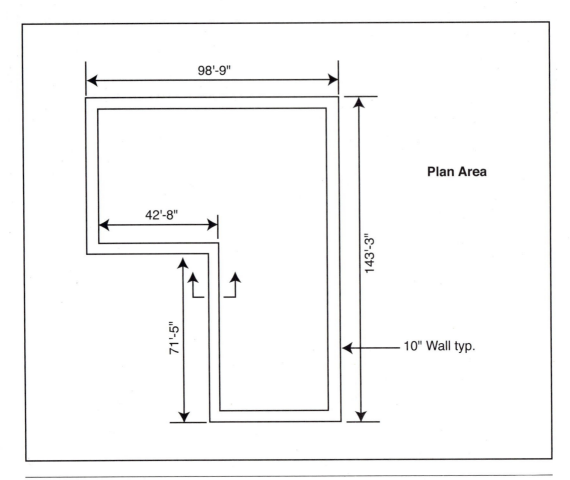

Figure 2.19 Plan of Wall

3

MEASURING EXCAVATION AND SITEWORK

OBJECTIVES

After reading this chapter and completing the review questions, you should be able to:

- Describe how excavation and other sitework are measured generally.
- Describe bank measure, and explain swell factors and compaction factors.
- Explain the potential hazards of excavation work, and describe the safety features adopted to deal with these dangers.
- Calculate volumes of excavation and backfill for basements, pits, and trenches.
- Describe the use of an electronic digitizer and its function in sitework takeoffs.
- Describe and use the "grid method" of calculating cut and fill volumes.
- Complete a takeoff of sitework for a residential project.

KEY TERMS

bank measure   plot plan

compaction factor   swell factor

digitizers

Introduction

Most homebuilders subcontract the excavation and sitework tasks required to construct a house so their estimators generally do not have to measure the work involved, but it is useful to be able to calculate the quantities of excavation and fill involved for a number of reasons:

1. In order to determine the amount of fill required, some or all of which may need to be imported if native excavated material is not suitable for backfill
2. To determine the amount of surplus soil to be removed from the site
3. To determine the duration of sitework and excavation tasks for a schedule
4. To compile an accurate cost estimate if you are the builder or subtrade performing this work

Measuring Sitework and Excavation Work Generally

Measuring sitework and excavation work is different from measuring most of the other work of a construction project because drawings show very little detail about the specific requirements of sitework or excavation operations. The drawings do provide details of the new construction required for the project, but information about what is currently to be found at the site of the proposed work is usually not indicated. On larger projects, builders are often advised in the "Instructions to Bidders" to satisfy themselves as to the present condition of the site. Furthermore, the size and depth of foundations may be defined in detail on the drawings, but there is typically nothing disclosed about the dimensions and shape of the excavations required to accommodate these constructions.

Before the estimator can measure the site work, she/he has to make an assessment of excavation requirements from whatever data there is on the drawings and from information gathered on visits to the site. Since the site conditions will also impact the pricing of other items in the estimate, the estimator should make use of a checklist of items to consider on a site visit so that important information is not overlooked. This is discussed in Chapter 1 under "Site Visits."

Bank Measure Swell and Compaction Factors

The soil that is extracted from an excavation is less dense than before it was excavated, so it will occupy more space than it did when it was in the ground. For example, if a hole of 10 cubic yards volume is excavated, the pile of soil removed from the hole might occupy 13 cubic yards; therefore, 13 cubic yards of material will have to be transported if it is required to be removed from the site. The difference between the volume of the hole and the volume of the material once it has been dug out is known as the **swell factor.**

A similar adjustment factor is required regarding filling operations. If an excavation of 10 cubic yards capacity is to be filled with gravel, as much as 14 cubic yards of loose gravel may be required because the material will have to be compacted. That is, it will be more dense after it is deposited and there will also be some waste of gravel to consider. Here the difference between the volume of the hole to be filled and the volume of fill material is referred to as the **compaction factor.**

In accordance with the general principle of measuring net quantities, excavation and backfill quantities are calculated using **bank measure.** Bank measure amounts are obtained by using the dimensions of the holes excavated or filled with no adjustment made to the quantities obtained for swell or compaction of materials. Swell factors and compaction factors will be accounted for later in the estimating process when items are priced.

Excavation Safety Considerations

The potential danger to workers in trenches and by the sides of excavations due to cave-ins of the earth embankments is a safety hazard that must be considered in every quantity takeoff of excavation work. The federal Occupational Safety and Health Administration (OSHA) Construction Safety and Health Regulations require that the sides of all earth embankments and trenches be adequately protected by a shoring system or by cutting back the sides of the excavations to a safe angle. As a consequence, the estimator must allow extra excavation for cutting back the face of excavations to a suitable angle wherever this is possible. Where restricted space or other circumstances prohibit cutbacks, the estimator may have to include a system of shoring and bracing, but this is usually an inferior choice because it is a far more expensive alternative. It is important that the estimator carefully study the

OSHA requirements for excavations since excavation rules are so strongly enforced by OSHA. The estimator will also need to consider particular state regulations regarding excavation safety requirements since specific details may vary from state to state. For instance, some states require that shoring systems over a certain depth be designed by and be constructed under the supervision of a professional engineer. This will have cost implications that cannot be ignored in the estimate.

Calculating Volumes of Excavation and Backfill for Trenches

A trench is an elongated excavation that is distinguished from a pit or a basement because it is excavated in a different way and usually with different equipment from that used for other types of excavation. Trenches are generally excavated using an excavator called a "backhoe" that sits astride the route of the trench. The machine digs material and casts it to one side as it moves backwards along the line of the trench.

As discussed in Chapter 2, because both the width and the depth of a trench may vary along its length, the volume of the trench is obtained by multiplying the centerline length of the trench by the average width and by the average depth:

Volume of Trench = Centerline Length × Average Width × Average Depth

Examples of Excavation and Backfill Calculations for Trenches

See Figure 3.1 for the details of a trench for a garden wall.

EXAMPLE 3.1

Average depth of trench = (5.0' + 7.0")/2
= 6.0'

Average width of trench:

Width of footing	3.0'
Add for work space	2.0' (this is 2 times 1'-0")
Add for cutback	6.0' (this is 2 times ½ depth of trench)
Total Width	= 11.0'
Volume of Excavation	= Centerline Length × Average Width × Average Depth
	= 100.0' × 11.0' × 6.0'
	= 6,600 cu. ft.
	= 6,600/27 cu. yds.
	= 244.4 cu. yds.

The best way to calculate the volume of backfill material required is to simply deduct the space taken up in the trench by the footing and the wall from the volume of excavation:

EXAMPLE 3.2

Volume of Backfill = Volume of Excavation − Volume of Footing − Volume of Wall
= 6,600 − (100.0 × 3.0 × 2.0) − (100.0 × 1.0 × 4.0)

(Note that only the portion of the wall that is in the trench is deducted, so the height of wall in the trench is the average trench depth minus 2'-0" for the footing: 6'-0"− 2'-0" = 4'-0".)
= 6,600 − 600 − 400
= 5,600 cu. ft.
= 5,600/27 cu. yds.
= 207.4 cu. yds.

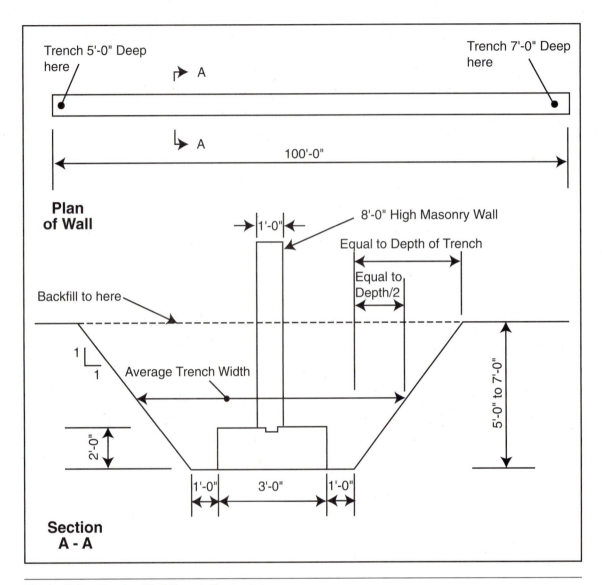

Figure 3.1   Garden Wall Trench Excavation

See Figure 3.2 for the details of a trench for a foundation wall.

EXAMPLE 3.3

Length of centerline	= (2 × 40.00') + (2 × 20.00') − (4 × 0.83')
	= 116.68'
Width of trench	= (2 × 1.00') + 2.00'
	= 4.00'
Average depth of trench	= (4.75' + 4.50' + 6.08' + 6.67')/4
	= 5.50'
Volume of Excavation	= Centerline Length × Average Width × Average Depth
	= 116.68 × 4.00 × 5.50

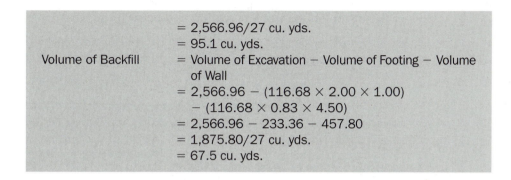

	= 2,566.96/27 cu. yds.
	= 95.1 cu. yds.
Volume of Backfill	= Volume of Excavation − Volume of Footing − Volume of Wall
	= 2,566.96 − (116.68 × 2.00 × 1.00) − (116.68 × 0.83 × 4.50)
	= 2,566.96 − 233.36 − 457.80
	= 1,875.80/27 cu. yds.
	= 67.5 cu. yds.

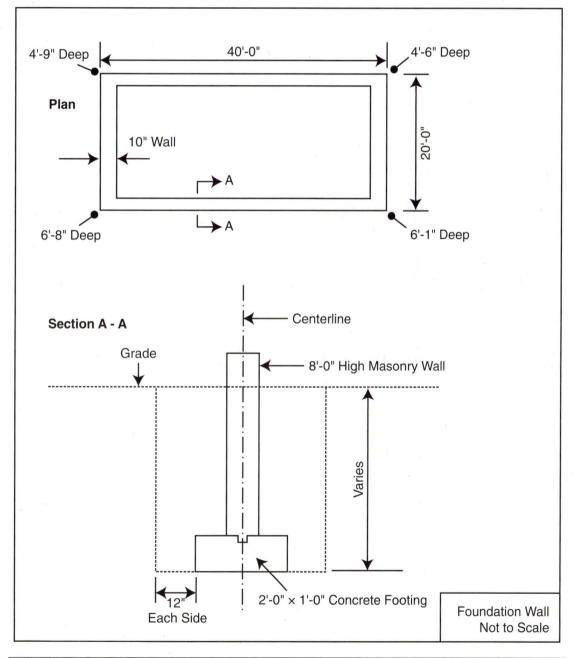

Figure 3.2   Foundation Wall Excavation

Calculating Volumes of Excavation and Backfill for Basements

A backhoe may be used to excavate small basements and pits, but larger holes are usually excavated using excavators called "front-end-loaders." These machines dig material in a scooping motion and may cast it to one side or load the soil onto trucks that remove it from the site.

If the ground elevation varies over the building area, the depth of the basement excavation will also vary. In this case, the estimator's first task is to determine the average depth from the existing grade down to the bottom of the excavation. Since most basements have a concrete slab-on-grade with a layer of gravel under this slab, basements are usually excavated down to the underside of this gravel. To obtain the average depth, measure the depth from the ground surface to the bottom of the gravel at each corner of the basement, and then calculate the average of these depths.

Similar to trenches, the sides of pits and basement excavations are usually cut back to a safe angle so that workers in the basement are not exposed to the dangers associated with the collapse of the sides of excavations. The easiest way to compute, at least roughly, the volume of this complicated shape of excavation is to use the average length and the average width in the calculations; this results in the following formula for basement excavations:

Volume of Basement = Average Length × Average Width × Average Depth

Examples of Excavation and Backfill Calculations for Basements

See Figure 3.3 for the details of a basement excavation for a house.

EXAMPLE 3.4

Average depth of excavation = (5.83' + 4.17' + 3.33' + 4.67')/4
                                                = 4.5'

Average length of excavation:
Length of building	40.0'	
Add for work space	2.0' (this is 2 times 1'-0")	
Add for cutback	4.5' (this is 2 times ½ depth of excavation)	
Total Length	= 46.5'	

Average width of excavation:
Width of building	28.0'	
Add for work space	2.0' (this is 2 times 1'-0")	
Add for cutback	4.5' (this is 2 times ½ depth of excavation)	
Total Width	= 34.5'	

Volume of Excavation   = Average Length × Average Width × Average Depth
                                     = 46.5' × 34.5' × 4.5'
                                     = 7,219 cu. ft.
                                     = 7,219/27 cu. yds.
                                     = 267.4 cu. yds.

To calculate the volume of backfill around the building, we once again deduct the space taken up by the building from the total volume of excavation.

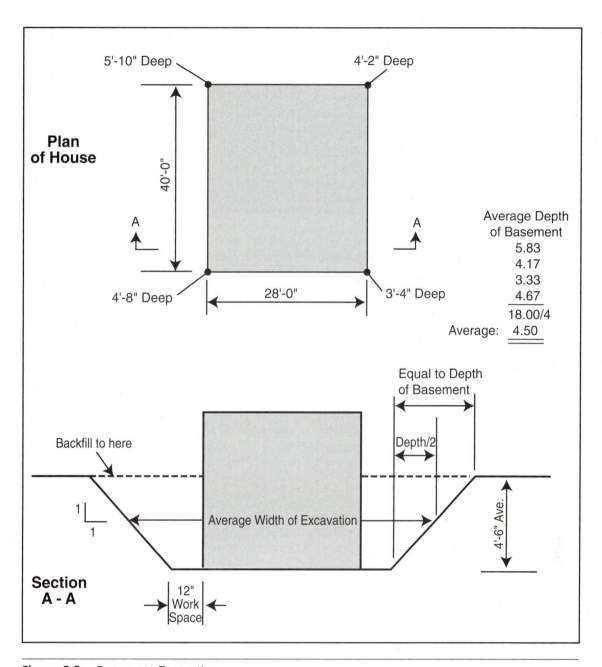

Figure 3.3   Basement Excavation

EXAMPLE 3.5

Volume of Backfill   = Volume of Excavation − Volume of Building
                      = 7,219 − (40.0 × 28.0 × 4.5)

(Note that only the portion of the building that is in the excavation is deducted, so the depth is 4'-6" which is the same depth as the excavation.)
                      = 7,219 − 5,040
                      = 2,179 cu. ft.
                      = 2,179/27 cu. yds.
                      = 80.7 cu. yds.

See Figure 3.4 for the details of a trench for a foundation wall.

EXAMPLE 3.6

Depth of excavation	$= (8.00' + 0.67') - 1.33'$
	$= 7.34'$
Average length of excavation	$= 50.00' + (2 \times 2.00') + 7.34'$ (lay-back)
	$= 61.34'$
Average width of excavation	$= 32.50' + (2 \times 2.00') + 7.34'$ (lay-back)
	$= 43.84'$
Volume of excavation	$= 61.34' \times 43.84' \times 7.34'$
(ignoring cutout)	$= 19,738.33$ cu. ft.
Volume of cutout	$= 24.50' \times 8.00' \times 7.34'$
	$= 1,438.64$ cu. ft.

(Note that lay-backs do not have to be added to the cutout dimensions because these have already been accounted for in the length and width of the overall excavation.)

Volume of excavation	$= 19,738.33 - 1,438.64$ cu. ft.
(less cutout)	$= 18,299.69/27$ cu. yds.
	$= 677.8$ cu. yds.
Volume of Backfill	$=$ Volume of Excavation $-$ Volume of Building
	$= 18,299.69 - (32.50' \times 25.50' \times 7.34')$
	$\quad - (24.50' \times 24.50' \times 7.34')$
	$= 18,299.69 - 6,083.03 - 4,405.84$
	$= 7,810.82/27$ cu. yds.
	$= 289.3$ cu. yds.

Use of Digitizers

Digitizers are electronic devices that enable the user to take measurements from drawings and input the data directly into a computer program. There are two main types of digitizer: sonic and tablet. Both of these types of digitizer employ a pointer or cursor to locate points and lines. With the sonic digitizer, the cursor emits a sonic code that is identified by two receivers that are used to calculate the precise location of the cursor. Using this system, any drawing can be scanned regardless of its size or the type of surface it is placed upon. The sonic receivers have be set up so that there is no obstruction between them and the cursor as it travels across the drawing, and the estimator has to ensure that measurements recorded are in accordance with the scale of the drawing. Clearly, drawings have to be drawn to scale for any type of digitizer to provide accurate data.

With tablet digitizers, drawings have to be laid out on an electronic tablet which functions to identify the specific location of the cursor as it moves over the surface of the tablet. Because of the size of construction project drawings, tablets as large as 42" × 60" and larger may have to be used if all the information on a drawing is to be accessible at one time.

The information gathered by the digitizer is then available for processing in computer programs so that lengths, areas, volumes, item counts, and sophisticated calculations such as cut and fill volumes can be determined very swiftly. Digitizers can also be linked to estimating software systems and used as an alternative to the keyboard for inputting the data into the system. Some estimating systems can

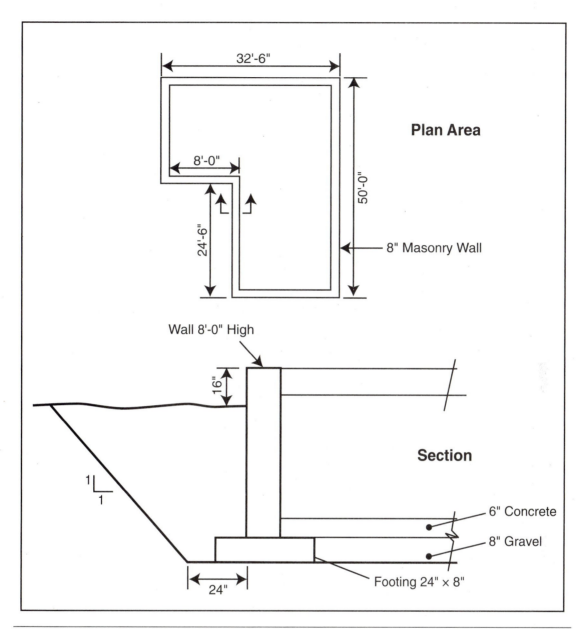

Figure 3.4   Basement Excavation 2

be operated directly from the digitizer, thereby eliminating any need to handle the keyboard. This setup provides a powerful tool for estimators because it allows large amounts of data to be quickly and accurately inputted into the computer without having to "key in" long lists of numbers.

One of the main disadvantages of using digitizers is that the accuracy of the system depends entirely upon the accuracy of the drawings that are scanned. A fundamental takeoff principle is that the estimator should only rely on scaled dimensions as a last resort, the reason being that when sizes are changed in the design process, these changes are often carried out by modifying the figured dimensions without changing the actual size of objects shown on the drawings to comply with their new dimensions. The result is that drawings wind up out of scale, and the digitizer system that scans these drawings generates erroneous output.

Cut and Fill Calculations

Certainly the fastest and probably the most accurate way to calculate volumes of cut and fill over a site when true scale drawings are available is to use an electronic digitizer in conjunction with a software program specifically for this type of application discussed previously. Alternatively, there are a number of "manual" methods of obtaining the quantities of cut and of fill. One such method uses a grid of elevations to provide the data necessary to make the calculations.

The Grid Method of Calculating Cut and Fill Quantities

Calculation by the "grid method" requires a survey of the site showing the elevation of the existing grade at each intersecting point on the grid. The elevation of the required new grade is also plotted at each intersection point. From these two elevations, the depth of cut or fill can then be obtained at each point. From here on, cut calculations are separate from fill calculations. To figure the volume of cut at an intersection point, the depth of cut at this point is multiplied by the area the intersection point applies to. By then adding together all of the individual cut volumes computed in this way, you will have the total volume of cut on the site. Following the same process using the fill depths will establish the individual and total fill volumes for the site. Figure 3.5 describes how grids are set up and how grid areas are used in this method of calculating cut and fill volumes.

The accuracy of this method of calculating cut and fill volumes depends upon the grid spacing. Generally the closer the grid spacing, the more accurate the results are. However, a closer grid spacing leads to more calculations and a longer processing time. Since this process is very repetitive, the processing time can be greatly reduced by using a computer program to perform the calculations.

For a complete calculation of volumes of cut and fill over a site using this "grid method" with a computer spread sheet program, see Figure 3.6.

Measuring Excavation and Sitework in a Takeoff

Here and in the sections that follow, rules of measurement are offered in order to provide a consistent method of measurement for estimators. This not only allows estimators to use and check each other's takeoffs more easily, but it also helps estimators work close together effectively on larger projects since following the same rules enables one estimator to know more readily what and how her/his colleagues are measuring in the estimating process.

Having a shared set of measurement rules also provides a more objective means of tracking productivities on a project. For instance, if different analysts were to track costs per cubic yard of excavations but each was measuring the work with different allowances for swell factors and compaction factors, there would be no way to compare the results of each analyst's work.

Landscaping and other site improvement work are sometimes required; this is most common on multi-unit projects or where the builder is responsible for an overall housing development. Landscaping is often subcontracted to specialized companies, but the homebuilder may occasionally perform the work using its own workforce, so we have included notes on how to measure site improvement operations, again to ensure uniformity in the preparation of takeoffs.

Setting Up the Grid:
— The grid lines labeled A, B, and C below run north-south and they are equally spaced.
— The grid lines labeled 1, 2, and 3 run east-west and they are equally spaced.
— Thus, the distance from line A to line B has to equal the distance from line B to line C.
— And, the distance from line 1 to line 2 has to equal the distance from line 2 to line 3.
— But, the spacing of the grid lines running north-south does not have to be the same spacing as the east-west grid lines.

Grid Areas:
— The "grid area" is the area bounded by four grid lines.
— Thus, the area bounded by lines 1, 2, A, and B is one "grid area."
— Note that there are four "grid areas" 1-2-A-B, 1-2-B-C, 2-3-A-B, and 2-3-B-C.
— Also note that each of these "grid areas" will have the same size.
— Thus, if the north-south grid spacing is 10 feet and the east-west grid spacing is 12 feet, then each "grid area" will be 120 square feet.

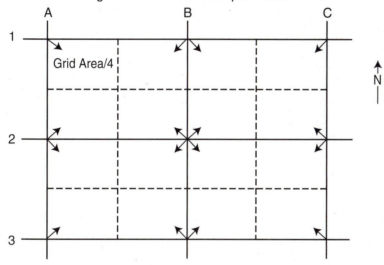

Intersection Points (Stations):
— The point where two grid lines intersect is called a "station."
— The arrows at each station indicate the area or areas that the station applies to.
— Thus, station 1-A applies to one area and the value of this area = "grid area"/4
— And, station 1-B applies to two areas, each of which has a value of "grid area"/4 etc.
— The term "frequency" is used in the calculations that follow; this refers to the number of areas a particular station applies to.
— So, in this diagram, these are the stations, their respective frequencies, and the area constant which is the grid area (120 square feet) divided by 4:

Station	Frequency	Area Constant
1-A	1	30
1-B	2	30
1-C	1	30
2-A	2	30
2-B	4	30
2-C	2	30
3-A	1	30
3-B	2	30
3-C	1	30

Figure 3.5 Grid Stations and Frequencies

The following data is shown at each intersection point (station) on the grid:

Elevation of New Grade	Elevation of Existing Grade
Depth of Cut	Depth of Fill

Plan of Site (Grid 20' x 16')

	A		B		C		D	
1	4.2	6.5	4.4	5.0	4.6	3.0	4.8	1.9
	2.3			0.6		1.6		2.9
2	4.4	5.1	4.6	3.2	4.8	2.8	5.0	4.5
	0.7			1.4		2.0		0.5
3	4.6	3.6	4.8	2.0	5.0	5.3	5.2	7.1
		1.0		2.8	0.3			1.9
4	4.8	1.9	5.0	4.0	5.2	8.2	5.4	10.0
		2.9		1.0	3.0			4.6
5	5.0	3.0	5.2	3.8	5.4	6.4	5.6	7.0
		2.0		1.4	1.0			1.4

Tabulation of Results:

Station	New Elev.	Existing Elev.	Depth Cut	Depth Fill	Frequency	Area Constant	Volume Cut	Volume Fill
1A	4.2	6.5	2.3	0.0	1	80	184	0
1B	4.4	5.0	0.6	0.0	2	80	96	0
1C	4.6	3.0	0.0	1.6	2	80	0	256
1D	4.8	1.9	0.0	2.9	1	80	0	232
2A	4.4	5.1	0.7	0.0	2	80	112	0
2B	4.6	3.2	0.0	1.4	4	80	0	448
2C	4.8	2.8	0.0	2.0	4	80	0	640
2D	5.0	4.5	0.0	0.5	2	80	0	80
3A	4.6	3.6	0.0	1.0	2	80	0	160
3B	4.8	2.0	0.0	2.8	4	80	0	896
3C	5.0	5.3	0.3	0.0	4	80	96	0
3D	5.2	7.1	1.9	0.0	2	80	304	0
4A	4.8	1.9	0.0	2.9	2	80	0	464
4B	5.0	4.0	0.0	1.0	4	80	0	320
4C	5.2	8.2	3.0	0.0	4	80	960	0
4D	5.4	10.0	4.6	0.0	2	80	736	0
5A	5.0	3.0	0.0	2.0	1	80	0	160
5B	5.2	3.8	0.0	1.4	2	80	0	224
5C	5.4	6.4	1.0	0.0	2	80	160	0
5D	5.6	7.0	1.4	0.0	1	80	112	0
							2,760 cu. ft.	3,880 cu. ft.
							102.2 cu. yd.	143.7 cu. yd.

Summary:
Bulk Cut	-	102.2 cu. yds.
Bulk Fill	-	143.7 cu. yds.
Import Fill Material	-	41.5 cu. yds.
Dispose Surplus	-	0 cu. yds.

Figure 3.6 Cut and Fill Excavation by the Grid Method

Measuring Notes

Excavation and Backfill

1. Measure excavations, backfill, and fill material in cubic yards "bank measure," that is, using the actual dimensions with no allowance for swell factors or compaction factors of materials.
2. Classify and measure separately excavations in the following categories:
   a. Site clearing
   b. Excavation over site to reduced levels
   c. Basement excavations
   d. Trench excavations
   e. Pit excavations
3. If different types of materials are to be excavated in any of the above categories, describe and measure separately each particular material.
4. Measure hand excavation separately.
5. Describe and measure separately different types of fill and backfill materials within each of the above categories.
6. Classify and measure separately fill and backfill materials in the following categories:
   a. Fill over site to raised levels
   b. Backfill to basements
   c. Backfill to trenches
   d. Backfill to pits
   e. Gravel under slabs on grade
   f. Gravel base course

Site Improvements

7. Measure concrete paving, sidewalks, curbs, and gutters in accordance with concrete work requirements (see Chapter 4).
8. Describe asphalt paving and measure in square feet.

Fences

9. Describe fences stating height and measure in linear feet.
10. Describe and enumerate gates.

Mulching, Seeding, and Sodding

11. Describe mulching material stating required thickness and measure in square yards.
12. Describe method of seeding and measure in square yards.
13. Describe sodding and measure in square yards.

Plants, Shrubs, and Trees

14. Describe and enumerate plants, shrubs, and trees.

Site Utilities

Water Lines

15. Enumerate connecting water lines to the main.
16. Describe water pipes stating the size and measure in linear feet.
17. Describe and enumerate valves and fittings.

Sewer Lines and Septic Tanks

18. Enumerate connecting the sanitary sewer to the main.
19. Describe sewer pipes stating the size and measure in linear feet.
20. Describe stating the required capacity and enumerate septic tanks.

Sample Drawings

Figure 3.7 is a set of drawings for a project comprising a single-family home. The takeoff examples in this chapter and in the chapters that follow are taken from these sample drawings. These drawings can also be found in Appendix D for quick reference.

The first of these drawings is a **plot plan** showing the size and shape of the property upon which the house is to be constructed. Land surveyors prepare plot plans to establish the location of property lines around the building, then, once the building is constructed, these plans are used to confirm the position of the building on the site. Plot plans are usually required when applying for a building permit and may also be required by the lender when setting up a mortgage on the property.

In addition to the dimensions of the site, the plot plan also shows:

- The property lines
- A North arrow
- The location of the building
- The outside dimensions of the building
- The setback of the building from the property lines
- The elevation of the main floor of the building relative to the ground elevations around the building
- The location of the underground utilities to the building

The remaining drawings show the design details of the house to be constructed.

Example of Sitework Takeoff

Figure 3.8a and Figure 3.8b show the takeoff for the sitework on the sample house project described on drawings shown in Figure 3.7.

Comments on the Sitework Takeoff Shown in Figure 3.8a and Figure 3.8b

1. This particular site has no topsoil to be stripped, so the first calculation is for the depth of basement to be excavated.
2. The average ground level over the area of the house is calculated from the elevations at the four corners of the house.
3. The depth of the basement excavation will extend from the ground level over the area of the house down to the level of the underside (u/s) of the gravel below the basement slab. The elevation of the underside of the gravel is not provided on the drawings, so we have to determine it from the information given:

The elevation of top of the main floor (this is indicated on the site plan)	100'-0"
Less: top of foundation wall to top of sub-floor	(1'-9¾")
Less: top of footing to top of foundation wall	(7'-1¾")
Less: depth of gravel	(-8")
	90'-4½"

This is equal to 90.38 feet.

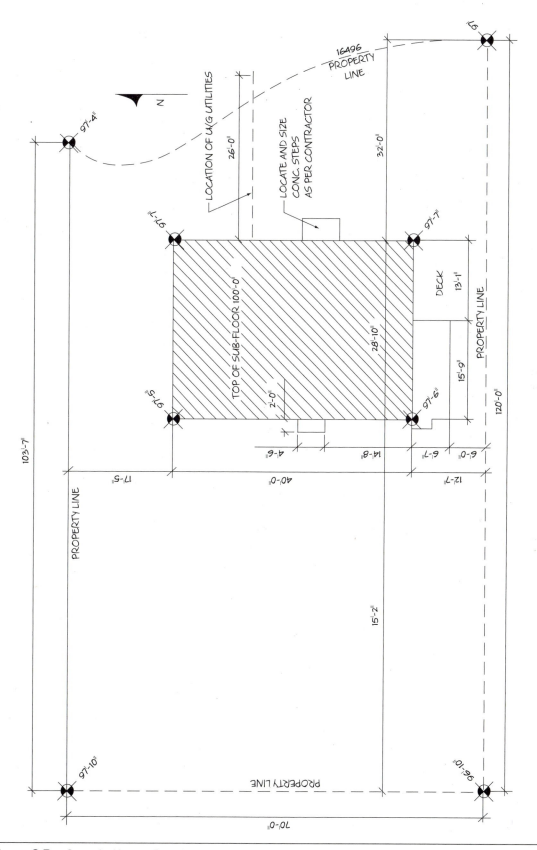

Figure 3.7   Sample House Drawings

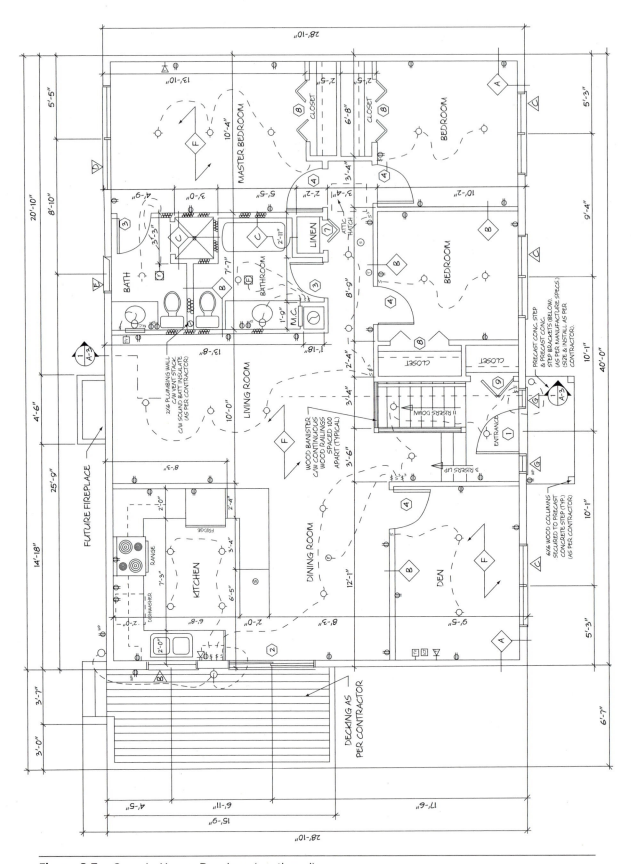

Figure 3.7   Sample House Drawings (continued)

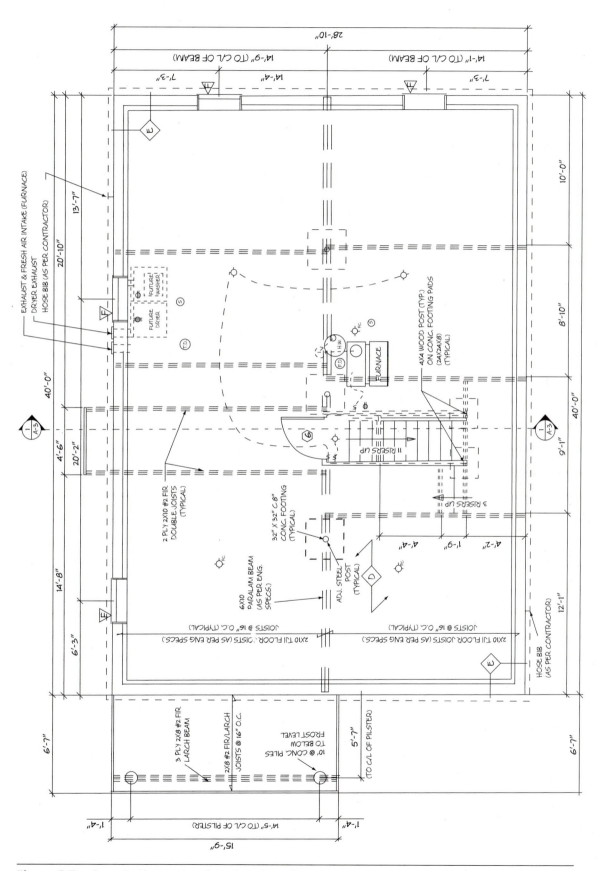

Figure 3.7   Sample House Drawings (continued)

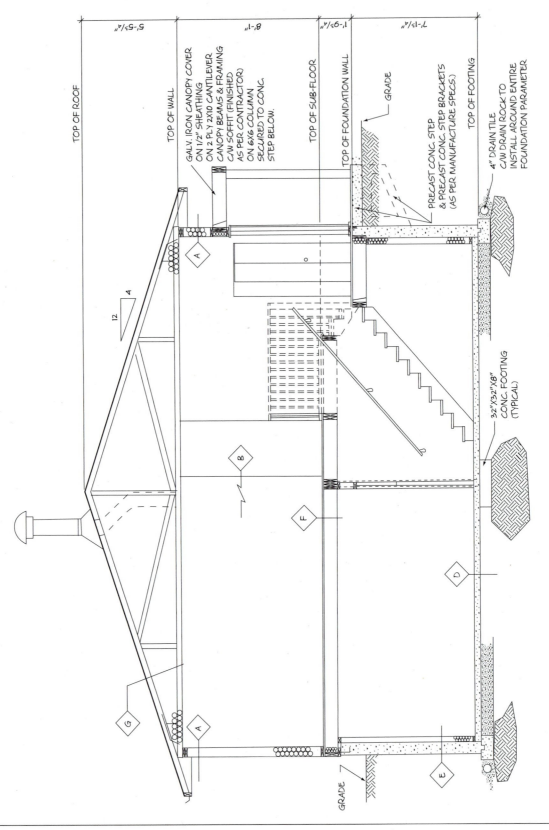

Figure 3.7 Sample House Drawings (continued)

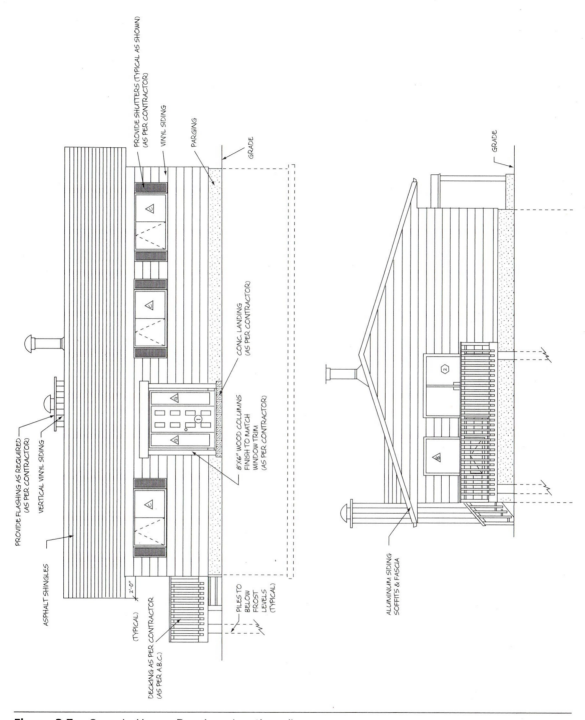

Figure 3.7   Sample House Drawings (continued)

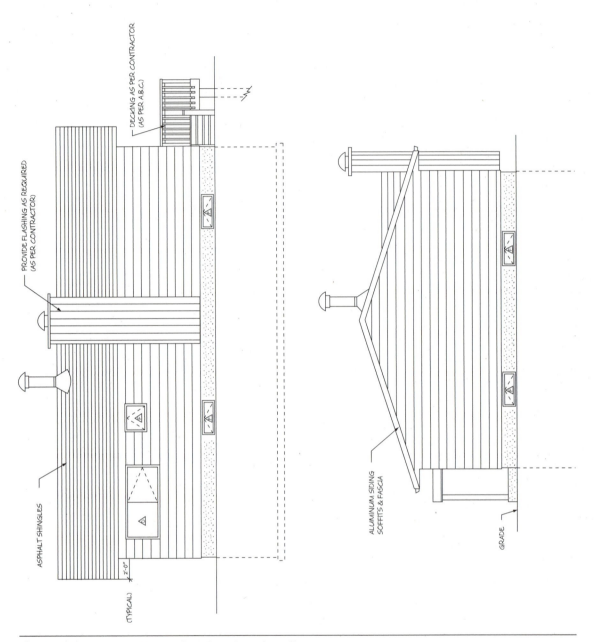

Figure 3.7   Sample House Drawings (continued)

WINDOW SCHEDULE

UNIT:	UNIT SIZING:	R.O. (LXH):
A	NOT IN CONTRACT	
B	36"X50" (F)	42"X72"
	36"X15" (A)	
C X3	27"X30" (A-F)	64"X37"
D	36"X30" (A-F)	83"X37"
E	27"X22" (A)	33"X29"
F X4	36"X24" (A)	37"X13"
G X2	12"X6'4" (SEALED) LIGHTS	67"X84" WHOLE DOOR FRAME

UNIT:	DOOR SCHEDULE	LOCATION:
1	3'X6'8"X1" (C/W SIDE LIGHTS)	F-ENTRANCE
2	5'X6'8" PATIO DOOR	NOOK
3	2'4"X6'8"X1"	ENSUITE & BATH
4	2'6"X6'8"X1"	BEDROOMS
5	NOT IN CONTRACT	
6	2'10"X6'8"X1"	BASEMENT
7	2'0"X6'8"	BI-FOLD
8	4'0"X6'8"	BI-FOLD
9	3'0"X6'8"	BI-FOLD

ELEC. LEGEND

S	SINGLE POLE SWITCH
S³	3-WAY POLE SWITCH
S_F	FURNACE SWITCH
⊖	DUPLEX CONVENIENCE OUTLET
⊖_R	ELECTRIC RANGE
⊖_D	ELECTRIC CLOTHES DRYER
⊖_R	RAZOR OUTLET
⊖_WP	WEATHER PROOF OUTLET
⊘	CEILING LIGHT
⊘_PC	PULL CHAIN LIGHT
◯	WALL LIGHT
F	CEILING FAN
F	BATHROOM EXHAUST FAN
T	THERMOSTAT
S	SMOKE ALARM
ICT	INTERNET CABLE
TV	TELEVISION OUTLET
▷	TELEPHONE JACK

G ROOF CONSTRUCTION
ASPHALT SHINGLES ON
1/2" O.S.B. SHEATHING c/w "H" CLIPS
ENG. APPROVED FINK TRUSSES @ 16" O.C.
R12 BATT INSUL. OR BETTER
6MIL. POLY. VAP. BARR.
1/2" DRYWALL CEILING
PROVIDE INSULATION STOPS AT WALL (TYP.)

F FLOOR CONSTRUCTION
FLOOR FINISH (AS PER CONTRACTOR)
UNDERLAY (WHERE REQUIRED AS PER CONTRACTOR)
3/4" T&G SUB FLOOR (GLUED & SECURED TO JOISTS)
2X10 TJI FLOOR JOISTS @ 16" O.C.
(OR AS PER ENG. SPECS).
NOTE: INSTALL R12 INSULATION BATT.
     @ TOP OF FOUNDATION WALLS
     ALONG INSIDE FACE OF RIM JOISTS
     C/W 6MIL. VAPOUR BARRIER
ALL STRAPPING & CROSS BRIDGING
(AS PER ENG. SPECS).
W/ 1/2" GYPSUM BOARD (BASEMENT CEILING)

A EXTERIOR WALL CONST (NEW & EXISTING)
(SIDING BY CONTRACTOR)
(BUILDING PAPER/SHEATHING MEMBRANE)
1/2" O.S.B. SHEATHING
2X6 WALL STUDS @ 16" O.C.
R20 BATT. INSUL. OR BETTER
6 MIL. POLY. VAP. BARR.
1/2" DRYWALL

B INTERIOR WALL CONST.
2X4 STUDS @ 16" o.c.
1/2" DRYWALL BOTH SIDES

C INTERIOR BATHROOM WALL CONST.
2X4 STUDS @ 16" o.c.
1/2" WATER PROOF DRYWALL.
(ALL AROUND TUB AREA
INCLUDING ANY MARKED
EXTERIOR WALLS).

D BASEMENT FLOOR SLAB
(AS PER CONTRACTOR, APPROVED BY ENG.)
4" MIN. CONC. SLAB
C/W 6MIL. POLY VAP. BAR.
ON 8" MIN. COMPACTED GRAVEL FILL
(REBAR OR WIRE MESH AS REQUIRED)

E FOUNDATION WALL CONST.
2X6 SILL PLATE (SECURED BY
LAG BOLTS INTO CONC. WALL,
C/W GASKETS TO MATCH) ON
8" WIDE CONCRETE WALL STRUCTURE:
C/W DAMPROOFING FROM ABOVE GRADE
ON OUTSIDE OF WALL
& PARGING ABOVE GRADE TO UNDER WALL SIDING
2-10MM REBAR TOP & BOTTOM.
ON 16"X8" FOOTING (C/W KEYWAY)
W/ #4 REBAR TOP & BOTTOM.
W/ WEEPING (DRAIN) TILE & ROCK
TO SURROUND EXTERIOR FACE OF FOOTINGS.
(ENTIRE PERIMETER OF STRUCTURE)
.10MIL POLY MOISTURE BARRIER
     (OUTSIDE FACE) TO ABOVE GRADE
C/W INTERIOR FURRING WALL CONSTRUCTION:
     2X4 WALL STUDS @ 16" O.C.
     (INCLUDING TOP & BOTTOM PLATES)
     W/ R12 INSULATION

Figure 3.7   Sample House Drawings (continued)

QUANTITY SHEET					SHEET No.	1 of 2	
JOB: House Example					DATE:		
ESTIMATOR: ABF			EXTENDED:		EXT. CHKD:		

DESCRIPTION	TIMES	Length	Width	Height			
Excavation and Backfill							
Elev of *Elev of*							
Ave Grade *U/S of Gravel*							
97.42 Main Floor: 100.00							
97.58 *Less*							
97.58 1'-9³/4"							
97.50 7'-1³/4"							
Ave: 97.52 -8" (9'-7¹/2")							
90'-4¹/2"							
(90.38) ← = 90.38							
7.14 Depth of Excav.							
40.00 x 28.83							
W/S 2 x 2.0 4.00 4.00							
C/B 2 x ¹/2 x 7.14 7.14 7.14							
51.14 39.97							
EXCAV BSMT		51.14	39.97	7.14	14,595		
					541	CY	
BACKFILL BSMT		as excav.			14,595		
DDT		40.00	28.83	7.14	(8,234)		
					6,361		
					236	CY	
4" Dia FOOTING DRAIN		145.66			146	LF	
2 x 40.00 80.00							
2 x 28.83 57.66							
137.66 Outside wall perimeter							
4 x 2 x 1.00 8.00							
145.66 Centerline of pipe							
DRAIN GRAVEL		145.66	2.00	2.00	583		
					22	CY	
DEDUCT BACKFILL BSMT		Ditto			(22)	CY	

Figure 3.8a   Excavation and Backfill Takeoff

QUANTITY SHEET   SHEET No. | 2 of 2 |

JOB: House Example   DATE:

ESTIMATOR: ABF   EXTENDED: EXT. CHKD:

DESCRIPTION	TIMES	Length	Width	Height			
Utility Services Trench							
Ave Depth _Ave Width_							
5.00 Bottom: 2.50							
6.00 Cutback: 5.50							
Ave: 5.50 8.00							
EXCAV TRENCH		26.00	8.00	5.50	1,144		
					42	CY	
SAND BEDDING		26.00	3.00	0.50	39		
Ave Width					1	CY	
Bottom: 2.50							
Cutback: 0.50							
3.00							
BACKFILL TRENCH		as excav.			1,144		
DDT		sand			(39)		
					1,105		
					41	CY	

4. Then the difference between the average ground level and the elevation of the bottom of the gravel gives the depth of the basement excavation:

$$97.52'$$
$$(90.38')$$
Depth of excavation: $7.14'$

5. The length and width of the basement excavation is based on the size of the house, so we start with the dimensions of the house to the outside of the walls shown on the basement plan.

6. A 2'-0" wide workspace is allowed outside of the foundation walls, and an allowance is also made for cutting back the sides of the excavation. (See Figure 3.9.)

7. The volume of backfill is equal to the volume of excavation less the space taken up by the building. (The space taken up by the footing projection is ignored here.)

    Note that the abbreviation "DDT" indicates a deduction, and the amount of the deduction is placed in brackets.

8. A footing drain is located 1'-0" outside of the foundation wall, so the length of this drain is calculated from the outside perimeter of the house:

Plus the adjustment for corners:	137.66'
4 corners $\times$ 2 $\times$ 1.0' at each corner gives:	8.00'
Centerline perimeter:	145.66'

9. The footing drain is required to be surrounded by drain gravel, so an area 2'-0" $\times$ 2'-0" is allowed around the pipe.

10. The space taken up by drain gravel reduces the amount of basement backfill required, so an adjustment is made to account for this.

11. A Utilities Trench is a trench that is excavated to accommodate building utilities such as sewer, water, and electrical lines. Figure 3.10 gives details of what the estimator allows for when measuring this trench work.

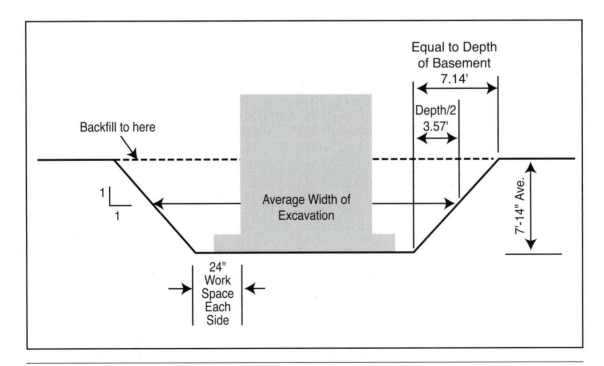

Figure 3.9   Cross-Section Through Basement Excavation

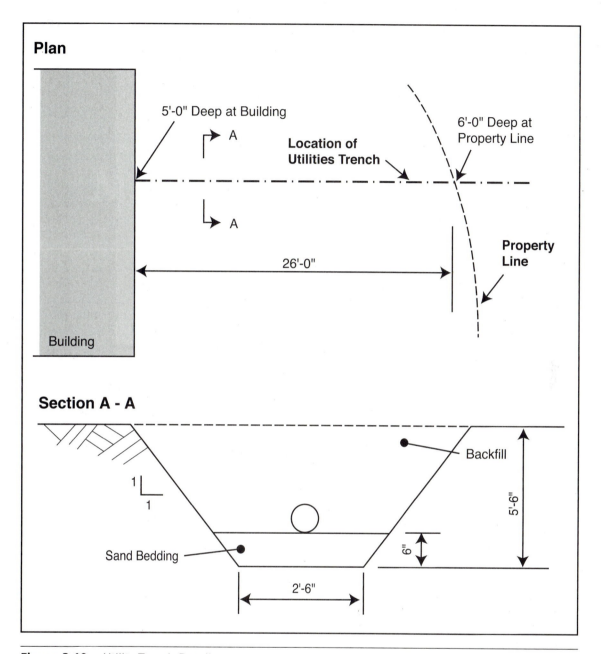

Figure 3.10   Utility Trench Details

SUMMARY

■ Even though most homebuilders subcontract the excavation work on a project, it is useful for the estimator to be able to take off quantities of excavation and sitework.

■ The estimator has to make an assessment of sitework requirements based on what is shown on the plans and in the specifications, but there is usually very little detail of excavation requirements provided on the bid drawings.

■ The estimator's site visit will be another useful source of information about sitework requirements for the project.

- Because soil expands when it is excavated, a swell factor is used to calculate the increased volume of material after it has been excavated.
- Similarly, a compaction factor is used to calculate the total amount of material required to backfill excavations with compacted soils and/or gravels.
- Excavation and backfill quantities are measured bank measure with no allowance for swell factors or compaction factors at the time of the takeoff.
- It is important that the estimator to be familiar with safety requirements for excavations to ensure that the sitework takeoff reflects these requirements.
- The following formulas are used to calculate the amount of excavations:
  - Volume of Trench = Centerline Length × Average Width × Average Depth
  - Volume of Basement = Average Length × Average Width × Average Depth
- The efficiency of the sitework takeoff, especially cut and fill operations, can be greatly increased by using a digitizer, an electronic device that enables you to take measurements from drawings and input this data directly into a computer program.
- The "grid method" may be used to calculate volumes of cut and fill over a site. This requires a survey showing the elevation of the existing grade at each intersection point on the grid. The elevation of the required new grade is also plotted at each intersection point and, from these two elevations, the depth of cut or fill can then be obtained at each point. From here on, cut calculations are separate from fill calculations. To figure the volume of cut at an intersection point, the depth of cut at this point is multiplied by the area "covered" by that intersection point. Then, by adding together all of the individual cut volumes computed in this way, you will have the total volume of cut on the site.
- In general, items of excavation, backfill, and fill materials are measured in cubic yards or cubic meters bank measure.
- Some of the information required to measure sitework is found on plot plans that show the size and shape of the property upon which the house is to be constructed.
- A takeoff of sitework for a house project is demonstrated.

RECOMMENDED RESOURCES

Information	Web Page Address
■ See the OSHA web site for information about construction safety.	http://www.osha.gov/
■ There is a great deal of information on the web about digitizers—Tenlinks is one useful site.	http://www.tenlinks.com/ Technology/Computers/ HARDWARE/INPUT/DIGIT.HTM
■ To see construction equipment in action, YouTube is a great resource. The link listed here shows a backhoe excavating a basement for a house.	http://www.youtube.com/watch?v= fH4lgQdQL_4&feature=related

REVIEW QUESTIONS

1. Explain why it is useful for an estimator of residential projects to have some skills in excavation takeoff.
2. Why is it generally more difficult to measure excavation work than it is to measure other work on a project?
3. How do swell factors and compaction factors affect the measurement of excavation and backfill?

4. What safety features does the estimator have to account for in connection with trench excavations?
5. How can digitizers help in an excavation takeoff?
6. How does the estimator account for surplus material left over from excavation and backfill operations?

PRACTICE PROBLEMS

1. Calculate the volume of excavation and backfill for the basement shown in Figure 3.11.
2. Calculate the volume of cut and fill for the site shown in Figure 3.12.
3. Calculate the volume of excavation and backfill for the trench shown in Figure 3.13.
4. Calculate the volume of cut and fill for the site area shown in Figure 3.14.

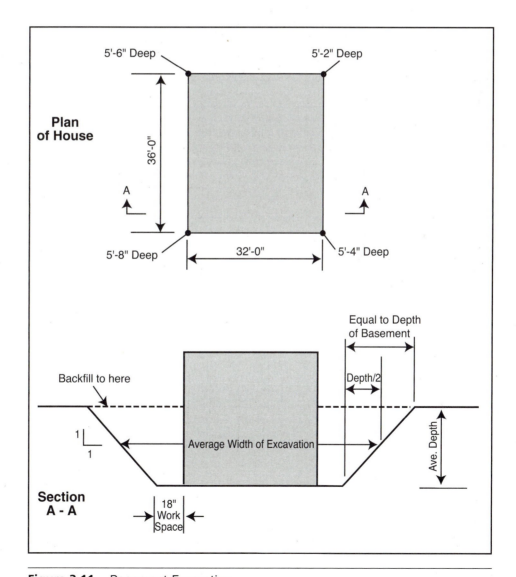

Figure 3.11   Basement Excavation

The following data is shown at each intersection point (station) on the grid:

Elevation of New Grade | Elevation of Existing Grade

Plan of Site (Grid 25' x 20')

	A		B		C		D	
1	5.0	6.5	5.5	5.0	6.0	3.0	6.5	1.9
2	5.2	5.1	5.7	3.2	6.2	2.8	6.7	4.5
3	5.4	3.6	5.9	2.0	6.4	5.3	6.9	7.1
4	5.6	1.9	6.1	4.0	6.6	8.2	7.1	10.0
5	5.8	3.0	6.3	3.8	6.8	6.4	7.3	7.0

Figure 3.12 Cut and Fill Excavation by the Grid Method

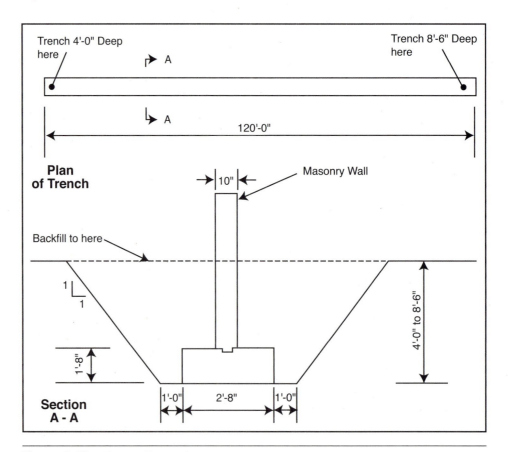

Figure 3.13 Trench Excavation

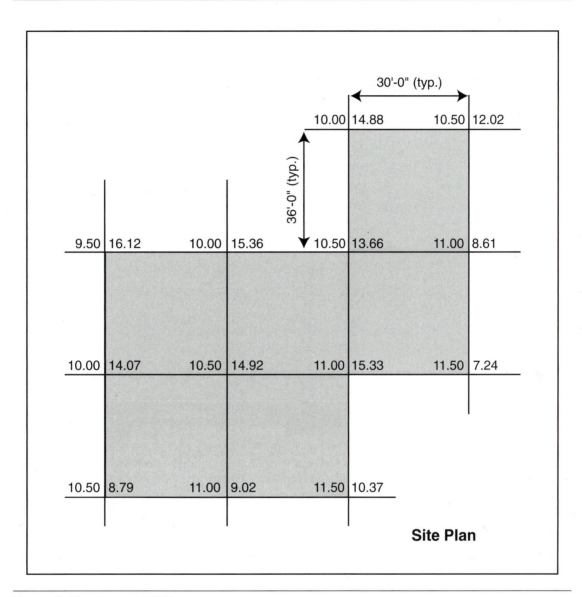

Figure 3.14   Cut and Fill Plan

4

MEASURING CONCRETE WORK

OBJECTIVES

After reading this chapter and completing the review questions, you should be able to:

- Describe how concrete work is measured in a takeoff by assembly.
- Measure concrete items from drawings and specifications.
- Make an assessment of formwork requirements from the details of concrete provided on drawings.
- Describe how concrete finishes are measured.
- Describe how welded wire mesh is measured.
- Describe how reinforcing steel is measured.
- Complete a takeoff of concrete work for a residential project.

KEY TERMS

blockouts	net in place	slab-on-grade
cutouts	pilasters	suspended slab
formwork	reinforcing steel (rebar)	welded wire mesh

Introduction

Preparing a quantity takeoff of concrete work requires the estimator to measure a combination of items, some of which are detailed on the drawings while others have to be inferred from what the drawings disclose. When drawings are well prepared, items of concrete such as footings, walls, and columns are clearly shown, but details of the **formwork** required for this work are not provided on the drawings. The assessment of formwork requirements will be based on the estimator's knowledge of what is required for each of the different concrete components. Similarly, concrete finishes and items such as curing slabs are not generally explicit on the drawings, so the estimator must add their knowledge of construction work to what is shown on the drawings in order to determine the extent of these implicit items.

Because the concrete is detailed on the drawings, it makes sense to begin the takeoff by measuring the volume of concrete in an item, then, after the concrete dimensions are defined, consider the formwork requirements followed by the finishes that are needed, etc. Also, in accordance with the basic principles previously discussed regarding taking off by assembly, a good practice is to measure all work associated with one concrete assembly before considering the next assembly.

For example, if there were concrete footings, walls, and columns to consider on a project, we could begin with the footings. First find the dimensions of footing concrete from the drawings, and then reuse these dimensions to calculate the area of formwork and the length of keyway required. Conclude this assembly by measuring anything else associated with the footings. After the work on the footings is measured, we would turn our attention to the walls and measure all the items associated with them. Finally, we would deal with the columns in the same way.

This approach allows the estimator to focus on one type of work item at a time and fully understand all of its requirements before moving on to the next item. Alternative approaches, such as measuring all the concrete for items then measuring all the formwork for these items, require the estimator to be constantly jumping from one type of item to another. Preparing a takeoff in this way requires the estimator to come back to an item such as footings three or four times, which is not an efficient way to proceed.

Measuring work by assembly also corresponds to the way most estimating softwares function. The estimator enters the dimensions of the concrete for an assembly such as a **slab-on-grade** and then the program, usually by means of menus, allows the estimator to select such things as the type of forms, the kind of reinforcing, the method of finishing, and so forth, required for this slab. As these selections will vary little from job to job, most softwares allow you to save the selections, so that the estimator on subsequent jobs needs only to enter the size of the concrete and merely by *pressing a button* can include all the additional requirements.

Measuring Notes—Concrete

1. Measure concrete in cubic yards **net in place.** Calculate the volume of concrete from the dimensions given on the drawings with no adjustment for "add on" factors. Additional material required because of spillage, expanding forms, and wastage is accounted for later in the pricing process by means of a waste factor added to concrete items.
2. Do not adjust the quantity of concrete for **reinforcing steel** and inserts that displace concrete. Also, do not deduct for openings in the concrete that are less than 1 cubic foot of volume.
3. Classify concrete and measure separately in the following categories:
   a. Pile caps
   b. Footings
   c. Retaining walls
   d. Grade beams
   e. Columns and pedestals
   f. Beams

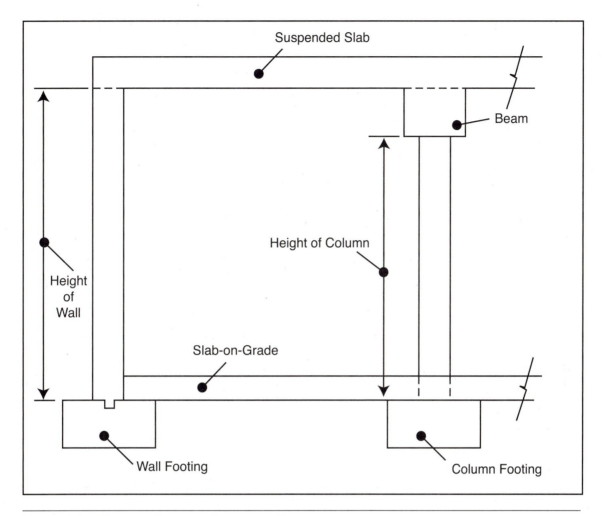

Figure 4.1   Height of Concrete Walls and Columns

       g. Slabs on grade
       h. **Suspended slabs**
       i. Floor toppings
       j. Stairs and landings
       k. Curbs
       l. Sidewalks

4. If different mixes of concrete are used in any of the categories listed in item 3, measure each mix separately. For instance, where different strengths of concrete are specified for an interior slab on grade and an exterior patio slab, separately calculate the quantity of concrete for each type of concrete in slabs.

5. Where columns and walls extend between the floors of a building, measure these components from the top surface of the slab below up to the undersurface of the slab or beam above. See Figure 4.1.

6. Beams may be measured separately from slabs. However, if they are to be poured monolithically with the slabs, the quantity of concrete in the beams should be added to the slab concrete for pricing.

Measuring Formwork Generally

The formwork operations involve a number of activities including fabricating and erecting the forms, stripping, moving, cleaning, and oiling the forms for reuse. All of these activities and the materials involved are allowed for in the pricing of the forms.

In general, the estimator measures the surface of the concrete that comes into contact with the forms. In the case of footing forms, this surface will usually be formed with lumber or plywood. Most footings on a housing project are formed with 2 × 8 lumber, in which case the length of this lumber will be measured. In the case where plywood forms are to be used, the area of plywood in contact with the concrete is measured.

Wall forms are usually measured in linear feet stating the height of the forms. Therefore, for an 8-foot high wall 100 feet long, the wall forms would be measured as 200 linear feet of 8" high wall forms.

The estimator does not have to be concerned about the design of the forms at the time of the takeoff; all that needs to be established is which surfaces of the concrete require forms. In the past, estimators have agonized over such things as whether the bottom of an opening or the sloped top surface of a wall needs to be formed. If discussion with your colleagues does not provide an answer, the prudent estimator will always exercise caution and allow the forms.

A recent innovation in home building, especially in the colder climate areas, is the introduction of insulated concrete foundations. These are specially made expanded polystyrene blocks that are filled with concrete to form the house foundations, thus eliminating the need for separate formwork. The Styrofoam acts as the forms and becomes the permanent insulation for the basement walls. To access more information see the Recommended Resources at the end of this chapter.

Measuring Notes—Formwork

1. Measure formwork in square feet of contact area, i.e., the actual surface of formwork that is in contact with the concrete.
2. Classify formwork into the same categories as listed for concrete. As an illustration, consider a project with concrete footings, walls, and columns: in this situation forms to footings, forms to walls, and forms to columns would each be described and measured separately. There are, however, a number of factors that may have no effect on the price of the concreting operations but do affect the price of formwork and, therefore, should be noted. For example, the volume of concrete in all walls, whether they are straight or curved, will have the same price, but the price of forms to curved walls will differ from that of straight walls, so the forms to curved surfaces must be kept separate.
3. Measure bulkhead and edge forms separately. For example, if there were openings in a wall, the area of **blockouts** (the edges around the openings) would be measured separately from the wall forms. Similarly, if there are **pilasters** projecting from the walls, the area of the pilasters would be calculated and noted separately from the wall forms. See Figure 4.2 for different categories of formwork in a wall system.
4. Measure forms to slab edges separately from forms to beams and forms to walls even where the edge forms may be extensions of beam or wall forms (Figure 4.3).
5. Forms are generally measured across openings for windows in walls. The estimator must distinguish between what are openings and what are **cutouts**.

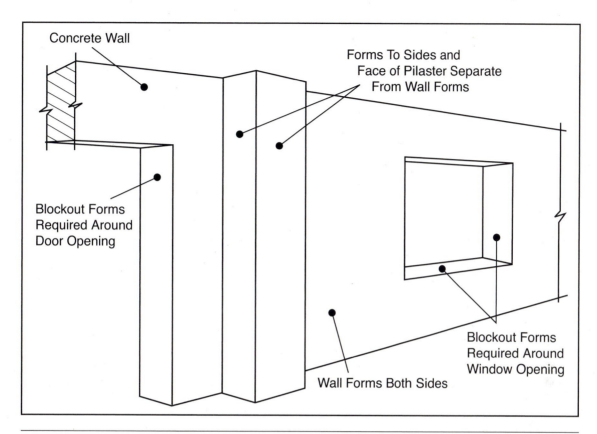

Figure 4.2   Formwork Categories

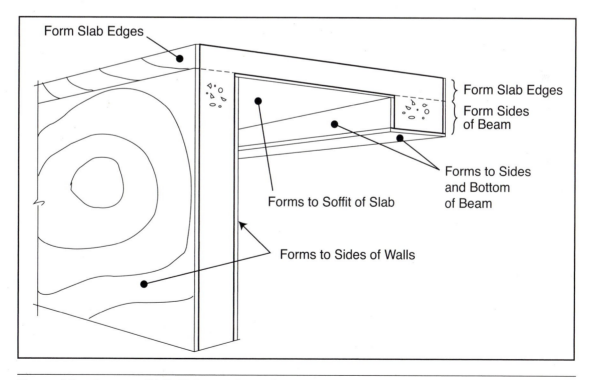

Figure 4.3   Concrete Wall, Slab, and Beam Forms

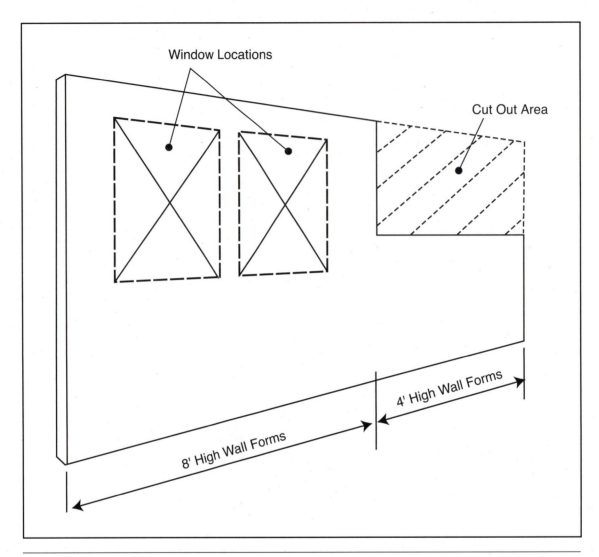

Figure 4.4 Openings and Cutouts in Formwork

Openings less than 100 square feet are not deducted, but all cutouts would be deducted (see Figure 4.4).

6. Describe items of formwork that are linear in nature stating their size and measure their length in feet. Grooves, chases, keyways, chamfers, and narrow strips of formwork, less than 1 foot wide, are measured in this fashion.

7. Describe forms to circular columns stating the diameter, and measure the height of the column in feet.

Measuring Notes—Finishes and Miscellaneous Concrete Work

1. Measure slab finishes in square feet of the plan area and vertical finishes in square feet of the exposed concrete surface.

2. Classify finishes first in terms of what is being finished: slabs, walls, columns, stairs, sidewalks, and so on. Then separate by type of finish: wood float, sack rubbed, steel trowel, broom finish, and such like.

REINFORCING BARS		
Bar Designation	Weight lbs/foot	Diameter
No. 2	0.167 lbs	0.250"
No. 3	0.376 lbs	0.375"
No. 4	0.668 lbs	0.500"
No. 5	1.043 lbs	0.625"
No. 6	1.502 lbs	0.750"
No. 7	2.044 lbs	0.875"
No. 8	2.670 lbs	1.000"

Figure 4.5   Reinforcing Bars

Curing Slabs and Vapor Barriers

3. Measure curing slabs in square feet.
4. Measure vapor barriers to slabs in square feet of the plan area; add 10 percent to this area to allow for overlaps.
5. Measure **welded wire mesh** in plan area with an additional 10 percent included for overlaps at the edges.

Reinforcing Steel

6. Measure bar reinforcing steel in linear feet and then convert these measurements to weights by multiplying the length of each bar by the unit weight per foot for that bar size. See Figure 4.5 for the unit weights of reinforcing bars of various sizes.

Miscellaneous Concrete Items

7. Describe and enumerate anchor bolts and other inserts into the concrete.
8. Measure inserts such as waterstops in linear feet.
9. Measure expansion joints in linear feet stating the size of joint and the type of material used. If a sealant is required in connection with the joint, measure it separately in linear feet stating the size and type of sealant. It is useful for the estimator to draw a small sketch on the margin of the takeoff to clarify this kind of complex detail.
10. Measure grout to anchor bolts and base plates in cubic feet separating different types of materials.
11. Measure saw cuts to concrete in linear feet stating the size of the cut. Here it is important to note whether old or new concrete is to be cut and what is to be cut: slabs, walls, or columns.

Example of Concrete Work Takeoff

Figure 4.6a and Figure 4.6b show the takeoff for the concrete work of the sample house project shown on drawings Figure 3.7 in Chapter 3.

QUANTITY SHEET						SHEET No.	1 of 2	

JOB: House Example DATE: _____

ESTIMATOR: ABF EXTENDED: _____ EXT. CHKD: _____

	DESCRIPTION	DIMENSIONS						
		TIMES	Length	Width	Height			
	Concrete Work							
2 x 40.00	*80.00*							
2 x 28.83	*57.66*							
	137.66	*Outside wall perimeter*						
4 x 0.67	*(2.66)*						*2 x 8*	*2 x 8*
	135.00	*Centerline of wall*				*Conc.*	*Forms*	*Forms*
						L x W x D	*2 x L*	*No. x 2 x (L+W)*
	CONC. FTGS.		*135.00*	*1.33*	*0.67*	*120*	*270*	*-*
	(Pads)	*3*	*2.67*	*2.67*	*0.67*	*14*	*-*	*32*
		2	*2.00*	*2.00*	*0.67*	*5*	*-*	*16*
						139	*270*	*48*
						5	*CY*	*270*
								318 LF
	FORM 2 x 4 KEYWAY		*135.00*			*135*	*LF*	
							8" High	
						Conc.	*Wall Forms*	*2 x 8 Blockouts*
						L x W x D	*2 x L*	*No. x 2 x (L + H)*
	CONC. WALLS		*135.00*	*0.67*	*8.00*	*724*	*270*	*-*
	(Windows) DDT	*4*	*3.08*	*0.67*	*1.08*	*(9)*	*-*	*33*
	(Drop Landing) DDT		*9.08*	*0.67*	*0.92*	*(6)*	*-*	*20*
						709	*270 LF*	*53 LF*
						26	*CY*	
1'-9³/₄"								
Less: *(10³/₄")* *Depth of joists and sheathing*								
Drop Landing: *11" = 0.92*								
	DAMP PROOF WALLS		*137.66*	*-*	*8.00*	*1,101*	*SF*	*-*
	#4 REBAR	*4*	*135.00*			*540*		
	Add 10% for Laps					*54*		
						594	*x 0.668*	
						397	*lbs*	

Figure 4.6a Concrete Work Takeoff

QUANTITY SHEET SHEET No. 2 of 2

JOB: House Example DATE: _____

ESTIMATOR: ABF EXTENDED: _____ EXT. CHKD: _____

DESCRIPTION	TIMES	Length	Width	Height	Conc. L x W x D	Finish and Curing L x W	
Concrete Work (Cont'd.)							
40.00 x 28.83 Outside dimensions							
(1.33) x (1.33)							
38.67 x 27.50 Inside							
CONC. S.O.G.		38.67	27.50	0.33	351	1,063	SF
					13	CY	
6mil POLY. V.B.	1.1	38.67	27.50		1,170	SF	
GRAVEL UNDER S.O.G.		38.67	27.50	0.67	712		
					26	CY	
2 x 6 SILL PLATE		137.66			138	LF	
137.66 / 6.00 = 23							
+4 Corners							
27 Bolts							
1/2" x 9" A.B.'s	27				27	No.	
Deck Supports							
DRILL 10" DIA. HOLES FOR PILES	2			6.00	12	LF	
CONC. PILES	2 x 3.14	0.42	0.42	8.00	9		
					1	CY	
10" DIA FIBER TUBES	2			8.00	16	LF	

Less 2 x 0.67

Figure 4.6b Concrete Work Takeoff

The design of the foundation system for this house consists of a concrete footing supporting a reinforced concrete basement wall. For information about alternative designs for foundation systems, see the *energy efficient construction details* listed in Recommended Resources at the end of this chapter.

Comments on the Concrete Work Takeoff Figure 4.6a and Figure 4.6b

1. First the centerline length of the perimeter footing is calculated; this is the same length as the centerline of the perimeter wall. The width and height of the footing are shown on the drawings, so these figures can be entered directly into the dimension columns.
2. The same strength concrete, 3,000 psi, is to be used for the footings, walls, and slab-on-grade so the strength is not mentioned in the takeoff description. This information is left until the pricing stage to consider.
3. The concrete for the three pad footings is added to the continuous footing concrete.
4. Footing forms are measured to each side of the continuous footing (2 × centerline length) and to the perimeter of each pad footing.
5. The keyway is the same length as the centerline of the wall.
6. The basement walls are measured as 8'-0" high, which is a standard dimension for a full-height basement.
7. Deductions are made from the wall concrete for the window openings because each opening exceeds 1 cubic foot in volume.
8. There is also a deduction from the wall concrete at the landing since the wall only comes up to the landing level at this location (drop landing).
9. Blockout forms are measured to the perimeter of the window openings and the opening for the drop landing.
10. The damp proofing (tar) to the exterior of the concrete wall is measured here. Note the exterior wall perimeter is used.
11. The drawing notes indicate two rows of #4 rebar is required at the top and also at the bottom of the wall. This is measured first in length plus 10 percent added for laps in the bars, and then the quantity is converted to pounds using the conversion factor of 0.668 pounds per foot for this bar size.
12. The basement slab-on-grade extends to the inside of the exterior walls so the thickness of the walls (2 × 8") is deducted from the outside dimensions to obtain the dimensions of the slab.
13. The plan area of the slab is measured for concrete finish and curing.
14. The vapor barrier under the slab is also measured as the plan area but the quantity is increased by 10 percent to allow for laps in the material.
15. A convenient place to measure the gravel under the slab-on-grade is directly after the concrete since it is the same area as the concrete—only the thickness differs from the concrete.
16. It is also convenient to measure the sill plate at this stage, but this item will be recapped and priced in the rough carpentry section.
17. The sill plate is secured by anchor bolts spaced at 6'-0" on center. The number of bolts required is calculated by dividing the exterior perimeter of the wall by the spacing (6'-0") then adding an extra bolt for each corner in the wall.
18. The work in the pile supports to the deck is measured. This consists of drilling holes, then the concrete and forms for the piles. Fiber tubes are the forms used for the circular piles.

SUMMARY

- Preparing a quantity takeoff of concrete work requires the estimator to measure a combination of items such as concrete, formwork, concrete finishes, etc.
- Whereas concrete is clearly shown on the project drawings, few details of formwork are provided. The estimator, therefore, needs to make an assessment of formwork requirements from the details of the concrete given.
- In a takeoff of concrete work, the estimator should work through the project dealing with one assembly at a time measuring all the work associated with one assembly before passing on to the next.
- Concrete is measured in cubic yards net in place.
- Concrete is classified in terms of the use to which the concrete is put.
- Different mixes of concrete are measured separately.
- Formwork is generally measured in linear feet or square feet of contact area.
- Formwork is classified in terms of use, just as concrete is. Forms to curved surfaces are measured separately.
- Because only the area of forms in contact with the concrete is measured, the estimator does not have to be concerned about details of the design of the forms at the time of the takeoff.
- Formwork to pilasters, bulkheads, edges, etc. are measured separately.
- Items of formwork such as grooves, keyways, notches, and chamfers are described stating their size and are measured in linear feet.
- Forms to circular columns are described giving the diameter of the column and are measured in linear feet or meters to the height of the column.
- Concrete finishes are generally measured in square feet.
- Slab finishes and curing slabs are all measured in square feet of the plan area.
- Welded wire mesh reinforcing is also measured in square feet or square meters of the plan area, with further percentage added to allow for the laps in the mesh.
- Builders often subcontract fabricating and placing reinforcing steel, in which case the subcontractors would take off and price this trade.
- A takeoff of concrete work for a house project is demonstrated.

RECOMMENDED RESOURCES

Information	Web Page Address
Energy-efficient construction details including many tips on how to build in order to save energy are found on many web sites. These are some examples of what is available:	
US Department of Energy	http://www1.eere.energy.gov/buildings/building_america/
Green Home Guide	http://greenhomeguide.com/askapro/question/are-skylights-a-good-idea
Earth Share	http://www.earthshare.org/2008/09/button-up-your.html?gclid=CPKehKiQnqMCFQVvbAod_355ng

Green construction   http://www.energyefficienthomes.ca/
                                   icf.html

■ Insulated concrete foundations   http://www.epsmolders.org/

REVIEW QUESTIONS

1. Preparing a takeoff of concrete work requires the estimator to measure a number of different items. Which of these items are usually shown on the project drawings and which are not shown?
2. Why is it recommended that the estimator takeoff assembly by assembly rather than measuring all the concrete together then all the forms together?
3. What size of opening is deducted when measuring wall forms?
4. If an opening for a door is required in a concrete wall, what is measured in the takeoff?
5. Why are pilaster forms measured separately from wall forms?
6. How is the concrete and formwork measured for circular columns?
7. Describe the procedure for measuring rebar.

PRACTICE PROBLEMS

1. Take off the excavation and concrete work required for the construction of the driveway shown on Figure 4.7.
2. Take off all the concrete work for a basement consisting of a 10" concrete wall, 7'-6" high. The wall is supported by a 2'-3" × 1'-0" concrete footing. Also include a 5"-thick concrete slab-on-grade placed on a 7"-thick bed of gravel. The size of the basement is 30'-0" × 32'-5" measured to the outside of the basement walls.
3. Take off all the concrete work for the foundations and slab-on-grade shown in Figure 4.8.

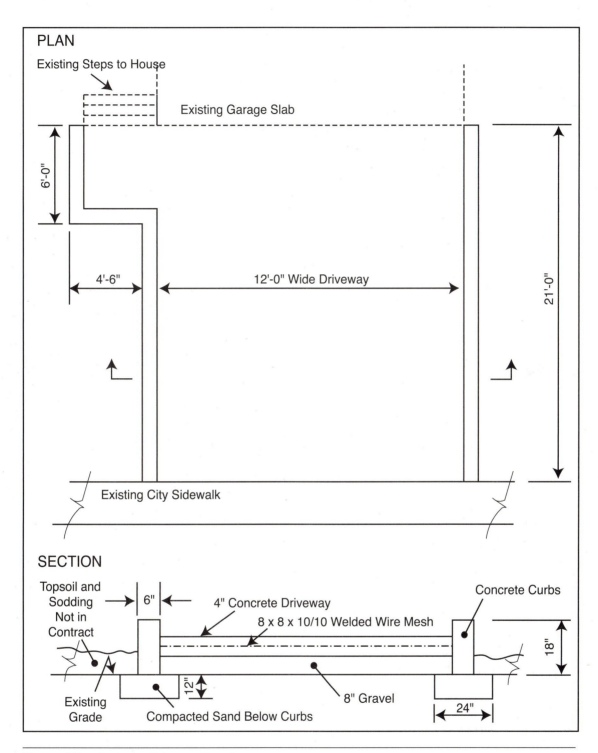

Figure 4.7   Concrete Driveway Details

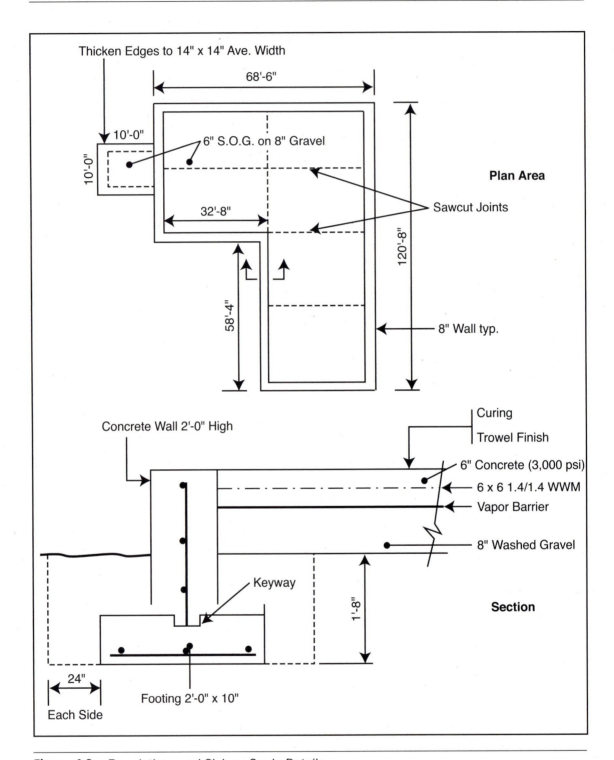

Figure 4.8 Foundations and Slab-on-Grade Details

5

MEASURING CARPENTRY WORK

OBJECTIVES

After reading this chapter and completing the review questions, you should be able to:

- Explain why a detailed knowledge of carpentry construction is useful in order to prepare a carpentry takeoff for a residential building.
- Calculate quantities of lumber using board measure units.
- Describe how to measure rough carpentry work in a takeoff.
- Describe how to measure finish carpentry and miscellaneous item work in a takeoff.
- Use quick methods to take off joists and studs.
- Complete a takeoff of rough carpentry, finish carpentry, and miscellaneous items for a residential project.

KEY TERMS

board measure	oriented strand board (OSB)	sheathing
engineered joist	parallam beam	soffits
joist hangers	pre-hung doors	subfloors
nominal sizes	rough hardware	teleposts

Introduction

In order to prepare a detailed, realistic quantity takeoff of the carpentry work the estimator must have a comprehensive knowledge of carpentry details and practices. This is particularly true for estimating housing and other buildings of wood frame construction because, on these types of projects, few specific details of framing requirements are shown on the design drawings. Instead, the builder is left to construct the framing in accordance with standard practices and code requirements.

As a consequence, in order to assess rough carpentry requirements adequately, the estimator has to be familiar with framing methods and must also be alert to design requirements such as the need for extra joists or additional studs in certain locations.

Students who are not experienced in carpentry work are advised to study one or more of the many reference books available on this subject which will help them to better understand the essentials of a rough carpentry takeoff.[1]

Measuring Rough Carpentry

Board Measure

Most lumber used in construction has been dressed; this means that surfaces are sanded to a smooth finish. This process reduces the size of the cross section of the lumber; however, even though the true dimensions of an item may be $1\frac{1}{2}$" by $3\frac{1}{2}$", it is still referred to by its **nominal size,** which is 2 by 4. See Figure 5.1 for the nominal and dressed sizes of lumber.

The unit of measurement of lumber is generally the **board measure** (BM), which is sometimes referred to as board foot (BF) and a 1000-board feet is written MBF. Board measure is a cubic measure where one unit of BM is equivalent to a 1×12 board 1 foot long. To calculate the BM of lumber, multiply the length in feet by the nominal width and thickness of the pieces in inches and divide the product by 12. Quantities are rounded off to the nearest whole board foot. Pieces of lumber, beams, joists, plates, etc., are first measured by length in the takeoff process and are then converted to board measure. See Figure 5.2 for examples of BM calculations.

Nominal	Dressed
1 x 2	$\frac{3}{4}$ x $1\frac{1}{2}$
1 x 3	$\frac{3}{4}$ x $2\frac{1}{2}$
1 x 4	$\frac{3}{4}$ x $3\frac{1}{2}$
2 x 4	$1\frac{1}{2}$ x $3\frac{1}{2}$
2 x 6	$1\frac{1}{2}$ x $5\frac{1}{2}$
2 x 8	$1\frac{1}{2}$ x $7\frac{1}{4}$
2 x 10	$1\frac{1}{2}$ x $9\frac{1}{4}$
2 x 12	$1\frac{1}{2}$ x $11\frac{1}{4}$
2 x 14	$1\frac{1}{2}$ x $13\frac{1}{4}$
3 x 3	$2\frac{1}{2}$ x $2\frac{1}{2}$
4 x 4	$3\frac{1}{2}$ x $3\frac{1}{2}$
6 x 6	$5\frac{1}{2}$ x $5\frac{1}{2}$

Figure 5.1 Nominal and Dressed Lumber Sizes

1. See the Web Page Help Box at the end of this chapter for information about carpentry construction details on the Internet

Takeoff Item	Calculation	Result
12 pieces of 1 x 2 - 10 feet long	12 x 1 x 2 x 10 / 12	20.00 BM
20 pieces of 1 x 4 - 8 feet long	20 x 1 x 4 x 8 / 12	53.33 BM
8 pieces of 1 x 12 - 12 feet long	8 x 1 x 12 x 12 / 12	96.00 BM
60 pieces of 2 x 4 - 8 feet long	60 x 2 x 4 x 8 / 12	320.00 BM
14 pieces of 2 x 12 - 16 feet long	14 x 2 x 12 x 16 / 12	448.00 BM
6 pieces of 3 x 3 - 8 feet long	6 x 3 x 3 x 8 / 12	36.00 BM
9 pieces of 3 x 4 - 12 feet long	9 x 3 x 4 x 12 / 12	108.00 BM
4 pieces of 4 x 6 - 14 feet long	4 x 4 x 6 x 14 / 12	112.00 BM
2 pieces of 4 x 8 - 10 feet long	2 x 4 x 8 x 10 / 12	53.33 BM

Figure 5.2   Board Measure Calculations

Measuring Notes—Rough Carpentry

Rough Carpentry Generally

1. Measure lumber by piece, by length in feet, or in board measure as described previously.
2. Measure items of lumber separately on the basis of the following categories:
   a. Dimensions: $2 \times 4, 2 \times 6, 2 \times 8$, etc.
   b. Dressing: "Rough sawn" and S4S (surfaced on four sides) are the usual alternatives.
   c. Grade: Construction lumber is graded in a two-step process. First, the grade category is identified, e.g., structural light framing; then the grade within that category, e.g., select structural. See Figure 5.3 for a list of the National Grading Rule lumber classes for construction lumber.
   d. Species: Most lumber for rough carpentry is soft wood. Common types include Douglas fir, larch, eastern hemlock, western hemlock, sitka spruce, cedar, and pine.
3. Lumber required to have a special treatment (i.e., kiln dried, pressure treated, etc.) shall be kept separate and described.
4. Measure wallboards in square feet.
5. Do not deduct for openings less than 40 square feet in area.
6. Classify wallboards and measure separately in the following categories:
   a. Type of material
   b. Thickness

Framing Work

7. Measure items of lumber for framing separately in the following categories:
   a. Plates   h. Rafters
   b. Studs   i. Ridges
   c. Joists   j. Hip and valley rafters
   d. Bridging   k. Lookouts and overhangs
   e. Lintels   l. Gussets and scabs
   f. Solid beams   m. Purlins
   g. Built-up beams n. Other items of framing

Lumber Class	Grade
Light Framing	Construction Standard Utility
Structural Light Framing	Select 1 2
Studs (2 to 4 inches thick, 2 to 4 inches wide)	Stud Economy Stud
Structural Joist and Planks	Select Structural 1 2 3
Appearance Framing	Appearance

Figure 5.3   Soft Wood Lumber Grades

Trusses, Truss Joists, and Truss Rafters

8. Enumerate and fully describe prefabricated trusses, truss joists, and truss rafters.

Manufactured Beams, Joists, and Rafters

9. Measure and fully describe manufactured beams, joists, and rafters in linear feet.

Sheathing

10. Measure **sheathing** in square feet; describe and measure separately wall, floor, and roof sheathing.
11. Measure diagonal work separately from other sheathing.
12. Measure common boards, ship-lap, tongued and grooved, plywood, and other types of sheathing separately.
13. Describe and measure separately work to sloping surfaces.

Copings, Cant Strips, and Fascias

14. Measure copings, cant strips, fascias, and other similar items in linear feet.

Soffits

15. Measure **soffits** in square feet; measure different materials separately.

Sidings

16. Measure sidings in square feet; describe the type of material used and state whether vertical, horizontal, or diagonal.

Vapor Barriers and Air Barriers

17. Measure vapor barriers and air barriers in square feet describing the type of material used. Increase quantities to allow for overlaps in these materials.

Underlay and Subfloors

18. Measure underlay and **subfloors** in square feet stating the type of material used.

Blocking and Furring

19. Measure blocking and furring in linear feet or board measure and classify in the following categories:
    a. Blocking stating purpose and location
    b. Furring stating purpose and location
    c. Nailing trips
    d. Strapping
    e. Grounds
    f. Rough bucks
    g. Sleepers

Rough Hardware

20. Include an allowance for **rough hardware** based on the value of the carpentry material in the estimate.

Measuring Finish Carpentry and Millwork

Finish carpentry includes trim, baseboard, stairs, paneling, doors, windows, cabinets, and all the other visible woodwork on a project. Some items of rough carpentry are measured at the same time as the finish carpentry items. For example, grounds or strapping may be required to support paneling, or blocking may be necessary for cabinets.

Measuring Notes—Finish Carpentry

Finish Carpentry Generally

1. Classify items of finish carpentry and measure separately according to materials, size, and method of installation, i.e., wall-mounted items would be measured separately from floor-mounted or ceiling-mounted items.
2. Measure grounds, rough bucks, backing, etc., in the Rough Carpentry section.
3. Include an allowance for rough hardware for carpentry work.

Trim

4. Measure trim in linear feet stating size.
5. Keep built-up items, such as valance boxes, false beams, etc., separate and fully describe them.
6. Measure baseboard and carpet strip through door openings, i.e., do not deduct for these openings.

Shelving

7. Measure shelving in linear feet stating width.

Stairs

8. Measure and fully describe prefabricated stairs stating the width of stair and the number of risers.
9. Describe balusters and handrails and measure in linear feet.

Cabinets, Countertops, and Cupboards

10. Enumerate and fully describe cabinets, countertops, and cupboards. Alternatively, units can be measured in linear feet describing the type and size of the unit and method of installation.

Paneling

11. Describe paneling and measure in square feet.

Measuring Notes—Doors and Frames

Doors and door frames are usually obtained from vendors who quote prices to supply and deliver the specified goods to the site. The builder needs to add the cost of handling and installing doors and frames. For this purpose, doors and door frames shall be enumerated stating their size and type of material.

Some doors may arrive at the site **pre-hung,** complete with finish hardware, while other doors must be hung at the site and fitted with necessary hardware. These should be enumerated separately.

On most projects, a door schedule that provides information about the type and size of doors and frames will be included with the drawings. Details of the hardware are also usually specified for each door. The estimator will find these schedules invaluable in preparing door takeoffs because all that remains to be done is to count the number of doors required of each type listed on the schedule. Sometimes even this information is provided on the schedule.

Measuring Notes—Windows

On some projects, subtrades may supply and install complete window units. However, on most housing projects, window and door components will be obtained from suppliers and then be installed on site by the general builder. To measure this work, individual windows shall be enumerated and fully described, including the rough opening sizes of the units.

Measuring Notes—Bathroom Accessories

Enumerate and fully describe the following items:

a. Shower curtain rods	f. Toilet roll holders
b. Soap dispenser units	g. Towel dispensers
c. Grab bars and towel bars	h. Waste receptacles
d. Mirrors	i. Medicine cabinets
e. Napkin dispensers	j. Coat hooks

Measuring Notes—Finish Hardware

Finish hardware includes the following items that shall be enumerated and fully described:

a. Hinges (in sets of two hinges)   g. Latch sets
b. Flush bolts   h. Kick plates
c. Bumper plates   i. Panic hardware
d. Deadlocks   j. Push plates
e. Doorstops   k. Pull bars
f. Lock sets   l. Door closers

Example of Rough Carpentry Floor System Takeoff

Figure 5.4a shows the takeoff for the rough carpentry floor system on this project.

Comments on the Floor System Takeoff Shown in Figure 5.4a

1. The rough carpentry takeoff begins at the foundations with the supports for the floor beam. The beam is supported by the concrete walls at the ends and by adjustable steel posts in the middle. These posts are commonly known as **teleposts.**

2. The floor beam consists of a 6" × 10" **parallam beam.** This is a manufactured wood component[2] that passes across the building. These beams are made by gluing together a number of thin boards to form a structural member bigger than the trees from which the boards were sawn.

3. **Engineered joists**[3] are used to support the floor sheathing. See Figure 5.5 for an illustration of a parallam beam and engineered joists.

4. The full length of the building is used for length of this beam even though the beam would stop short of the outside of the basement walls.

5. Floor joists are required at 16" on center. To determine the number of joists, divide the length of the floor by 1.33', round up to the nearest whole number, and then add the extra end joist.

6. Double joists are required as shown on the floor-framing plan.

7. Header joists are needed at the outside of the building running perpendicular to the joists (40' long). They are also required next to the stairs and landing. Again, this is shown on the floor-framing plan.

8. All lumber will be classified as *Structural Grade Light Framing* on this project so this does not have to be included in the takeoff descriptions.

9. A 4" × 4" wood post is shown at each side of the basement stairs.

10. Floor sheathing extends across the whole floor area plus the future fireplace area. Note that there are no deductions for openings where the stairs are located.

11. The landing at the front entrance is to be constructed of the same materials as the floor system. There are sufficient joist and sheathing material measured here for the landing.

Deck Construction

12. The 2 × 8 joists for the deck are supported on one side by a 3-ply 2 × 8 beam and on the other side by a header attached to the outside of the basement wall.

---

2. See the Web Page Help Box at the end of this chapter for information about "parallam" beams on the Internet

3. See the Web Page Help Box at the end of this chapter for information about engineered joists on the Internet

QUANTITY SHEET   SHEET No. | 1 of 10 |

JOB: _____House Example_____   DATE: _____

ESTIMATOR: _ABF_____   EXTENDED: _____   EXT. CHKD: _____

DESCRIPTION	DIMENSIONS						
	TIMES	Length	Width	Height			
Rough Carpentry							
- Floor System							
3" Dia. TELEPOSTS	3	-	-	-	3	No.	
6 x 10 PARALLAM BEAM		40.00	-	-	40	LF	
40.00 / 1.33 = 30 + 1							
2 X 10 ENG. JOISTS	31	28.83	-	-	894		
(Extension)	5	2.00	-	-	10		
Fireplace: 4.50 / 1.33 = 4 +1 (Doubles)	2	16.75	-	-	34		
	2	14.75	-	-	30		
By Stairs: 14'-1"	3	14.08	-	-	42		
Less (4'-2")	2	9.92	-	-	20		
9'-11" (Headers)	2	40.00	-	-	80		
	2	9.08	-	-	18		
	6	3.33	-	-	20		
					1147	LF	
4 x 4 WOOD POSTS	2	8.00	-	-	16		
					21	BM	
¾" T & G FLOOR SHEATHING		40.00	28.92	-	1157		
(Fireplace)		4.50	2.00	-	9		
					1166	SF	
Deck							
2 x 8 IN BEAM	3	15.75	-	-	47		
					63	BM	
2 x 8 JOISTS	13	8.00	-	-	104		
(Headers)	2	16.00	-	-	32		
15.75 / 1.33 = 12 + 1					136		
					181	BM	
15.75							
5.58 *2 x 4 CEDAR DECKING*		15.75	5.58	-	88	SF / 0.33 =	266 LF
3.00							
1.67 *40" HIGH CEDAR RAILING*		27.67	-	-	28	LF	
1.67							
27.67 *3'-7" WIDE CEDAR STEPS*	1	-	-	-	1	No.	
COMPRISING 2-TREADS							

Figure 5.4a   Floor System Takeoff

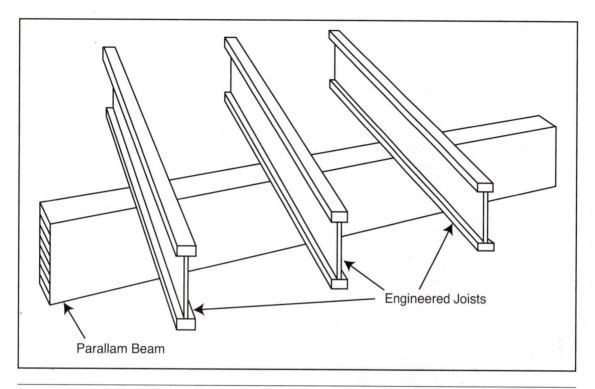

Figure 5.5   Parallam Beam and Engineered Joists

13. The length of 2 × 4 boards required for decking is easily calculated by dividing the area of the deck (88 SF) by the 4" center-to-center spacing of the boards (0.33 feet).

Canopy Over Entrance

14. The height of the 6 × 6 posts is not clear from the drawings, but we can see that it is going to exceed the height of windows (G) that are 84" high. So 8'-0" high posts are allowed for.
15. The canopy is taken to be the same size as the precast concrete step at the entrance. Although this is not dimensioned, the wall dimensions suggest that it is about 7'-0" long and its width is about half this (3'-6").
16. The joists are attached to the building with **joist hangers**—these are enumerated.

Measuring Joists—Quick Method

It does not matter which way you run the joists for the takeoff of a floor system; you will generate roughly the same quantity whichever way they run. In fact, you can calculate the total length of joists required by simply dividing the floor area by the joist spacing and adding for headers and double joists. For example, in the house takeoff above the floor area is 1,162.2 square feet; dividing this value by the joist spacing (1.33') gives 874 feet of joists. Adding for double joists, headers, etc., takes this up to 1,127 feet, which is not too far off the more accurately calculated quantity of 1,147 feet. When waste factors are added, the difference may not be significant.

Example of Wall System Takeoff

Figure 5.4b, Figure 5.4c, and Figure 5.4d show the takeoff for the rough carpentry wall system on this project.

Comments on the Wall System Takeoff Shown in Figure 5.4b, Figure 5.4c, and Figure 5.4d

17. The perimeter of the outside face of the building is used to calculate the length of the exterior stud wall. This wall comprises three 2 × 6 plates, with studs spaced at 16" on center.

18. Two extra studs are allowed at each corner in the wall, at the intersection of two walls and at each opening.

19. One-half inch **oriented strand board (OSB)** is specified as wall sheathing on this job. Note that there are no deductions from the total area for window or door openings.

20. 2 × 10 lintels are specified for openings in the exterior walls. The length of these lintels is equal to the width of the rough opening plus 3".

21. A 2 × 4 stud wall is shown on the inside of the concrete basement walls and to the sides of the basement stairs. Three plates are measured for this wall. The wall length is obtained by dividing the total length of the plates by 3; this length is then divided by 16" to arrive at the number of studs. Again, extra studs are added for corners and openings in this wall.

22. *Interior main floor:* non-load–bearing partitions may have just two plates but three plates have been allowed because many builders like partitions to match the exterior walls; this allows them to use the same standard stud length for all walls.

23. Plates for partitions are measured without deductions for door openings.

24. There are so many interior partitions on the main floor that a systematic approach is needed to ensure that all walls are included in the takeoff. The technique used here is to consider first the walls that run left to right (L-R) starting at the top of the drawing, then the walls that run from top to bottom (T-B) starting at the left of the drawing. It is a good idea to highlight the lengths of the walls as they are measured to keep track of progress and ensure all partition walls have been accounted for.

25. The number of studs for these partitions is calculated as before with extra studs added for corners/intersections and openings.

26. A 2 × 6 stud wall is called for between the two bathrooms.

27. A single 2 × 4 lintel is measured for each of the openings in these non-load–bearing partitions.

Counting Studs—Quick Method

Calculating the number of studs required for wall is always an approximation because there are many places where an extra stud may be added and places where one can be eliminated. A quick way to calculate, at least roughly, the number of studs required for a wall without assessing the number of extras required is to allow one stud every foot when the stud spacing is 16". In house takeoff above this method would generate 142 of the 2 × 6 studs, where the *accurate* calculation came to a total of 167. Here again, when waste factors are added, the difference may not be significant.

QUANTITY SHEET　　　　　　　　　　SHEET No. ☐ 2 of 10

JOB: _____ **House Example** _____　　DATE: _____

ESTIMATOR: __ **ABF** _____　EXTENDED: _____　EXT. CHKD: _____

| DESCRIPTION | DIMENSIONS | | | | | | |
	TIMES	Length	Width	Height			
Canopy Over Entrance							
6 x 6 8' POSTS	2	-	-	-	2	No.	
2 x 10 JOISTS	7	3.50	-	-	25		
(Headers)	2	7.00	-	-	14		
7.00 / 1.33 = 6 + 1 = 7					39		
					65	BM	
2 x 10 JOIST HANGERS	7	-	-	-	7	No.	
Wall System							
- Main Floor - Exterior							
2 x 40.00　*80.00*							
2 x 28.83　*57.66*							
137.66　Outside Wall Perimeter							
2 x 6 PLATES	3	137.66	-	-	413		
(Chimney)	3	8.50	-	-	26		
2 x 2.00　4.00					438	BM	
4.50							
8.50							
137.66 / 1.33 = 106							
8.5 / 1.33 = 6 + 1							
2 x 6 STUDS	106	-	-	-	106		
(Extras)	2 x 25	-	-	-	50		
(Chimney)	7	-	-	-	7		
(Extras)	2 x 2	-	-	-	4		
					167	No.	
1/2" OSB WALL SHEATHING		137.66	-	8.00	1101		
		8.50	-	8.00	68		
					1169	SF	

Figure 5.4b　Wall System Takeoff

QUANTITY SHEET   SHEET No. | 3 of 10 |

JOB: _____ **House Example** _____   DATE: _____

ESTIMATOR: __ABF_____   EXTENDED: _____   EXT. CHKD: _____

DESCRIPTION		TIMES	Length	Width	Height			
			DIMENSIONS					
Wall System (Cont'd)								
2 x 10 LINTELS	(Door 1)	2	5.83			12		
	(Door 2)	2	5.25			11		
5'-7"	5'-4"	(Windows B)	2	3.92			8	
+ 3"	+ 3"	(Windows C)	3 x 2	5.58			33	
5'-10"	5'-7"	(Windows D)	2	7.17			14	
5'-0"	6'-11"	(Windows E)	2	3.00			6	
+ 3"	+ 3"					72		
5'-3"	7'-2"					120	BM	
3'-8"	2'-9"							
+ 3"	+ 3"							
3'-11"	3'-0"							
Interior Basement								
2 x 4 PLATES		3	135.68			407		
	(Side of Stairs)	2 x 3	9.91			59		
2 x 39.67 =	79.34 14.08	3	3.33			10		
2 x 28.17 =	56.34 −4.17					476		
	135.68 9.91					318	BM	
2 x 4 STUDS		-	-	-	-	141	No.	
476 / 3 =	158.83 / 1.33 = 119.4							
	corners 2 x 6 = 12							
	openings 2 x 5 = 10							
	141.4							
2 x 10 LINTELS	(Windows F)	4	3.43			14		
3.08						23	BM	
0.25								
3.43								
2 x 4 LINTELS	(Door 6)	2	3.08			6		
2.83						4	BM	
0.25								
3.08								

Figure 5.4c   Wall System Takeoff

QUANTITY SHEET   SHEET No. | 4 of 10 |

JOB: _____ House Example _____   DATE: _____

ESTIMATOR: ABF _____   EXTENDED: _____   EXT. CHKD: _____

DESCRIPTION		DIMENSIONS					
	TIMES	Length	Width	Height			
Wall System (Cont'd)							
Interior Main Floor							
2 x 4 PLATES (L - R)		12.08			12		
	2	3.25			7		
7.58 2.33		2.92			3		
10.34 8.75		1.75			2		
17.92 3.33		17.92			18		
6.67		6.67			7		
13.67 21.08		21.08			21		
1.67		2.33			2		
15.34 (T - B)		8.25			8		
	2	15.34			31		
		3.00			3		
		1.67			2		
		2.17			2		
	2	2.42			5		
		9.42			9		
	3	10.17			31		
162.0 / 1.33 = 122				3 x	162		
Corners 23 x 2 = 46				=	485		
Openings 12 x 2 = 24					324	BM	
192							
2 x 4 STUDS	-	-	-	-	192	No.	
2 x 6 PLATES	3	4.67			14	BM	
1.75							
2.92							
4.67 *2 x 6 STUDS*	-	-	-	-	5	No.	
4.67 / 1.33 = 4 + 1 = 5							
2 x 4 LINTELS (Door 3)	2	2.58			5		
2.33 (Door 4)	4	2.75			11		
0.25 (Bifold 7)		2.25			2		
2.58 (Bifold 8)	3	4.25			13		
(Bifold 9)		3.25			3		
					34	LF = 23 BM	

Figure 5.4d   Wall System Takeoff

QUANTITY SHEET SHEET No. 5 of 10

JOB: House Example   DATE: _____

ESTIMATOR: ABF   EXTENDED: _____ EXT. CHKD: _____

DESCRIPTION	DIMENSIONS						
	TIMES	Length	Width	Height			
Roof System							
40.0 / 2 = 20 + 1 − 2							
= 19							
"KING POST" TRUSSES	-	-	-	-	19	No.	
28'-10" SPAN							
GABLE ENDS	-	-	-	-	2	No.	
½" WALL SHEATHING	2 x ½	32.83	-	5.50	181	SF	
28.83							
Overhangs 4.00							
32.83							
1 x 3 RIBBONS	5	40.00			200	LF	
2 x 4 RIDGE BLOCKING		44.00			29	BM	
28.83 / 2 = 14.42							
Overhang 2.00							
16.42							
/ 3 = 5.47							
5.47 x √10 = 17.30							
2 x 4 BARGE RAFTERS	4	17.30			69		
					46	BM	
17.3 / 2.0 = 9 + 1 = 10							
2 x 4 LOOKOUTS		4 x 10	4.00		160		
					107	BM	
2 x 6 ROUGH FASCIA	2	44.00			88	BM	
ROOF SADDLE 2'-3" LONG	-	-	-	-	1	No.	
(Behind Chimney)							

Figure 5.4e   Roof System Takeoff

QUANTITY SHEET   SHEET No. 6 of 10

JOB: House Example   DATE:

ESTIMATOR: ABF   EXTENDED: EXT. CHKD:

DESCRIPTION	TIMES	Length	Width	Height			
Roof System (Cont'd)							
2 x 4 CEILING BLOCKING		137.66			92	BM	
¹/₂" OSB ROOF SHEATHING	2	44.00	17.30		1522	SF	
Eaves							
VENTED ALUM. SOFFIT	2	40.00	2.00		160		
(Sloped)	4	17.30	2.00		138		
					298	SF	
ALUM. "J" MOULD	2	40.00			80		
	4	17.30			69		
					149	LF	
ALUM. FASCIA 6" HIGH	2	44.00			88		
	4	17.30			69		
					157	LF	
Stairs							
3'-4" WIDE STAIR w. 11-RISERS	-	-	-	-	1	No.	
DITTO w. 3-RISERS	-	-	-	-	1	No.	
BASEMENT HANDRAIL		12.00			12.00	LF	
WOOD RAILINGS 40" HIGH		9.33			9.00	LF	

3.33
6.00
9.33

Figure 5.4f   Roof System Takeoff

Example of Roof System Takeoff

Figure 5.4e and Figure 5.4f show the takeoff for the rough carpentry roof system on this project.

Comments on the Roof System Takeoff Shown in Figure 5.4e and Figure 5.4f

28. The number of trusses required is calculated in a similar manner to joists; the length of the building is divided by the truss spacing plus one for the end truss, but two trusses are deducted since gable ends will be substituted for these.
29. Wall sheathing is measured to the gable ends.
30. The 1 × 3 ribbons are not shown on the drawings; they are attached, along the length of the roof, to the truss members that exceed 6'-0" long to perform the same function as bridging.
31. Ridge blocking is measured along the ridge of the trusses plus the width of the overhang outside of the gables.
32. The length of the barge rafter is equal to the slope length of the roof. This length can be determined using the slope ratio of the roof. With a 1:3 slope, the slope length (hypoteneuse) is the square root of 10, times the roof height; see the sketch in Figure 5.6.
33. See Figure 5.6 and Figure 5.7 for details of the ridge blocking, barge rafters, lookouts, and ceiling blocking.
34. The rough fascia is attached to the tails of the trusses and the bottom ends of the barge rafters.
35. See Figure 5.8 for details of the eaves.

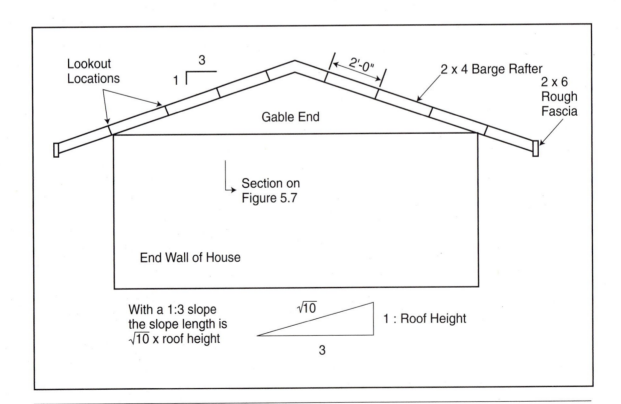

Figure 5.6   Gable End to Roof

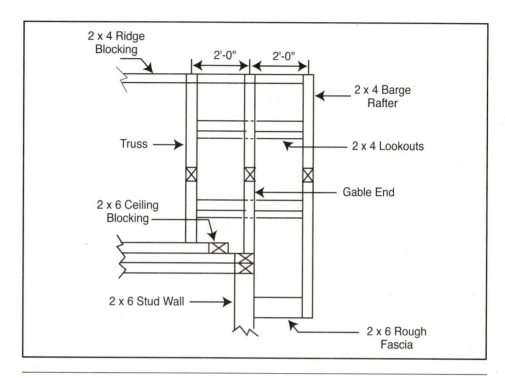

Figure 5.7 Section Through Roof Gable

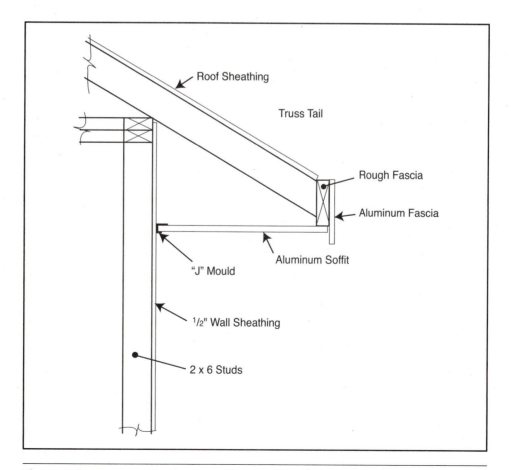

Figure 5.8 Detail at Eaves

Example of Finish Carpentry Takeoff

Figure 5.4g, Figure 5.4h, Figure 5.4i, and Figure 5.4j show the takeoff for the rough carpentry wall system on this project.

Comments on the Finish Carpentry Takeoff Shown in Figure 5.4g, Figure 5.4h, Figure 5.4i, and Figure 5.4j

36. The stairs do not have to be measured in detail; they will be obtained prefabricated for installation.
37. The length of the 12-foot handrail is scaled from the plans.
38. Minimal details are included in the takeoff descriptions of doors, hardware, and windows; this allows the takeoff to proceed without having to constantly check on the different item specifications. At this stage there is no need, for instance, to ascertain the precise type of interior doors required when they are all to be of the same type. Specific requirements can be verified when these items are priced.
39. Doors will be priced as complete units so it is not necessary to take off doorframes, trim, and door stops since the full door price will include all of these items.
40. Note that the quantity of hinges is given in pairs while the other items of hardware are enumerated.
41. The length of baseboard is determined by taking two times the overall length plus two times the overall width of all rooms regardless of shape. Note that closets are considered to be separate rooms, and no deductions are made for door openings.
42. In the baseboard length calculation, the dimensions indicated on the drawings are used without adjustments for wall thicknesses, etc. It is not necessary to calculate the precise length of each piece of baseboard because the time taken to do this is not justified by the cost of the item.

Example of a Dormer Roof System Takeoff

Figure 5.9a and Figure 5.9b show the drawings of a roof system that incorporates two dormer windows.

Figure 5.10a and Figure 5.10b show the takeoff for the dormer roof system.

Comments on the Dormer Roof System Takeoff

1. The number of rafters required on one side of this roof is calculated in a similar manner to joists: the length of the building is divided by the rafter spacing (16") plus one for the end rafter, but two rafters are deducted as gable ends will replace the end rafters.
2. Given that the roof has a 12:12 slope, the rise and the run are both equal to 10'-0" plus 2'-0" for the overhang that gives 12'-0". You can then use the Pythagoras formula to calculate the diagonal joist length.
3. Double rafters are added to the sides of the dormers, but there is no adjustment for the six rafters that have to be cut at the dormers. The cut pieces are considered to be just waste.
4. We are assuming the use of prefabricated gage ends for this roof, but sheathing to the gables is still measured.
5. Measuring the ridge, ceiling joists, collar ties, and rough fascia is straightforward.
6. The width of the roof sheathing is equal to the length of the joists that was calculated earlier.

QUANTITY SHEET   SHEET No. 7 of 10

JOB: ____House Example_____   DATE: _____

ESTIMATOR: __ABF_____   EXTENDED: _____   EXT. CHKD: _____

DESCRIPTION	DIMENSIONS						
	TIMES	Length	Width	Height			
Finish Carpentry							
Doors							
3'-0" x 6'-8" EXTR. DOOR (1)	-	-	-	-	1	No.	
C/W 12" WIDE SIDELIGHTS							
5'-0" x 6'-8" PATIO DOOR (2)	-	-	-	-	1	No.	
2'-4" x 6'-8" INTR. DOOR (3)	-	-	-	-	2	No.	
2'-6" x 6'-8" INTR. DOOR (4)	-	-	-	-	4	No.	
2'-10" x 6'-8" INTR. DOOR (6)	-	-	-	-	1	No.	
2'-0" x 6'-8" BIFOLD (7)	-	-	-	-	1	No.	
4'-0" x 6'-8" BIFOLD (8)	-	-	-	-	3	No.	
3'-0" x 6'-8" BIFOLD (9)	-	-	-	-	1	No.	
ATTIC ACCESS HATCH	-	-	-	-	1	No.	
Door Hardware							
4" BUTT HINGES (1)	-	-	-	-	1½	Pr	
3½" BUTT HINGES (3, 4, 6)	7	-	-	-	7	Pr	
KEY-IN-KNOB LOCKSET (1)	-	-	-	-	1	No.	
DEAD BOLTS (1)	-	-	-	-	2	No.	
PASSAGE SETS (4)	-	-	-	-	4	No.	
PRIVACY SETS (3, 6)	-	-	-	-	3	No.	

Figure 5.4g   Finish Carpentry Takeoff

7. Stud walls are located about half way down the rafter, so they will be approximately half the height of the roof: this gives 5'-0" for their height.
8. 2 × 8 Headers are required at the top and bottom of the dormer openings.
9. There are also 2 × 4 stud walls to the sides of the dormers.
10. The 2 × 4 rafters are to the dormer roof.
11. Sheathing is required to the dormer roofs, the dormer sides, and the triangles (gables) outside the dormer rafters.

QUANTITY SHEET SHEET No. 8 of 10

JOB: House Example DATE:

ESTIMATOR: ABF EXTENDED: EXT. CHKD:

DESCRIPTION		DIMENSIONS					
	TIMES	Length	Width	Height			
Finish Carpentry (Cont'd)							
Windows							
36" x 50" F / 36" x 15" A (B)	-	-	-	-	1	No.	
27" x 30" A - F (C)	-	-	-	-	3	No.	
36" x 30" A - F (D)	-	-	-	-	1	No.	
27" x 22" A (E)	-	-	-	-	1	No.	
36" x 24" A (F)	-	-	-	-	4	No.	
1" x 2" WINDOW TRIM (B)	2	3.50			7		
	2	6.00			12		
(C)	2	5.33			11		
	2	3.08			6		
(D)	2	6.92			14		
	2	3.08			6		
(E)	2	2.75			6		
	2	2.42			5		
(Front Door)		5.58			6		
	2	7.00			14		
(Patio Door)		5.25			5		
	2	6.92			14		
					105	LF	
Cabinets							
2' x 3' FLOOR MOUNTED		8.00			8.00		
C/W COUNTERTOP		4.00			4.00		
		3.33			3.33		
	2	3.00			6.00		
		3.33			3.33		
					24.66	LF	

Figure 5.4h Finish Carpentry Takeoff

QUANTITY SHEET   SHEET No. | 9 of 10

JOB: _____ House Example _____   DATE: _____

ESTIMATOR: __ABF_____   EXTENDED: _____   EXT. CHKD: _____

DESCRIPTION	DIMENSIONS						
	TIMES	Length	Width	Height			
Cabinets (Cont'd)							
1' x 2'-8" WALL MOUNTED		1.75					
		2.00					
		4.75					
		3.25					
		4.00					
					15.75	LF	
2' x 2'-6" BATHROOM VANITY		5.33					
		3.33					
					8.66	LF	
1'-3" WIDE CLOSET SHELVES	3	6.00	-	-	18.00		
		4.00	-	-	4.00		
(Linen)	4	2.25	-	-	9.00		
					31.00	LF	
ADJUSTABLE CLOSET RODS	-	-	-	-	4	No.	
Bathroom Accessories							
TOILET ROLL HOLDER	-	-	-	-	2	No.	
5'-4" x 3' MIRROR	-	-	-	-	1	No.	
MEDICINE CABINET	-	-	-	-	2	No.	
SHOWER ROD	-	-	-	-	1	No.	
Appliances							
DISHWASHER	-	-	-	-	1	No.	
FRIDGE	-	-	-	-	1	No.	
RANGE	-	-	-	-	1	No.	

Figure 5.4i   Finish Carpentry Takeoff

QUANTITY SHEET   SHEET No. 10 of 10

JOB: _____ House Example _____   DATE: _____

ESTIMATOR: ABF _____ EXTENDED: _____ EXT. CHKD: _____

DESCRIPTION	TIMES	Length	Width	Height			
		DIMENSIONS					
Finish Carpentry (Cont'd)							
1" x 3" BASEBOARD	2	33.33	-	-	67		
(Living/Dining/Entrance)	2	28.00	-	-	56		
	2	8.25	-	-	17		
(Lin Closet)	2	2.92	-	-	6		
	2	2.17	-	-	4		
(Front Ent Closet)	2	2.33	-	-	5		
	2	4.00	-	-	8		
(Ensuite)	2	7.58	-	-	15		
	2	5.25	-	-	11		
(Bathroom)	2	7.58	-	-	15		
	2	9.33	-	-	19		
(M Bed)	2	10.33	-	-	21		
	2	13.82	-	-	28		
(3-Closets)	6	6.67	-	-	40		
	6	2.42	-	-	15		
(Bed 2)	2	10.00	-	-	20		
	2	10.17	-	-	20		
(Bed 3)	2	8.75	-	-	18		
	2	10.17	-	-	20		
(Den)	2	12.08	-	-	24		
	2	9.42	-	-	19		
					446	LF	
PLUMBING Rough in and Finish	-	-	-	-	1	No.	
HEATING Rough in and Finish	-	-	-	-	1	No.	
ELECTRICAL Rough in and Finish	-	-	-	-	1	No.	

Left margin notes beside baseboard entries: 40.00, (6.67), 33.33

Figure 5.4j Finish Carpentry Takeoff

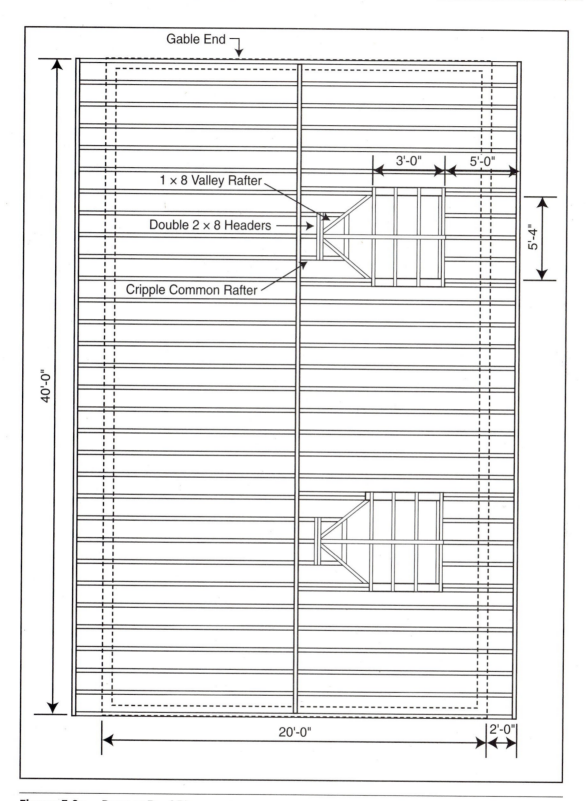

Figure 5.9a   Dormer Roof Plan

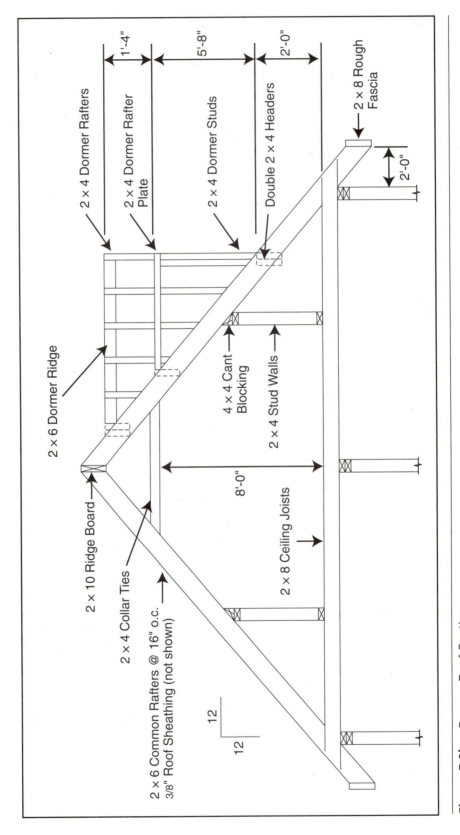

Figure 5.9b Dormer Roof Section

QUANTITY SHEET   SHEET No. | 1 of 2 |

JOB: Dormer Roof Example   DATE: _____

ESTIMATOR: DJP _____   EXTENDED: _____   EXT. CHKD: _____

DESCRIPTION	DIMENSIONS						
	TIMES	Length	Width	Height			
Roof System							
40.0 / 1.33 = 30 + 1 - 2							
= 29							
2 x 6 RAFTERS	2 x 29	16.97	-	-	984		
(Doubles)	4	16.97	-	-	68		
2 X 12^2 = 288					1052	BM	
$\sqrt{288}$ = 16.97							
GABLE ENDS - 20' SPAN	-	-	-	-	2	No.	
$^3/_8$" WALL SHEATHING	2 x ½	24.00	-	10.00	240	SF	
(To Gables)							
2 x 10 RIDGE		40.00	-	-	40		
					67	BM	
2 x 8 CEILING JOISTS	29	22.00	-	-	638		
					851	BM	
2 x 4 COLLAR TIES	29	6.00	-	-	174		
					232	BM	
2 x 8 ROUGH FASCIA	2	40.00	-	-	80		
					107	BM	
$^3/_8$" ROOF SHEATHING	2	40.00	16.97	-	1358	SF	
2 x 4 PLATES	2 x 2	40.00	-	-	160		
(Stud Walls)					107	BM	
4 x 4 CANT BLOCKING	2	40.00	-	-	80		
					53	BM	

Figure 5.10a Dormer Roof System Takeoff

QUANTITY SHEET   SHEET No. [2 of 2]

JOB: Dormer Roof Example   DATE: _____

ESTIMATOR: DJP _____   EXTENDED: _____   EXT. CHKD: _____

DESCRIPTION	DIMENSIONS						
	TIMES	Length	Width	Height			
2 x 4 STUDS	2 x 31	5.00	-	-	310		
					207	BM	
40'/1.33' = 30 + 1 = 31							
DORMER FRAMING							
2 x 8 HEADERS	2 x 2	5.33	-	-	21		
	2 x 2	2.67	-	-	11		
$2 \times 2.67^2 = 14.26$					32		
$\sqrt{14.26} = 3.78$ on plan					43	BM	
$1.33^2 + 3.78^2 = 16.07$							
$\sqrt{16.07} = 4$							
2 x 8 VALLEY	2 x 2	4.00	-	-	16		
					21	BM	
2 x 4 STUDS	2 x 2	5.67	-	-	23		
	2 x 2	4.67	-	-	19		
	2 x 2	3.67	-	-	15		
					56		
					37	BM	
2 x 4 PLATES	2 x 2	3.00			12		
					8	BM	
2 x 4 RAFTERS	2 x 8	3.00			48		
$1.33^2 + 2.67^2 = 8.91$	2 x 2	1.50			6		
$\sqrt{8.91} = 3$					54		
					36	BM	
2 x 6 RIDGE	2	5.67			11	BM	
³/₈" ROOF SHEATHING	2 x 2	5.67	3.00		68	SF	
³/₈" WALL SHEATHING	4 x ½	3.00	-	5.67	34		
(Dormer Gable)	2 x ½	5.33	-	1.33	7		
					41	SF	

Figure 5.10b   Dormer Roof System Takeoff

SUMMARY

- A thorough knowledge of carpentry construction is required to prepare a detailed estimate of this trade. This is particularly true for estimating wood frame structures where carpentry details are often lacking.
- Lumber is usually described according to its *nominal size,* which is larger than the actual *dressed* size of lumber.
- The unit of measure of lumber is generally the board measure (BM).
- To calculate the BM of a piece of lumber, multiply the length of the piece in feet by the nominal width and thickness of the piece in inches, and then divide by 12.
- Items of lumber are classified in terms of:
  - Dimension
  - Dressing
  - Grade
  - Species
- Framing lumber is further classified in terms of use, for example, plates, joists, lintels, etc.
- Trusses and rafters are described and enumerated.
- Sheathing and siding is measured in square feet or square meters.
- Items of finish carpentry are generally classified and measured separately according to materials, size, and method of fixing.
- Cabinets are described and enumerated; alternatively, units are measured in linear feet describing type and size of unit.
- Doors and windows are enumerated stating type and size of each. This information is usually obtainable from door and window schedules provided on the drawings.
- Bathroom accessories are enumerated and fully described in the takeoff process.
- Finish hardware is enumerated and fully described in the takeoff.
- Exterior and interior finishes are usually subcontracted, but the builder's estimator should be able to takeoff and price this work if required.
- A takeoff of rough carpentry and finish carpentry work for a house project is demonstrated.

RECOMMENDED RESOURCES

Information	Web Page Address
■ Carpentry details—There are many books and online resources illustrating carpentry details; this is just one of them.	http://openlibrary.org/books/ OL2876275M/Basic_carpentry_ illustrated
■ Parallam Beams	http://www.usglu-lam.com/ products/parallambeams/
■ Engineered Joists	http://www.usglu-lam.com/ products/ijoist/

REVIEW QUESTIONS

1. Why is it desirable for the estimator of carpentry work to have a thorough knowledge of carpentry construction?
2. Describe how lumber is differentiated in the takeoff.

3. How are wallboards measured?
4. List six framing classifications that have to be measured separately in a takeoff.
5. Describe how air barriers are measured.
6. What is a "parallam" beam?
7. Describe a quick method of calculating the length of decking required for a given area of deck.

PRACTICE PROBLEMS

1. Calculate the board measure amounts in Figure 5.11.
2. If a floor measured 20'-0" × 31'-0", how many joists at 16" on center would be required if they spanned the 20'-0" dimension?
3. Calculate the length of barge rafters from the following data:

	Case 1	Case 2	Case 3	Case 4
Roof slope	2 : 12	3 : 12	4 : 12	6 : 12
Building width	24'-0"	29'-0"	30'-0"	36'-0"
Eaves overhang	2'-0"	2'-6"	3'-0"	4'-0"

4. Take off all the roof system components (except for dormers) shown in Figure 5.9a and Figure 5.9b for a building of dimensions 50'-0" × 24'-0" measured to the outside of the exterior walls.

Takeoff Item	Board Measure
100 pieces of 1 x 3 - feet long	
25 pieces of 1 x 6 - 6 feet long	
11 pieces of 1 x 8 - 10 feet long	
60 pieces of 2 x 3 - 8 feet long	
36 pieces of 2 x 4 - 12 feet long	
19 pieces of 2 x 6 - 10 feet long	
90 pieces of 3 x 3 - 12 feet long	
14 pieces of 4 x 4 - 18 feet long	
22 pieces of 4 x 10 - 16 feet long	

Figure 5.11   Board Measure Calculations

6

MEASURING MASONRY AND FINISHES

OBJECTIVES

After reading this chapter and completing the review questions, you should be able to:

- Explain how masonry work and finishes are measured in a takeoff.
- Measure masonry items, exterior, and interior finishes from drawings and specifications.
- Explain how masonry items and finishes are classified in the takeoff process.
- Describe the main factors affecting the measurement of brick masonry.
- Use conversion factors to calculate quantities of bricks, blocks, and masonry mortar.
- Given the size of masonry units and the thickness of the mortar joints between them, calculate the value of conversion factors to determine the number of bricks/blocks and the volume of mortar.
- Complete a takeoff of masonry work, exterior, and interior finishes for a residential project.

KEY TERMS

brick-on-edge course	R-value	standard bricks
ladder reinforcement	scope of work	
net in place	soldier course	

Introduction

Masonry work includes construction with clay bricks, concrete bricks and blocks, clay tiles, natural and artificial stone. Subcontractors usually undertake masonry work; they supply materials, equipment, and the labor required to complete the work of this trade. Here we will deal with the measurement of standard clay brick and concrete block masonry. While masonry work that involves products other than **standard bricks** and concrete blocks may be measured in a similar fashion to that described herein, this specialized work is beyond the scope of this text.

As we have previously stated regarding other trades, a thorough knowledge of masonry construction is a definite asset when taking off the work in this trade. Particular attention has to be paid to items such as masonry bond and mortar joint treatment that have significant effect upon material usage, waste factors, and mortar consumption. A minimum knowledge of terminology is required before masonry requirements can be properly interpreted to determine such items as the type of units required and their requisite features, together with aspects of the numerous accessories associated with masonry work. We have not attempted to explore any of the many features of masonry construction practices in this text. The student is advised to consult and study one or more of the many texts that deal with this specific subject.

Measuring Masonry Work

In accordance with the general principles followed in the measurement of the work of other trades, masonry work is measured **net in place**, and the necessary allowances for waste and breakage are considered later in the estimating process when this work is priced. The units of measurement for masonry are generally number of masonry pieces such as concrete blocks or, in the case of bricks, the number of thousands of clay bricks. Calculating the number of masonry units involves a two-stage process:

1. The area of masonry is measured.
2. A standard factor is applied to determine the number of masonry units required for area measured.

A number of other items associated with unit masonry which may or may not be detailed or noted on the drawings also have to be measured, including mortar, wire **ladder reinforcement** to joints, metal ties used between walls, and extra work and materials involved in control joists, sills, lintels, and arches. All of these items are considered in the masonry takeoff as detailed in the following text. Loose and rigid insulation to masonry is also included in this work, and in some places the masonry trade includes membrane barriers, insulation, and similar materials behind masonry where the masonry is the last work installed.

Estimators need to be familiar with the **scope of work** in their geographical location. This defines what items the masonry contractor has to include and what items of work are excluded from their trade. The masonry estimator will then be able to takeoff and price everything that is included in the masonry trade on the project. The knowledge of scope is essential to the person estimating the masonry work for a specialized masonry trade contractor; it is also valuable to the builder's estimator whose job it is to ensure that all the work of the project is included in the total price, whether the work is to be performed by the masonry trade or someone else.

The builder estimator's role in handling problems associated with a subtrade's scope of work is discussed in Chapter 11. The builder's estimator will, however, need to address items of work that are connected with the masonry trade but which are specifically excluded from the subtrade quote. If these items of work are not to be included as part of the subtrade's work, their cost will have to be estimated with the builder's work.

A typical example of the kind of work item that is related to masonry work but may not be included in the quote from a subtrade is insulation material. Even if rigid insulation is attached to the blockwork, the masonry subtrade may not supply or install it. Therefore, it would have to be measured and priced separately.

Brick Masonry

Bricks are generally made in large batches in a manufacturing plant where a mixture of clay and water is first dried, then molded, and lastly *fired* to produce the finished product. They can also be made of different materials, manufactured by different methods, and also, they can be used in many different ways in the construction process. All of these factors will influence the price of the masonry, but the main factors affecting the measurement of brick masonry are the size of the brick units, the size of the joints between bricks, the wall thickness, and the pattern of brick bond utilized. Bricks are available in a large number of different sizes, but the Common Brick Manufacturers Association has adopted a standard size with the nominal dimensions of 2¼" by 3¾" by 8". Bricks of this size are referred to as standard bricks, and this is the brick size used in the takeoff examples that follow.

All of the variables previously listed will also impact the measurement of the mortar needed for brick masonry. From the size of the bricks and the thickness of the mortar required, the amount of mortar per 100 square feet of wall can be readily calculated; examples are shown below. To avoid having to perform these calculations for all of the combinations of sizes involved, many estimators refer to texts that offer tables of factors to use in calculating numbers of masonry units and volumes of mortar per 100-square foot of wall area. For example, from Walker's *The Building Estimators Reference Book*[1], the amount of mortar required for the bricks previously described is specified as 5.7 cubic feet per thousand bricks. Information about quantities of bricks per square foot and mortar requirements over a wide range of brick sizes is also available from brick manufacturers and from organizations such as the Brick Institute of America (BIA).

Sometimes designated courses of bricks are required to be laid in a different way than the rest of the courses in a wall. For example, a **soldier course** may be required above openings or a **brick-on-edge course** may be required at the sill of openings. These features are measured separately because the number of bricks per square foot of the special course may be different from the number of bricks per square foot of wall area generally.

Concrete Blocks

There are fewer variables to consider with concrete blocks than with clay bricks, but blocks do come in different sizes, and the thickness of joints between blocks can vary. See Figure 6.1 for standard block and lintel block sizes together with some architectural blocks for an 8-inch thick wall.

Conversion Factors

As mentioned, tables are published in reference works that list quantities of bricks or blocks per square foot area of masonry. The number of "standard bricks" per square foot of wall for different wall thicknesses, joint thicknesses, and various bonds is given in Walker's reference book which states, for instance, that a 4½-inch wall with ⅜-inch brick joints and laid in running bond will have 6.67 bricks per square foot of wall area.

Tables also list the number of standard-sized concrete blocks and the amount of mortar required per 100 square feet of wall. From the same reference source cited,

---

1. *The Building Estimator's Reference Book*, published by Frank R. Walker Company, 28th. Edition, Chicago, 2006.

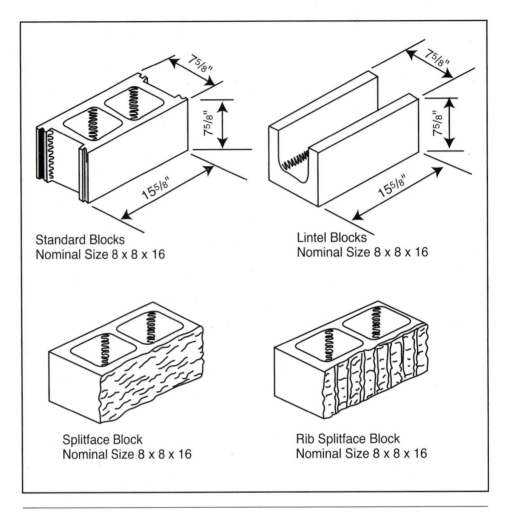

Figure 6.1 Concrete Blocks

for a wall of 8-inch nominal thickness built with standard blocks of size 7⅝ by 7⅝ by 15⅝ inches and a joint thickness of ⅜ inches, the tables indicate that 112.5 blocks together with 6 cubic feet of mortar are required for 100 square feet of wall area.

Where non-standard size brick or block units are to be used, the estimator may have to calculate the conversion factors. So, if bricks of size 3⅝ by 3⅝ by 15⅝ inches with ⅜-inch wide mortar joints, the conversion factor may be calculated in the following way:

The length of a brick is 15⅝ plus a mortar joint (³/₈) = 16 inches

The height of a brick is 3⅝ plus a mortar joint (³/₈) = 4 inches

So, the area of the face of a brick, including mortar joints, is 16 × 4 = 64 square inches

1 square foot (144 square inches) / 64 square inches = 2.25

(Note: Only the dimensions of the face of the brick are required; the brick thickness does not affect the calculation.)

Therefore, there are 2¹/₄ bricks for each square foot of wall face.

For the volume of mortar per square foot of wall area, the volume of brick plus joints is calculated less the volume of brick per square foot of wall area. (Note: Here the brick thickness—3⅝—does have to be taken into account.)

EXAMPLE 6.1

Volume of brick + joints	= 3⅝ × (3⅝ + ⅜) × (15⅝ + ⅜)
	= 232 cubic inches/brick
And: since 1 cu. foot	= 1728 cubic inches
Volume of brick + joints	= 232 / 1728
	= 0.134 cubic feet/brick
Volume of brick alone	= 3⅝ × 3⅝ × 15⅝
	= 205.23 cubic inches
Thus, 205.23 / 1728	= 0.119 cubic feet/brick
Therefore, the volume of mortar	= 0.134 − 0.119
	= 0.015 cubic feet/brick
Hence, the volume of mortar per square foot of wall area	
	= 0.015 × 2.25
	= 0.035 cubic feet/square foot

This is equivalent to 0.129 cubic yards of mortar per 100 square feet of wall area. In practice, however, this mortar quantity will be too low because, in addition to mortar wasted and spilled, in the process of laying bricks some mortar is squeezed into the voids in the bricks and, sometimes, mortar is left to bulge from the joints. Therefore, the mortar factor could be increased by as much as 40 percent to allow for the additional quantity required.

Measuring Notes—Masonry

1. Measure all masonry quantities "net in place" and do not deduct for openings that are less than 10 square feet.
2. Masonry work includes scaffolding and hoisting.
3. Measure masonry work in the following categories:
   a. Facings
   b. Backing to facings
   c. Walls and partitions
   d. Furring to walls
   e. Fire protection
4. Masonry work that is circular on plan is measured separately.
5. Describe cleaning of exposed masonry surfaces and measure in square feet.
6. Measure silicone treatment of masonry surfaces in square feet stating the number of coats required.
7. Measure expansion joints or control joints in masonry in linear feet and fully describe these joints. Specify any required joint filler material (caulking) in the description of the joint system.
8. Measure mortar in cubic feet and provide details of any admixtures required.
9. Separately measure colored mortar where it is required.

10. Measure wire reinforcement in masonry joints in linear feet and fully describe. "Ladder" or "truss" reinforcement may be specified; indicate which is required.
11. Enumerate anchor bolts, sleeves, brackets, and similar items that are built into masonry and fully describe these items.
    (Note: Built-in metal components, such as anchor bolts, lintels, and similar items may be supplied and installed by the Masonry Trade; alternatively, they could be supplied by the Miscellaneous Metals subtrade and installed by the Masonry Contractor. Clarify which is the case in the takeoff.)
12. Measure building in lintels, sills, copings, flashings, and similar items in linear feet and fully describe the items involved.
13. Enumerate weep holes where they are required to be formed using plastic inserts and suchlike.
14. Measure rigid insulation to masonry work in square feet describing the type and thickness of material.

Brick Masonry

15. Enumerate bricks describing the type and dimensions of the bricks.
16. Separately measure bricks required to be laid in any other pattern than running bond, examples: soldier courses or brick-on-edge courses.
17. Enumerate and fully describe brick ties.

Concrete Block Masonry

18. Enumerate concrete block masonry units stating the type and size of blocks.
19. Separately enumerate special units required at corners, jambs, heads, sills, and other similar locations.
20. Measure in cubic yards and fully describe loose fill or foam insulation to blockwork cores.
21. Measure in cubic yards concrete in core fills and bond beams stating the strength and type of concrete to be used.
22. Measure in linear feet reinforcing steel to core fills and bond beams stating the size and type of rebar to be used.

Trade rules in some locations require the subcontractor to supply reinforcing steel to the project to supply the rebar for the masonry work. The masonry subtrade or the builder, whomever is completing the masonry work, then installs this rebar.

Example of Concrete Block Masonry Takeoff

See Figure 6.2 for details of an alternative concrete block basement for the house, and see Figure 6.3 for the takeoff of an alternative concrete block basement for the house.

Comments on the Concrete Block Masonry Takeoff

1. The concrete footing is the same as that for the original concrete foundation design, so the footing concrete and forms takeoff remains the same as before.
2. For the number of concrete blocks for the wall, first calculate the area of the wall, then multiply this area by 112.5 blocks per 100 square feet.
3. There are no deductions for windows or the drop landing since none of these openings exceeds 10 square feet.
4. Multiply the area of the wall by 6/100 to determine the cubic feet of mortar required for this concrete block wall.

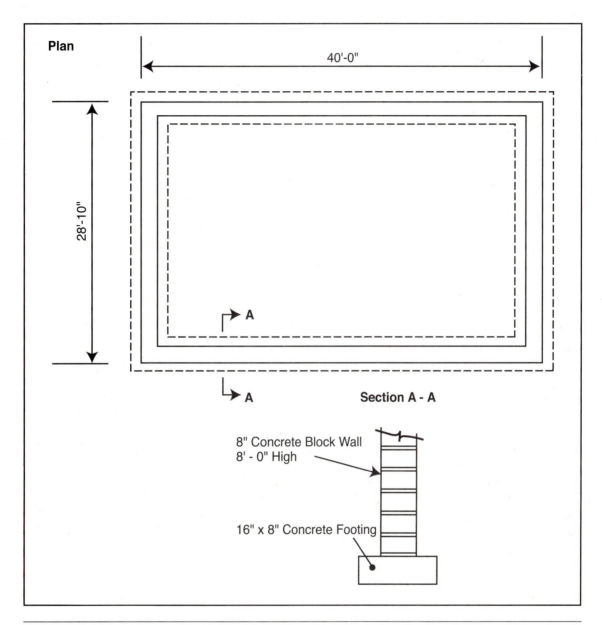

Figure 6.2   Alternative Concrete Block Foundation

5. Six-inch wide truss type reinforcing is required in every second course of the blockwork. The length of this material can be quickly determined by dividing the area of the wall by the spacing of the reinforced courses, which is 16".

6. Lintel blocks are needed over the window openings. The width of each window is 3'-0", and the lintel extends 28" wider than the opening (14" each side). This extension of the lintel provides support, and the length 28" was used to give a total of 64", which is exactly the length of four blocks ($4 \times 16" = 64"$).

7. The lintel blocks are filled with concrete and reinforced with a single #4 size reinforcing rod.

QUANTITY SHEET						SHEET No.	1 of 1

JOB: _____ House Example _____ DATE: _____

ESTIMATOR: ABF _____ EXTENDED: _____ EXT. CHKD: _____

	DESCRIPTION		DIMENSIONS					
		TIMES	Length	Width	Height			
	Concrete and Masonry Work							
2 x 40.00	80.00							
2 x 28.83	57.66							
	137.66 Outside Wall Perimeter							
4 x 0.67	(2.66)						2 x 8	2 x 8
	135.00 Centerline of Wall					Conc.	Forms	Forms
						L x W x D	2 x L	No. x 2 x (L+W)
	CONC. FTGS.		135.00	1.33	0.67	120	270	-
	(Pads)	3	2.67	2.67	0.67	14	-	32
		2	2.00	2.00	0.67	5	-	16
						139	270	48
						5	CY	270
								318 LF
	8" CONC. BLOCK WALL		135.00		8.00	1,080	x 112.5/100 =	1215 Blocks
	MORTAR		-	-	-	1,080	x 6.00/100 =	65 cu. ft.
								2 CY
	JOINT REINFORCING		-	-	-	1,080	/ 1.33 =	812 LF
	(6" wide truss every 2nd course)							
	LINTEL BLOCKS	4	5.33			21	/ 1.33 =	16 Blocks
	Deduct 8" CONC. BLOCKS		-	-	-	< 16 >	Blocks	
	CONC. IN BOND BEAMS	4	5.33	0.50	0.58	6	cu. ft.	
	#4 REBAR	4	5.33			21	LF	

Figure 6.3   Alternative Masonry Foundation Takeoff

Example of Brickwork Takeoff

Figure 6.4 and Figure 6.5 give details of an alternative exterior finish of brick facings to the front of the house. See Figure 6.6 for the takeoff of this brick facings exterior finish to the front of the house.

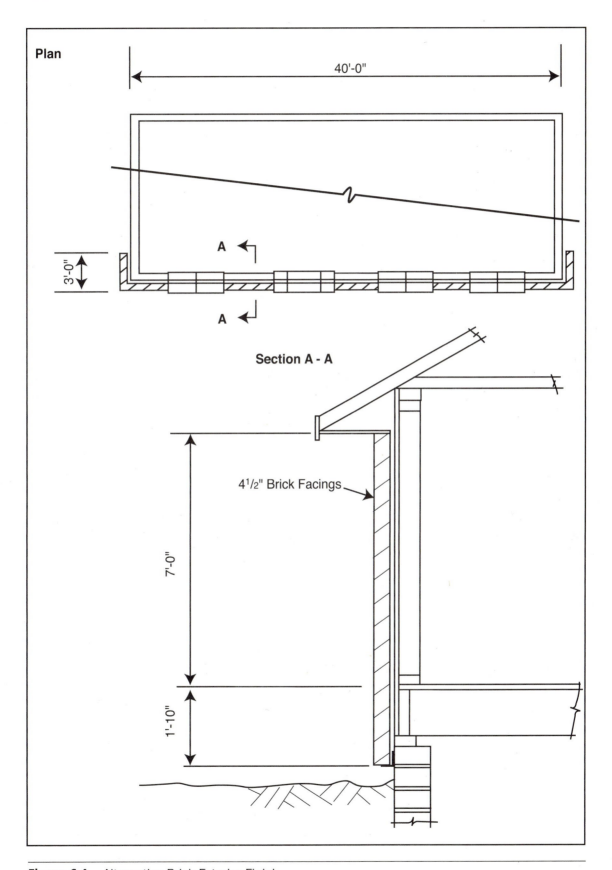

Plan

40'-0"

3'-0"

A

A

Section A - A

4¹/₂" Brick Facings

7'-0"

1'-10"

Figure 6.4   Alternative Brick Exterior Finish

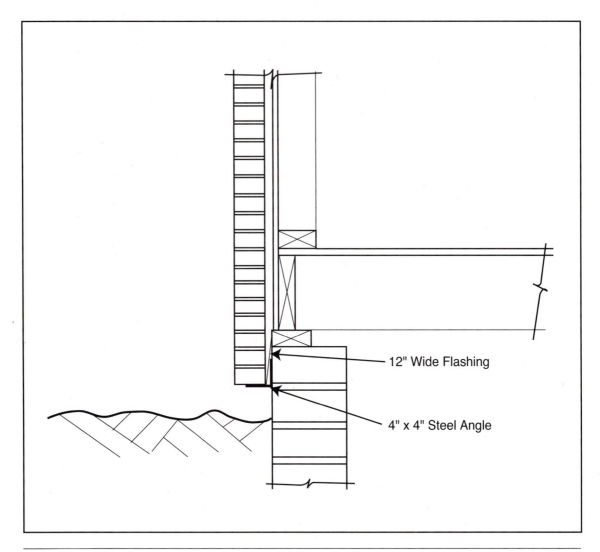

12" Wide Flashing

4" x 4" Steel Angle

Figure 6.5 Masonry Wall Details

Comments on Brickwork Takeoff

1. Figure 6.4 shows facing bricks are to be applied to the front left side of the house as shown on the sketches. The front width plus the two returns gives an overall brickwork length of 46'-0".
2. The *net* area of brickwork is required so all window openings are deducted (but any opening less than 10 square feet is ignored).
3. To calculate the number of bricks with a ⅜" mortar joint, multiply the wall area by the standard factors of 6.67 bricks per square foot.
4. The amount of mortar required for bricks with this joint thickness is 8.7 cubic feet per 1,000-bricks.
5. Brick ties are spaced 2'-0" horizontally and 1'-4" vertically; the number of ties required is obtained by dividing the wall area by the product of the spacing (2 × 1.33).

QUANTITY SHEET SHEET No. 1 of 1

JOB: House Example DATE: _____

ESTIMATOR: ABF EXTENDED: _____ EXT. CHKD: _____

DESCRIPTION	TIMES	Length	Width	Height			
Brick Facings Alternate							
40.00 7.00							
6.00 1.83							
46.00 8.83							
FACING BRICKS		46.00	-	8.83	406		
(Windows) DDT	3	5.33	-	3.08	(49)		
(Front Door) DDT		9.17	-	6.67	(61)		
					296	x 6.67 =	1973 No.
MORTAR		-	-	-	1,973	x 8.7 / 1000	17 cu. ft.
BRICK TIES		-	-	-	296 /(2 x 1.33)	=	111 No.
(Spacing 24" x 16")							
4" x 4" x 1/4" ANGLE		46.00			46		
	3	7.33			22		
5.33 9.17		11.17			11		
2.00 2.00					79	LF	
7.33 11.17							
1/2" DRILLED ANCHORS	24	-	-	-	24	No.	
46.00 / 2.00 = 23 + 1 = 24							
12" WIDE FLASHING		46.00			46 LF		
WEEP HOLES	17	-	-	-	17	No.	
46.00 / 3.00 = 16 + 1 = 17							

2 x 3.00 (against Brick Facings Alternate rows)

2 x 1.00 (against ANGLE rows)

Figure 6.6 Alternative Brick Facings Takeoff

6. The steel angle lintel over the windows and the front entrance extends 1 foot on each side of the opening, so 2 feet is added to the opening width to give the length of lintel required.

7. The shelf angle supporting the full length of the brick wall is attached to the footing by means of drilled anchors into the concrete foundation wall. The number of these inserts is found by dividing the length of the angle by the spacing of the inserts. See figure 6.5.

8. A 12" wide flashing is installed at the foot of the wall; this is the same length as the angle support at this location.

9. Weep holes at 4'-0" spacing are built into the lowest brick course which sits on top of the shelf angle. These allow water that seeps behind the facings to drain to the outside of the brickwork.

Measuring Exterior and Interior Finishes

Because there can be a large number of different items to take into account in the finishes to a building, the estimator needs to follow a systematic method of measuring this work in order to prepare an accurate takeoff. The method adopted here for exterior finishes is to begin by considering what is required for the foundations, then for the exterior walls, followed by the roof. All of the foundation treatments and finishes are measured before moving on to the items related to the walls above grade; then, once this is completed, the roofing items are dealt with.

Measuring interior finishes begins by considering all the work associated with the floors, next the walls finishes and associated items are measured, and finally, finishes to ceilings are measured.

Measuring Notes—Exterior Finishes

Finishes Generally

1. Measure exterior and interior finishes in square feet.
2. Do not deduct for openings less than 40 square feet.

Damp Proofing and Waterproofing

3. Measure damp proofing and waterproofing separately in the following categories stating the material to which it is applied:
   a. Damp proofing
   b. Built-up bituminous waterproofing
   c. Sheet waterproofing
   d. Fluid applied waterproofing
   e. Water repellants
4. Separately measure the protective covering.
5. Measure grooves, chases, and similar items in linear feet.

Thermal Protection

6. Measure thermal protection in the following categories stating the resistance of the material to heat flow (the **R-value** of the material):
   a. Building insulation
   b. Roof and deck insulation
   c. Exterior insulation and finish systems

       d. Vapor barriers

       e. Air barriers (building wrap)

  7. Describe and measure separately insulation installed by blowing into place.

Siding

  8. Measure siding in the following categories stating the material to which it is applied:

       a. Aluminum siding

       b. Wood and plywood siding

       c. Vinyl siding

       d. Composite wall panels

       e. Fiber-reinforced cementitious panels

       f. Exterior wall assemblies

Roofing

  9. Measure roofing in the following categories stating the material to which it is applied:

       a. Manufactured roof panels

       b. Plastic roof panels

       c. Manufactured composite roof panels

       d. Asphalt singles

       e. Wood shakes

       f. Roof tiles

       g. Built-up bituminous roofing

       h. Single membrane roofing

 10. Separately describe the component parts of roof systems including: sheathing, paper, primer, vapor barrier, underlayment, insulation, felt, asphalt, gravel, etc.

Flashings and Sheet Metal

 11. Measure flashings and sheet metal in linear feet stating the width in the following categories:

       a. Sheet metal roofing

       b. Sheet metal flashings

       c. Plastic flashings

 12. Separately measure gutters, valleys, hips, and underlayment strips.

Roof Specialties and Accessories

 13. Measure roof specialties and accessories in linear feet in the following categories:

       a. Eaves trough and gutters

       b. Down spouts

 14. Describe and enumerate roof scuppers.

 15. Describe and enumerate plash pads.

 16. Measure roof pavers in square feet.

 17. Describe and enumerate roof hatches.

Example of Exterior Finishes Takeoff

Figure 6.7a and Figure 6.7b show the takeoff for the exterior finishes takeoff.

QUANTITY SHEET					SHEET No.	1 of 2	

JOB: **House Example** DATE: _____

ESTIMATOR: **ABF** EXTENDED: _____ EXT. CHKD: _____

	DESCRIPTION	TIMES	Length	Width	Height			
				DIMENSIONS				
	Exterior Finishes							
2 x 40.00 =	*80.00*							
2 x 28.83 =	*57.66*							
	137.66							
	ASPHALT D/PROOFING		*137.66*	*-*	*8.00*	*1,101*	*SF*	
	97.42							
	97.58							
	97.58							
	97.50							
	390.08 *100.00*							
Ave. Elev.	*97.52* → *(97.52)*							
	2.48 – 0.83 + 0.50 = 2.15							
	1/2" PARGING		*137.66*	*-*	*2.15*	*296*	*SF*	
	8.00							
	(1.00)							
	1.08							
	8.08							
	VINYL SIDING		*137.66*	*-*	*8.08*	*1,112*		
	(Windows C) DDT	*3*	*5.33*	*-*	*3.08*	*(49)*		
	7.00 (Entrance)		*5.58*	*-*	*6.15*	*(34)*		
Add	*0.96* (Patio Door 2) DDT		*5.00*	*-*	*6.67*	*(33)*		
Less	*(1.81)* (Window B) DDT		*3.50*	*-*	*6.00*	*(21)*		
	6.15 (Window E) DDT		*2.75*	*-*	*2.42*	*(7)*		
	(Window D) DDT		*6.92*	*-*	*3.08*	*(21)*		
						946	*SF*	
	BUILDING WRAP		*1,112 sf + 10%*			*1,223*	*SF*	
	PAINT EXTERIOR WINDOWS (B)		*3.50*	*-*	*6.00*	*21*		
	(C)	*3*	*5.33*	*-*	*3.08*	*49*		
	(D)		*6.92*	*-*	*3.08*	*21*		
	(E)		*2.75*	*-*	*2.42*	*7*		
	(F)	*4*	*3.08*	*-*	*1.08*	*13*		
						112	*SF*	
	PAINT EXTERIOR DOORS (1 + G)		*5.58*	*-*	*6.67*	*37*		
	(2)		*5.00*	*-*	*6.67*	*33*		
						71	*SF*	

Figure 6.7a   Exterior Finishes Takeoff

QUANTITY SHEET SHEET No. 2 of 2

JOB: House Example DATE:

ESTIMATOR: ABF EXTENDED: EXT. CHKD:

DESCRIPTION	DIMENSIONS						
	TIMES	Length	Width	Height			
Exterior Finishes (Cont'd)							
40.00							
2 x 2.00 = 4.00							
44.00							
4" EAVES GUTTER	2	44.00	-	-	88	LF	
2.50							
8.00							
3.00							
13.50							
3" DOWN SPOUTS	2	13.50	-	-	27	LF	
40.00 x 28.83							
4.00 x 4.00							
44.00 x 32.83							
210 lbs ASPHALT SHINGLES	1.054	44.00	-	32.83	1,523		
(Starter Strip)	2	44.00	-	1.00	88		
					1,611	SF	
RIDGE CAP		44.00	-	-	44	LF	
4" WIDE DRIP-EDGE FLASHING	2	44.00	-	-	88	LF	
7'-0" x 3'-6" PRECAST CONC. STEP	1	-	-	-	1	No.	
7'-0" x 3'-6" GALV. IRON CANOPY COVER	1	-	-	-	1	No.	
STAIN CEDAR DECK		15.75	5.58	-	88	SF	
STAIN CEDAR RAILINGS		27.67	-	3.33	92	SF	

Figure 6.7b Exterior Finishes Takeoff

Comments on the Exterior Finishes Takeoff

1. The measurement of exterior finishes begins at the foundations and proceeds up to the roof finish.
2. On this house, the finishes consist of asphalt damp proofing on the foundations below grade, parging on the exposed foundations, and vinyl siding on the framed walls, all of which are measured in square feet. No deductions are made for openings less than 40 square feet of damp proofing and parging or for openings less than 1 square foot of siding.
3. The damp proofing extends around the perimeter of the basement, so the takeoff begins with a calculation of the outside perimeter.
4. Sand/cement parging is to be applied to the exposed top of the concrete foundation wall. Parging will extend from the bottom of the siding (which lines up with the bottom of the floor joists) down to 6" below grade. The grade is not constant, though, so first we calculate the average elevation of grade around the building that is found to be 97.52'. The elevation of the top of the floor is 100.00', so the distance from grade to the floor is 2.48'. Next, we deduct the height of joist and floor deck, which is 0.83', and finish the calculation by adding 0.5'. This is the depth that the parging extends below grade, so the full height of the parging works out to be 2.15'.
5. The siding is applied over the joist headers but only up to the eaves soffit, which gives a height of 8'-1".
6. The area of all openings (over 1 square foot) are deducted from the siding but not from the building wrap since only openings over 40 square feet are deducted from building wraps, vapor barriers, and the like.
7. The exterior of windows and doors is to be painted. Note that there are no deductions for the glass from the window areas measured.
8. Eave gutters and down spouts are not shown on the drawings but are allowed for here.
9. The quantity of roof shingles is left in square feet here because this is a small area. But roofing is usually measured in squares that are areas of 100 square feet with no deduction for openings of less than 40 square feet.
10. Precast concrete steps are taken off here although they are not strictly exterior finishes.
11. Eave soffits, fascias, and trim are all prefinished, otherwise painting would have to be measured here.

Measuring Notes—Interior Finishes

Flooring

1. Measure flooring in the following categories:
   a. Floor treatment
   b. Resilient flooring
   c. Carpet
   d. Fluid-applied flooring
   e. Specialty flooring
2. Describe base, treads, nosings, risers, feature strips, and the like, and measure these items in linear feet.
3. Describe and measure separately flooring with decorative patterns, boarders, game courts, etc.
4. Fixing in place, cutting, fitting around obstructions, edging, threshold strips, etc., are all deemed to be included.

Gypsum Wallboard and Plaster

5. Measure gypsum wallboard and plaster in the following categories stating whether applied to walls or ceilings:
   a. Gypsum wallboard (sheetrock)
   b. Furring and lathing
   c. Gypsum plaster
   d. Portland cement plaster
6. Describe and measure separately gypsum wallboard and plaster to columns, pilasters, beams, bulkheads, ducts, and the like.
7. Describe coves, cornices, bases, moldings, and suchlike and measure in linear feet.
8. Enumerate column bases and caps to columns.
9. Corner beads, joint and strip reinforcement, holes, notches, etc., in wallboard are all deemed to be included; that is, these items are not measured but they will be allowed for when the work is priced.

Tile Finishes

10. Measure tile finishes in the following categories stating whether applied to floors, walls, or ceilings:
    - Ceramic tile
    - Quarry tile
    - Stone tile
    - Glass mosaics
    - Plastic tile
    - Metal tile
    - Acoustic tile
    - Mirror tile
11. Do not deduct for areas more than 1 square foot.
12. Describe and measure separately tiles to columns, pilasters, beams, bulkheads, ducts, and the like.
13. Measure round-edged tiles, internal and external angle tiles, coved base tiles, and suchlike in linear feet.
14. Enumerate tiles forming letters and numbers.

Paint and Coatings

15. Measure paint and coatings in the following categories stating whether applied to floors, walls, or ceilings:
    a. Paint work
    b. Stains and transparent finishes
    c. Laminated plastic coatings
    d. High-performance coatings
16. Measure separately exterior and interior work in the above categories.
17. The contact area of the surface to be finished is the area that is measured.
18. Describe and measure separately special decorative work such as striped colors or scratched or patterned work.
19. Measure windows and doors overall, both sides; make no deductions for glass areas.
20. Measure balustrades, railings, fencing, and similar items overall, both sides.
21. Measure handrails and similar items in linear feet.
22. Measure open-wed steel joists overall, both sides.

23. Enumerate painted letters and numbers.
24. Preparation, covers, protection, and cleanup are deemed to be included.

Example of Interior Finishes Takeoff

Comments on Interior Finishes Takeoff, Figure 6.8a and Figure 6.8b

1. Flooring, drywall wallboards, insulation, vapor barriers, paint, and other finishes are measured in square feet with no deductions for openings less than 40 square feet.
2. The main floor of the house is carpeted except for the kitchen, bathrooms, and front entrance (including the closet by the front entrance). These areas are to be finished with vinyl flooring.
3. The stairs down to the basement are to be carpeted. Generally the area of the stairs is included in the main floor area, but $3.33' \times 2.00'$ has to be added because these stair treads show up in the basement and are not part of the main floor plan.
4. Also, the area of the stair risers is added because it too is not captured in the measurement of the main floor area.
5. A carpet-edging strip is required where the carpet meets the vinyl flooring by the kitchen and at the bathroom doorways.
6. The wallboard ceiling is treated with a textured finish.
7. Loose insulation is spread in the attic space above the ceiling with a vapor barrier between the drywall ceiling and the insulation.
8. Insulation stops are installed under the roof between the tails of the trusses to maintain a space for ventilation above the insulation over the eaves. Divide the length of the roof (40 feet) by the spacing of the trusses (2 feet) to determine the number of stops on each side of the roof.
9. The exterior wall of the basement is insulated and finished with pained wallboards.
10. The length of the drywall required for the main floor (446 feet) is obtained from the previous takeoff of the baseboard, and the length of the basement drywall corresponds to the interior perimeter of the basement.
11. Paint is measured on both sides of doors and on the top and bottom of closet shelves.
12. Painting railings and balustrades is measured in square feet calculated by multiplying the length of the railing by the height.

QUANTITY SHEET   SHEET No. | 1 of 2 |

JOB: House Example   DATE:

ESTIMATOR: ABF   EXTENDED: EXT. CHKD:

DESCRIPTION		DIMENSIONS					
	TIMES	Length	Width	Height			
Interior Finishes							
- Floors							
CARPET		40.00	28.92	-	1,157		
(Stairs)		3.33	2.00	-	7		
2.00 (Stair Risers)	14	3.33	0.58	-	27		
7.25 (Kitchen) DDT		11.25	8.67	-	(98)		
2.00 (Bathrooms) DDT		15.34	7.59	-	(116)		
11.25 (Front Entrance) DDT		9.16	4.17	-	(38)		
13.67 1.75 3.50					938	SF	
1.67 2.92 3.33							
15.34 2.92 2.33							
7.59 9.16							
VINYL FLOORING (Kitchen)		11.25	8.67		98		
(Bathrooms)		15.34	7.59		116		
(Front Entrance)		9.16	4.17		38		
					252	SF	
CARPET EDGING STRIP		6.00			6		
	2	2.33			5		
					11	LF	
1/2" DRYWALL CEILING		40.00	28.92		1,157	SF	
TEXTURED CEILING FINISH		-	-	-	1,157	SF	
R35 LOOSE INSULATION		-	-	-	1,157	SF	
6mil POLY V.B.		1,157 SF + 10%		-	1,273	SF	
INSULATION STOPS	2	20			40	No.	
40.00 / 2 = 20							

Figure 6.8a   Interior Finishes Takeoff

QUANTITY SHEET   SHEET No. | 2 of 2 |

JOB: __House Example_____   DATE: _____

ESTIMATOR: __ABF_____   EXTENDED: _____   EXT. CHKD: _____

DESCRIPTION		TIMES	Length	Width	Height			
Exterior Walls								
R20 BATT INSULATION			137.66	-	8.00	1,101	SF	
6mil POLY V.B.			1,101 SF + 10%			1,211	SF	
137.66 Outside Perimeter								
-4 x 2 x 0.67 = *(5.33)*								
132.33 Inside Perimeter								
R12 BATT INSULATION			132.33	-	8.00	1,059	SF	
(Basement)								
6mil POLY V.B.			1,059 SF + 10%			1,165	SF	
All Walls								
1/2" DRYWALL WALLS	(Main)		446.00	-	8.00	3,568		
	(Basement)		132.33	-	8.00	1,059		
						4,627	SF	
PAINT WALLS			as above			4,627	SF	
PAINT DOORS	(3)	2 x 2	2.33	-	6.67	62		
	(4)	2 x 4	2.50	-	6.67	133		
	(6)	2	2.83	-	6.67	38		
	Bifold (7)	2	2.00	-	6.67	27		
	Bifold (8)	2 x 3	4.00	-	6.67	160		
	Bifold (9)	2	3.00	-	6.67	40		
						460	SF	
PAINT WOOD RAILINGS			9.33	-	3.44	32	SF	
PAINT SHELVES		2 x 3	6.00	1.25		45		
		2	4.00	1.25		10		
		2 x 4	2.25	1.25		23		
						78	SF	

Figure 6.8b   Interior Finishes Takeoff

SUMMARY

- Masonry work includes construction with clay bricks, concrete bricks and blocks, clay tiles, and natural and artificial stone.
- A thorough knowledge of masonry construction is required to prepare a detailed estimate of this trade.
- First the surface area of the masonry is measured in square feet in the takeoff process, then a factor is applied to convert these measurements to number of pieces such as bricks or blocks required to complete the work.
- A number of other items associated with unit masonry are also measured in the takeoff including mortar, wire "ladder" reinforcement to joints, metal ties used between walls, and extra work and materials involved in control joists, sills, lintels, and arches.
- Masonry estimators need to be familiar with the definition of the scope of work for the masonry trade in their geographical location.
- The main factors affecting the measurement of brick masonry are the size of the brick units, the size of the joints between bricks, the wall thickness, and the pattern of brick bond utilized.
- Where bricks are required to be laid in a different pattern than the other bricks in a wall, such as a soldier course or a brick-on-edge course, they should be measured separately from the regular bricks.
- There are fewer variables to consider with concrete blocks than with clay bricks, but blocks do come in different sizes and the thickness of joints between blocks can vary, which will affect the value of the factor used to calculate the number of blocks per square foot of wall area.
- Different types of blocks such as those used for bond beams and copings, together with any other non-standard block, are measured separately from standard blocks.
- Various publications provide information about conversion factors to calculate such items as the number of blocks per square foot or square meter of wall.
- Conversion factors are also available to enable the estimator to calculate the volume of mortar required per 1,000 bricks or blocks of different sizes.
- Given the size of masonry units and the thickness of the mortar joints between them, estimators can calculate their own conversion factors.
- Mortar is measured in cubic yards and any add mixtures are described.
- Bricks shall be enumerated describing the type and dimensions of bricks.
- Concrete block masonry units shall be enumerated stating the type and size of blocks.
- A manual takeoff of masonry work for a house is demonstrated.
- The estimator needs to follow a systematic method of measuring exterior and interior finishes work in order to prepare an accurate takeoff.
- Exterior and interior finishes are generally measured in square feet.
- Exterior finishes include:
  - Damp proofing and waterproofing
  - Insulation and vapor barriers
  - Siding
  - Roofing
  - Flashings and sheet metal work
  - Roof specialties and accessories
- A manual takeoff of exterior finishes for a house is demonstrated.

- Interior finishes include:
  - Flooring
  - Gypsum wallboard and plaster
  - Tile finishes
  - Paint and coatings
- A manual takeoff of interior finishes for a house is demonstrated.

RECOMMENDED RESOURCES

Information	Web Page Address
■ National Concrete Masonry Association	http://www.ncma.org/
■ Taylor Concrete Products	http://www.taylorconcrete.com/default./asp
■ Brick Industry Association	http://www.brickinfo.org/
■ Western States Clay Products Association	http://brick-wscpa.org/
■ Architectural Engineering Information	http://www.aecinfo.com/index.html
■ Dow Building Solutions—Building Wrap	http://building.dow.com/na/en/products/housewrap/
■ Owens Corning • Siding • Roofing • Insulation	http://www.owenscorning.com/index.asp
■ FloorBiz—Flooring	http://www.floorbiz.com/

REVIEW QUESTIONS

1. Which two variables, apart from brick or block size, have an appreciable effect on masonry material usage and mortar consumption?
2. Describe the usual method of calculating quantities of bricks or blocks in a masonry takeoff.
3. State the standard brick size adopted by the Common Brick Manufacturer's Association.
4. List three sources that could be referred to in order to obtain information about the number of bricks and amount of mortar required for different brick sizes.
5. How are special brick courses (such as soldier or brick-on-edge courses) dealt with in a masonry takeoff?
6. Openings greater than what area are deducted from masonry wall areas?
7. How are scaffolding and hoisting operations accounted for in a masonry work estimate?
8. How is insulation to masonry work measured?
9. What are the units of measurement for each of the following items measured with finishes?
   a. Waterproofing
   b. Loose insulation material
   c. Plastic flashings
   d. Splash pads
   e. Roof hatches
   f. Rubber base

g. Carpet edging strip

h. Paint to railings

10. Describe a systematic method of measuring the exterior finishes of a house.

11. What is *parging,* and where is it often found?

12. What are *insulation stops,* and how are they measured in a takeoff?

PRACTICE PROBLEMS

1. How many custom bricks size 3" × 3" × 11" are there per 100 square feet of wall area when the mortar joint is ¼"?

2. How many custom bricks size 4" × 4" × 15" are there per 100 square feet of wall area when the mortar joint is ½"?

3. If brick ties are spaced 16" × 24", how many ties are required for a wall that is 40'-0" long × 8'-0" high?

4. If brick ties are spaced 18" × 20", how many ties are required for a wall that is 140'-0" long × 12'-0" high?

5. If decking consists of 2 × 6 boards laid out diagonally at 6" on center, how many linear feet are required for a deck that measures 26'-6" × 10'-4"?

6. If decking consists of 2 × 3 boards laid out diagonally at 3" on center, how many linear feet are required for a deck that measures 44'-8" × 12'-2"?

7. Take off the masonry work for the building shown in Figure 6.9.
   Additional Notes:
   • Walls are 10"-wide regular concrete blocks.
   • 6.5 cubic feet of mortar is required per 100 square feet of blocks.
   • Include 10" ladder reinforcement in every second course.
   • Fill block cavities with loose perlite insulation.
   • 39 cubic feet of insulation material is required per 100 square feet of blocks.
   • Include damp proofing to the exterior of these walls.

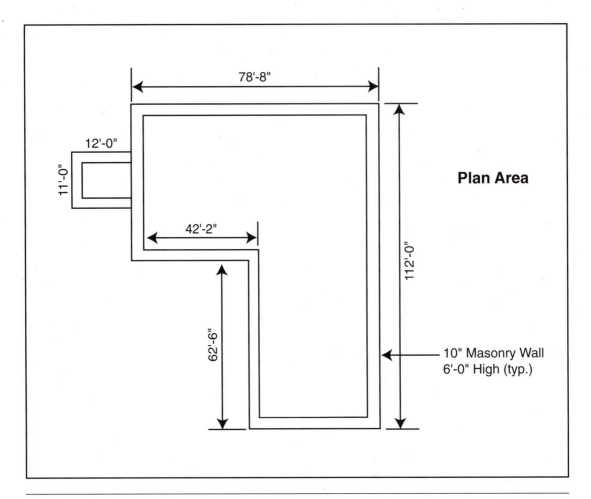

Figure 6.9   Foundation Plan

7

ESTIMATING PLUMBING, HEATING, VENTILATING, AND AIR CONDITIONING (HVAC), AND ELECTRICAL WORK

OBJECTIVES

After reading this chapter and completing the review questions, you should be able to:

- Explain why it is useful for the building estimator to be able to prepare a takeoff of plumbing, heating, ventilating, air conditioning (HVAC), and electrical work.
- Explain why it is useful for the building estimator to be able to make sketches of plumbing, HVAC, and electrical requirements for a project.
- Describe how plumbing, HVAC, and electrical work are measured in a takeoff.
- Explain how the builder's estimator can price the plumbing, HVAC, and electrical work.
- Explain the use of quick methods of estimating plumbing, HVAC, and electrical work.
- Describe the cost per fixture method of estimating plumbing.

KEY TERMS

backflow preventer	heat exchangers
duplex houses	HVAC
electrical estimating	plumbing

Introduction

Plumbing, heating, ventilating, air conditioning (HVAC), and electrical work are invariably performed by subcontractors on residential projects, but these trades represent a significant part of a project, usually accounting for more that 30 percent of the total project cost. It is, therefore, useful for the builder's estimator to be able to check trade quotes to ensure that they cover the full scope of work for that trade.

Certainly estimates of large and/or complicated projects call for the expertise of specialists in these trades. Like any other estimators, those working on plumbing, **HVAC,** and electrical work need a thorough knowledge of how the work is done in that trade before they can assemble accurate quotes for the work. To be accurate, the prices for these quotes will have to be estimated using the same meticulous method used by building estimators with their work. That is, first prepare an accurate takeoff of the work items, then, by means of unit prices, calculate the anticipated cost of these items of work.

The ability to compile a takeoff of plumbing, HVAC, and electrical work can also be a useful skill for the builder's estimator who sometimes needs to put together a quick price for these trades on a small project. Indeed, some builders like to develop their own cost estimate for each trade's scope of work to ensure that the quotes received are reasonable; takeoff skills are definitely an asset if this analysis is to be done properly. Below we have identified the main items of work and how they are measured in a takeoff for each of these trades.

Plumbing Work Generally

On small residential projects, such as single houses and **duplex houses,** it is unusual to see much information about the plumbing system on the project plans other than, perhaps, the location of the main plumbing vent and basement floor drains. Lacking details from the plans, the estimator often has to draw sketches to be able to assess the plumbing layout for a house. These sketches can then help the estimator evaluate measurements such as the length of pipes for water supply and sewers on the project.

Measuring Notes—Plumbing Work

Water Distribution

1. Enumerate connections of water lines to the main.
2. Describe and measure hot water and cold water pipes in linear feet.
3. Describe and enumerate fittings and valves.
4. Describe and enumerate hangers and supports for pipes.

Sanitary Sewer Pipes and Fittings

5. Describe and measure pipes in linear feet to 5 feet from the building.
6. Describe and enumerate **backflow preventers,** cleanouts, floor drains, roof drains, valves, and fittings.
7. Describe and enumerate hangers and supports for piping.

Fixtures and Equipment

8. Describe and enumerate fixtures and equipment, including:
   a. Water meters
   b. Gas meters
   c. Water pressure reducing valves
   d. Water heaters

     e. Pumps

     f. Lavatory basins

     g. Water closets

     h. Bidets

     i. Sinks

     j. Laundry tubs

     k. Bath tubs

9. Describe and enumerate separately equipment supplied by the builder and installed by the plumber, including:

     a. Refrigerators

     b. Garbage disposal equipment

     c. Dishwashers

Gas Lines

10. Describe and measure in linear feet gas line piping from the gas meter to all terminal points.
11. Describe and enumerate fittings and valves.
12. Describe and enumerate hangers and supports.

Miscellaneous Plumbing Items

13. Describe and enumerate permits for the work.
14. Describe pipe insulation and measure in linear feet.
15. Describe and enumerate pressure testing of lines.
16. Describe and enumerate cutting, drilling, and patching required for this work.
17. Measure cleanup resulting from this work as an item.

HVAC Work Generally

Often the only HVAC information provided on plans of small residential projects is only the location of the furnace. So, before the estimator is able to takeoff this work, she/he needs to determine the position of heating ducts, air outlets and inlets, and thermostats. An outline of the location of ductwork sketched on the floorplans is useful for this purpose.

Measuring Notes—HVAC Work

Ductwork and Venting

1. Describe ductwork, stating the size and measure in linear feet.
2. Describe range, cooktop, laundry, and bath venting stating the size and measure in linear feet.
3. Describe transitions and measure in linear feet stating the size at each end.
4. Describe elbows stating the size and enumerate.
5. Describe "Y" fittings stating the size and enumerate.
6. Describe fittings, hangers, and sleeve frames stating the size and enumerate these items.

Mechanical Equipment

7. Describe and enumerate mechanical equipment items, including:

     a. Air handling units

     b. Furnaces

   c. **Heat exchangers**
   d. Heat pumps
   e. Refrigeration compressors
   f. Refrigerant condensing units
   g. Water coolers
   h. Humidity control equipment
   i. Floor-heating and snow-melting equipment

HVAC Controls

8. Describe control wiring and measure in linear feet.
9. Describe low-voltage wiring and measure in linear feet.
10. Describe and enumerate electric and electronic controls.
11. Measure both testing and balancing as items.

Miscellaneous HVAC Items

12. Describe and enumerate permits for the work.
13. Describe duct insulation and measure in square feet.
14. Describe and enumerate fire stops and flashings.
15. Describe and enumerate cutting, drilling, and patching required for this work.
16. Measure cleanup resulting from this work as an item.

Electrical Work Generally

Residential plans generally show the location of light fixtures, switches, and at least some of the wiring. The location of duplex receptacles is also noted on the drawings, however, the layout of power wiring is seldom shown. Sketches, especially with colored pens, can here again be a great help to the estimator measuring the work of this trade.

Measuring Notes—Electrical Work

Wiring

1. Describe wire, cable, and conduits and measure in linear feet.
2. Describe cable tray and bus duct systems and measure in linear feet.
3. Describe and enumerate hangers, sleeves, and supports.

Electrical Equipment, Fittings, and Fixtures

4. Describe and enumerate electrical equipment, fittings, and fixtures, including:
   a. Fire alarm systems
   b. Security systems
   c. Communication systems
   d. Medium-voltage switching equipment
   e. Low-voltage switching equipment
   f. Electrical panels and sub-panels
   g. Junction boxes
   h. Outlet receptacles
   i. Switches
   j. Light fixtures
   k. Door bells
   l. Fans

m. Range hoods
n. Smoke detectors
5. Describe and enumerate installing appliances, including:
    a. Cooktops
    b. Microwave ovens
    c. Built-in ovens
    d. Range hoods

Miscellaneous Electrical Items

6. Describe and enumerate permits for the work.
7. Describe and enumerate fire stops and flashings.
8. Describe and enumerate cutting, drilling, and patching required for electrical work.
9. Measure testing and hot checking entire system as an item.
10. Measure cleanup required as a result of this work as an item.

Pricing the Work

Materials can be priced in the usual manner using supplier quotes for the items involved. However, because the builder does not perform this work with their own forces, there will be no in-house historic cost information available to use to estimate the cost of labor and equipment for this work. The builder's estimator will, therefore, have to rely on productivities and prices obtained from subtrades if they are willing to share this information. Otherwise, the estimator will have to resort to publications such as *RSMeans Plumbing Cost Data,*[1] *RSMeans Electrical Cost Data,* and so on for prices.

Estimating Shortcuts

Instead of compiling a full takeoff on the smaller projects, many plumbing, HVAC, and electrical estimators use shortcuts to compute their prices for this work. Building estimators who need quick budget prices for these trades can also make use of these time-saving methods.

For the plumbing, HVAC, and electrical trades, a price per square foot of gross floor area could be used to provide at least a rough estimate of the cost of that trade. Often, though, especially for bid estimates, something that gives a more accurate price is called for.

Cost Per Fixture Plumbing Estimate

A quick estimate for plumbing work can be obtained by means of a unit price for each plumbing fixture on the project. The cost of all necessary piping and any other required work is allowed for in each of these unit prices. So, enumerating the following fixtures and multiplying by the applicable price for each fixture will give the estimated price for the complete plumbing work:

a. Floor drains
b. Water heaters
c. Lavatory basins
d. Water closets

---

1. R. S. Means Company Inc., Kingston, MA

e. Sinks

f. Laundry tubs

g. Bath tubs

Additional prices may be required for installing appliances, special fixtures, and any out-of-the-ordinary plumbing work requirements.

Plumbing Example

Suppose you are considering adding a three-fixture bathroom to a home design and you would like to know roughly what the plumbing price would be. The plumber can quickly estimate the price based on the number and type of fittings involved and all-inclusive price for these fittings.

EXAMPLE 7.1

Estimate of plumbing price for a 3-fixture bathroom:

1 bathtub 72" × 36"	$2,500
1 water closet	$1,485
1 lavatory basin 19" × 16"	$1,160
	$5,145
Extra for pipe and fittings 30%	$1,544
Total	$6,689

Quick Estimates of HVAC Work

The cost of HVAC work can be estimated quite quickly by using a combination of prices per unit area and unit prices for items of equipment. The overall cost of ductwork including ducts and all fittings is proportional to the floor area of the house. A fairly accurate price per square foot, therefore, can be obtained from the analysis of projects that have been completed in the past. Thus, to estimate the total cost of ductwork for a house, multiply the unit price from a past project by the gross floor area of the new house.

The estimator can also compile all-inclusive equipment item prices for each item of equipment used in the HVAC system. An item price would include the cost of purchasing the item of equipment, installing the item, and any further costs necessary to complete its installation. So, to compile a full HVAC estimate, determine how many of the following equipment items are required, and multiply by the appropriate item price for each:

a. Furnaces

b. Heat exchangers

c. Refrigeration compressors

d. Refrigerant condensing units

e. Water coolers

f. Humidifiers

g. Kitchen and bathroom vents

HVAC Example

To quickly calculate the heating and ventilating price for a house, we can use a price per unit area and add the cost of non-standard fixtures. So, if the proposed house is

2,200 square feet of floor area and a special humidifier and exhaust fan are required, this would be the estimate:

EXAMPLE 7.2

Quick heating and ventilating price:

Basic HVAC 2,200 SF at $5.50/SF	$12,100
Humidifier	$2,500
Exhaust fan	$350
	$14,950

Quick Estimates of Electrical Work

The cost of electrical work can also be estimated quickly by means of a combination of prices per unit area and prices for items of equipment and fixtures:

1. To estimate the full price of wiring for lighting, multiply gross floor area by the inclusive unit price per square foot.
2. To estimate the full price of power wiring and fittings, multiply the number of power outlets by inclusive price per outlet.
3. To estimate the price of equipment and fixtures, multiply the number of each item by unit price for that item, including:
   a. Electrical panels
   b. Subpanels
   c. Light fixtures
   d. Doorbells
   e. Fans
   f. Range hoods
   g. Smoke detectors
   h. Cooktops
   i. Microwave ovens
   j. Built-in ovens
   k. Range hoods

Electrical Example

To quickly calculate the electrical price for a house, we can use a price per unit area and add the cost of non-standard fixtures. So, if the proposed house is 2,200 square feet of floor area and a security system with motion detectors is required, this would be the estimate:

EXAMPLE 7.3

Quick electrical price:

Basic electrical 2,200 SF at $8.50/SF	$18,700
Security system	$3,500
	$22,200

SUMMARY

- It is customary to perform plumbing, HVAC, and electrical work by subcontractors on residential projects.
- Because plumbing, HVAC, and electrical work accounts for so much of the total price of a project, it is useful for the building estimator to be able to check the subtrade quotes for these trades.
- Because building estimators sometimes estimate these trades on small projects, the ability to compile takeoff of this work can be useful to them.
- Very little information about plumbing, HVAC, and electrical work is found on drawings for small residential projects, so estimators often need to prepare sketches of this work.
- Plumbing pipes and gas lines are measured in linear feet in a takeoff.
- Plumbing fittings and equipment are enumerated.
- HVAC ductwork is measured in linear feet for each size of duct.
- HVAC fittings and items of equipment are enumerated.
- Electrical wiring, cables, and conduit are measured in linear feet.
- Electrical equipment, fittings, and fixtures are enumerated.
- Installation of appliances is enumerated.
- The builder's estimator will generally have to rely on publications such as *Means Plumbing Cost Data* in order to price plumbing, HVAC, and electrical work.
- Costs of plumbing, HVAC, and electrical work can be quickly estimated using the price per unit area estimating method.
- A quick estimate of plumbing work can be assembled using the cost per plumbing fixture method.
- A combination of price per unit area and unit prices per item of equipment can be used to generate a quick HVAC cost estimate.
- A combination of price per unit area and unit prices per item of equipment can also be used to generate a quick electrical work estimate.

RECOMMENDED RESOURCES

Information

- R.S. Means publishes specialized estimating texts for plumbing, HVAC, and electrical work.

- Information on the latest estimating software can be obtained with keywords: Plumbing, HVAC, and **Electrical Estimating** on your web search engine.

Web Page Address

http://www.rsmeans.com/

REVIEW QUESTIONS

1. What portion of the total cost of a residential project does plumbing, HVAC, and electrical work together account for?
2. Why it is useful for a building estimator to be able to prepare a takeoff of HVAC work?
3. How much information about the plumbing system is usually provided on housing plans?

4. To what distance outside the house are sanitary sewer lines measured in a plumbing takeoff?
5. How is piping insulation measured?
6. Describe how you would measure bathroom venting in an HVAC takeoff.
7. How is testing and balancing of an HVAC system measured?
8. If a cooktop is required for a kitchen, its cost is included in the builder's work, so why is it measured again as part of the electrical scope of work?
9. Wiring or conduit passing through walls sometimes requires fire stops or flashings. How are these items measured?
10. Describe a quick method for estimating the cost of HVAC work.

PRACTICE PROBLEMS

1. Calculate a quick plumbing price for a four-fixture bathroom based on the following data: Fixtures: 1 bathtub at $2,450; 1 shower unit at $1,000; 1 WC at $1,400; and 1 Lavatory basin at $850. Also include an additional 30 percent for pipe and fittings.
2. Calculate quick plumbing, HVAC, and electrical prices for a 3,100-square-foot home based on the following data: Plumbing is expected to be $6.50 per square foot; HVAC $4.25 per square foot; and electrical work $6.75 per square foot.

8

PRICING GENERAL EXPENSES

OBJECTIVES

After reading this chapter and completing the review questions, you should be able to:

- Define general expenses.
- Identify general expense requirements for a project and explain how general expense requirements are assessed.
- Describe how general expense prices are calculated.
- Calculate general expense items for a residential project.
- Identify general expense items that are calculated as add-ons.
- Complete the pricing of general expenses of a house project.

KEY TERMS

add-ons	hoardings	small tools and
bar-chart schedule	markup and fee	consumables
British Thermal Units	new home warranty	specifications
(BTUs)	payroll additive	
green building certification		

Introduction

The cost of labor, material, equipment, and subtrades expended on the items measured in the quantity takeoffs is usually referred to as the direct costs of the work. The general expenses of a project comprise all of the additional, indirect costs necessary to facilitate the construction of the project. These indirect costs are sometimes referred to as project overheads.

There are many possible general expense items on a project. The General Expenses Sheet presented in Figure 8.1 provides a summary of some of the more common items of general expenses associated with residential construction. On a small- to medium-sized project, this General Expenses Sheet can be used as a checklist that allows the estimator to choose those items that are required on the particular project under

	Quantity Unit	Unit Rate Labor	Labor $	Unit Rate Material	Material $	Unit Rate Equip.	Equip. $	Subtrade $	Total $
PROJECT: LOCATION: DATE: ESTIMATOR: **GENERAL EXPENSES**									
Design									
Schedule									
Project Superintendent									
Assistant Superintendent									
Stakeout									
Survey and Plot Plan									
Safety and First Aid									
Rentals: – office trailer									
– office equipment									
– office supplies									
– storage									
– tool lockups									
– toilets									
Permanent Connections:									
– Sewer Connection									
– Underground Wiring									
Municipal Charges									
Temporary Site Services:									
– Water services									
– Electrical services									
– Telephone									
Scaffolds									
Hoardings									
Temporary Heating									
Temporary Fire Protection									
Site Access									
Site Security									
Equipment Rentals									
Truck Rentals									
Dewatering Excavations									
Collecting/Sorting Recyclables									
Cleanup:									
– General site cleanup									
– Final site cleanup									
– Cleaning the furnace									
– Cleaning windows									
– Move-in cleanup									
Snow Removal									
Photographs									
Project Signs									
Soils Testing									
Green Building Certification									
New Home Warranty									
TOTALS TO SUMMARY:									

Figure 8.1   General Expenses Sheet

consideration. The selected items can then be priced either by inserting the prices and calculating the total costs manually, or by setting up the general expenses summary on a computer spreadsheet and allowing the computer to perform the applicable calculations. On larger projects, a more comprehensive list of items may be required.

Because most general expense items are not usually referred to in the project drawings and **specifications**, the estimator needs some construction knowledge and experience to first of all identify which general expense items are required for the project and, second, to accurately price these requirements. The price of general expenses can be a substantial component of the bid amounting to 15 percent more of the total price in some cases. Accurate assessment of general expenses can be critical to the competitiveness and financial success of the venture, so many estimators, including highly experienced veterans, consider it beneficial to consult with other staff about the general expense demands of a project.

Job superintendents, construction managers, and other senior company personnel can usually help with the appraisal and pricing of general expenses, especially when special site facilities and equipment are being considered. There are often several alternative ways of providing general expense facilities on a project, and discussions can sometimes yield solutions that are more competitive or more effective than those elicited by the estimator working alone. Some companies go so far as to make it a policy that a team of senior staff meet and consider the general expense needs of all major projects.

General expense requirements are assessed by examining the plans and specifications and by considering the amount of work involved in the project as measured in the takeoff process. Therefore, the task of estimating general expense prices can begin only when the quantity takeoffs are complete. However, some general expenses cannot be priced until the total price of the work is determined. These items are separated from the other general expense items and are priced at the end of the estimate as **add-ons**. Figure 8.2 contains a list of some general expense add-on items.

Project Schedule

It is useful to have a realistic schedule of the project work at the time of the estimate because one of the principal variables that determine the cost of general expense

GENERAL EXPENSE ADD-ONS		
DESCRIPTION	METHOD OF CALCULATION	BASED ON $
Small Tools and Consumables	Percentage	Total Labor
Payroll Additive	Percentage	Total Labor
Sales Tax	Percentage	Total Materials
Building Permit	Percentage or Step Rate	Total Bid
Insurance – Builder's risk	Percentage	Total Bid
– Liability	Percentage	Total Bid
Interest Charges	Percentage	Total Bid
Markup and Fee	Percentage or Lump Sum	Total Bid
Value Added Tax	Percentage	Total Bid

Figure 8.2   General Expense Add-ons

items is the duration of the project. Also, owners often take into account the project duration specified by homebuilders competing to build a custom home. If the owner is looking for the shortest possible project duration, the builder with a better schedule may win over a lower-priced bidder.

Because builders usually have only a short time to prepare their estimate and submit their proposal for the construction of a custom home, it is not practical to prepare a detailed schedule using such techniques as critical path method. Consequently, more approximate methods are employed for schedules at the time of the estimate.

One approach taken by many homebuilders is to simply search previous job records for a project of similar size and construction to the one being estimated and use the actual duration of the past project as a guide to the duration of the new project. When a homebuilder repeatedly builds projects of a similar nature, this technique can produce quite accurate results. If detailed information has been kept about the time taken to complete the work of each of the various trades involved in previous projects, a fairly sophisticated schedule of the future project can be produced.

Even when the new project is of a different type of construction or size from the homebuilder's previous projects, an approximate schedule in the form of a bar chart of the major activities can often still be drafted. From the information contained in the plans, together with details from the takeoffs of the amount work involved in the project, veteran supervisors can usually put together at least a rough schedule based on their experience on past jobs.

An example of a **bar-chart schedule** prepared for an estimate is depicted in Figure 8.3. This kind of schedule can be roughed out in just a few minutes using a computer spreadsheet template that has been previously set up for this purpose. The first draft of the estimate schedule can be based on mere guesswork, but the durations of key tasks such as foundations, framing, and finishes can be updated as information becomes available from subcontractors.

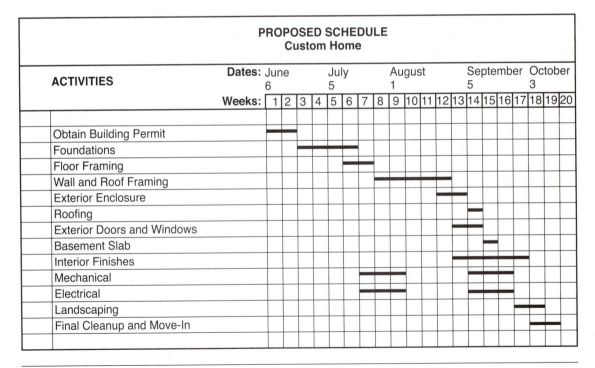

Figure 8.3 Sample Project Schedule

General Expenses

Project Supervision

The cost of project supervision is considered to be part of company overhead by some homebuilders, so they do not incorporate this expense into the general expenses of a project. Where a company does include the cost as a general expense, the price of this item is calculated by multiplying the monthly payroll cost of the supervisor(s) by the number of months they are required to be on the project. When a number of projects under construction at the same time are under the direction of a single supervisor, the cost of supervision is shared by all projects.

Other personnel may be required on the project in addition to a project supervisor. For example, a helper may be needed when the building is staked out or, when many homes are under construction, an additional *assistant* superintendent may be required.

The wages used to price project staff may or may not include payroll burden (the cost of fringe benefits). Estimators should follow a standard practice regarding payroll burden. If it is not firmly established whether net wages or gross wages are used in the pricing process, it will be very easy to make an overpricing or underpricing error. In the examples below, net wage amounts are used to price site personnel, and the total amount for site personnel is added to the rest of the labor cost of the estimate. An allowance for payroll burden is subsequently added to the total labor amount at the end of the estimate.

Safety and First Aid

In this item, we consider the cost of meeting general safety and first aid requirements at the site. Safety requirements for a project may include one or more of the following expenses:

1. The cost of worker's time when attending safety meetings
2. The cost of workers' time when engaged in other safety related activities
3. The provision of safety awards on the project
4. The provision of safety supplies at the site

The homebuilder often needs to complete additional work on a project in order to comply with safety regulations. This is usually measured in the takeoff process together with the rest of the work. Examples include cutting back excavation embankments and providing shoring to the sides of deep trenches which is taken off and priced with the excavation work, and using scaffold systems that meet safety codes which adds to the cost of the scaffold system. In addition, there may be activities associated with the project that are purely safety related and are not part of any particular work item. Workers' time spent attending safety meetings or taking safety courses are examples of this kind of safety activity. The estimator, generally by reviewing records of previous projects, has to anticipate how much labor time will be spent on these safety matters.

Safety supplies such as hard hats, lifelines, special safety clothing, and suchlike are expended during the course of construction activities. Some companies include the cost of these items in the "**Small Tools and Consumables**" item in the add-ons, whereas other companies prefer to estimate these costs separately. The estimator should determine company policy and work accordingly.

First aid expenses can be divided into two basic categories:

the cost of providing on-site first aid supplies, and

the cost of labor engaged in administering first aid services at the site.

OSHA regulations require construction sites to have first aid kits sufficient to meet the needs of the project. It is necessary to estimate the cost of supplies used up over the course of the work. This is usually done by investigating the amounts expended on first aid supplies on past jobs from project records.

Regulations also call for persons qualified to administer first aid to be present on construction sites. On projects up to a certain size, this requirement is met by having a number of the trade's people and site staff obtain first aid qualifications. When a carpenter with a first aid ticket spends an hour providing first aid to a fellow worker, the cost of that hour will rightly be charged to the project first aid budget, so an estimate of the total number of hours administering first aid is required in order to establish the budget.

In order to properly estimate the cost of these first aid provisions, the estimator needs to be familiar with the particular OSHA regulations that apply at the site of the project.

Temporary Offices, Site Storage, and Sanitary Facilities

Office accommodation is seldom required when only one or two homes are to be constructed. Supervisors are usually able to work out of a pickup truck in these situations. However, with larger developments where many units are to be constructed over a period of years, sales offices (and possibly also construction offices) may be required for such a development.

Facilities provided to meet the needs of a development vary in scope from a small rental trailer up to an elaborate building that is constructed at the beginning of construction and removed when work is finally completed and all units are sold. A large sales office, which may include mock-ups of suites, is sometimes constructed for multi-unit projects. In these cases, a full estimate of building, servicing, and later removing the structure will be required.

Rental of temporary site facilities usually includes one or more of the following items:

1. Rentals:
   - Office trailers
   - Storage trailers
   - Tool lockups
   - Toilets
2. Costs to move in and move out trailers
3. Computers, printers, and related office equipment
4. Office supplies
5. Heating costs

On renovation projects, the owner may be able to provide facilities that meet the storage space and sanitary facility needs of the homebuilder. It is important to determine whether the homebuilder or the owner will pay heating, utilities, and other such expenses in connection with the use of these spaces.

There will often be Health Department rules to comply with regarding temporary site sanitary facilities. Homebuilders frequently use portable chemical toilets to meet these needs, however, in some situations, toilets connected to the sewer system and equipped with adequate plumbing may be necessary.

Utilities and Site Services

Utilities and site services requirements are going to vary a great deal depending on the type of project to be constructed. Where a new home is to be built on a serviced lot in a city, site services will almost certainly be in place and will normally be available for temporary use during the construction of the project.

When the house is to be constructed is in a new subdivision, there may or may not be services available at the time of construction. The estimator will have to investigate this and may have to include in the estimate the cost of such items as temporary generators.

In contrast, there will be a far greater amount of work needed and, therefore, higher costs to build a house for the first time on a large acreage. Here utilities will probably have to come in from some distance, and special facilities such as a septic tank and sewage disposal field may have to be constructed.

Utilities—Permanent Connections

Here we allow for the municipal charges for the permanent connections to sewer systems, natural gas, and electrical systems. In some locations, the municipality performs the work for a fee. In other areas the homebuilder's subtrades make the connections but there may still be additional municipal charges to pay. The estimator needs to determine what is required for the particular project under consideration and obtain details of the fees likely to be charged by the municipality involved.

Utilities—Temporary Site Services

A number of temporary site services are often required in order to pursue construction operations on a project. Site services include the following items:

- Water services
- Electrical services for power and lighting
- Telephone services

In each case, there is usually an installation cost to extend the particular service to the site and a cost for the resources consumed during the course of construction operations. Installation costs vary widely according to how far the services have to be brought to reach the site. The information about which services are available on site and which have to be brought in from afar may be provided in the bid documents for custom homes. This data is usually only obtainable from a thorough investigation of the site. See Chapter 2 for a discussion of site investigations.

The largest consumption cost of all of these items will probably be for electrical power used on the project. Electricity consumption for items that are heavy users of electricity (such as temporary hoists) should be calculated separately from the general electrical consumption on the project.

The cost of the electricity used to meet the general power and lighting needs of the job is most often estimated by once again resorting to the technique that consists of using the historic costs of the item on previous projects as a basis for making a prediction about future expenses on this item.

This method of using cost records to predict future costs is also often used to assess the amount of phone charges on a project. Project supervisors generally use cell phones since they spend much of their time on the move rather than in a fixed office location. A phone line may, however, be required if a sales office is to be set up at the site of the development.

If a phone line is to be installed, contact the phone company to obtain information about installation charges and estimate monthly charges from the costs incurred on previous projects remembering to include for past and possible future rate increases.

Scaffolds, Hoardings, and Temporary Enclosures

Scaffolds, **hoardings**, and temporary enclosures are constructed on a project in response to specific project needs, or the owner may call for particular items such as dust curtains, which are often needed on renovation projects. This is another situation where the estimator is advised to consult with experienced construction personnel in her/his company to assess requirements since there are a wide variety of different temporary structures that could possibly be needed on a project, including:

1. Fences or barricades set up around the perimeter of work areas
2. Temporary scaffolds for work on exterior walls or roof
3. Temporary enclosures built to provide security and/or allow a space to be heated
4. Temporary covers to the exterior openings of buildings to improve security or to allow the building to be heated
5. Temporary partitions or dust curtains erected in or around a work area to confine the debris and noise of construction operations
6. General protection of existing structures on or adjacent to the work site

In each case, the total cost of work includes the cost of materials, labor, and any necessary equipment required to erect the structures involved, plus the cost of taking down and removing the materials from the site after their use.

Figure 8.4 shows the sketch of a temporary enclosure erected over a construction work area to provide a heated space in winter conditions. Lightweight joists are supported on 2 × 4 stud walls around the exterior all covered with reinforced plastic sheets. Materials will be used many times over, so unit costs will be low and the joists will be rented only for the time they are required for use on the enclosure.

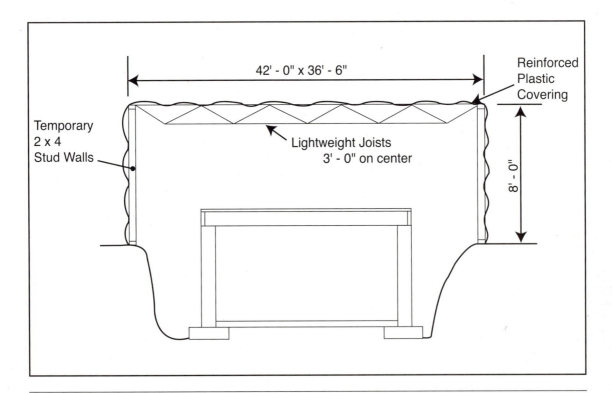

Figure 8.4 Temporary Enclosure

Takeoff

Lightweight Joists: (42.00/3.0) + 1 = 15 Joists   =   15   No.
      Stud Walls: 2 x 42.0 x 8.0   =   672
                    2 x 36.5 x 8.0   =   <u>584</u>   =   1,256   SF

Plastic Covering: Walls: 1,256
                    Roof: 42.0 x 36.5   =   <u>1,533</u>   =   2,789
                    Add 10% for laps: <u>279</u>
                                            3,068   SF

ITEMS	Quantity	Unit	U/Rate Lab.	Labor $	U/Rate Matl.	Materials $	Total $
Lightweight Joists	15	No.	$40.00	600	$50.00	750	1,350
Stud Walls	1,256	SF	$0.80	1,005	$0.30	377	1,382
Plastic Covering	3,068	SF	$0.06	184	$0.08	245	429
Total Price:				1,789	—	1,372	3,161

Figure 8.5   Temporary Enclosure Estimate

Figure 8.5 shows takeoff of the work involved in this enclosure together with prices for materials and labor to erect and dismantle the items involved.

Temporary Heating

Temporary heating can be divided into three categories:

1. Heating provided in localized areas to allow operations like masonry to continue in cold weather
2. Heating a temporary enclosure
3. Heating the house to allow the finishing trades to complete their work

Categories 1 and 2 are usually achieved by the use of portable space heaters that are oil, gas, or propane fuelled. This type of heater may also be used for item 3, or the homebuilder may be allowed to use the building's permanent heating system if it is operational.

Two costs need to be estimated when pricing these items: the rental rate or ownership costs of the portable space heaters used and the cost of fuel consumed, both of which depend upon the duration of the heating period and the amount of heat required. There will also be labor costs for maintenance and operation of heaters, but this cost should be considered with the other site personnel expenses considered previously.

If the homebuilder proposes to use the permanent heating system of a building as a source of temporary heating during construction, the estimator needs to calculate the fuel cost up to the time the owner takes possession of the house.

Notes:

1. A temporary enclosure of size: 42'-0" x 36'-6" x 12'-0" is to be heated for a period of 30-days.
2. Heating is required to be continuous 24-hours per day at the rate of 100 BTUs per CY per hour.
3. Heating is to be provided by propane heaters with an output of 70,000 BTUs per hour.
4. Heaters are rented for $75.00 per week, and propane is available for $1.05 per gallon.
5. A gallon of propane provides 90,000 BTUs of heat.

Calculation:

			Cost $
Volume to Be Heated: 42.0 x 36.5 x 12.0	=	18,396 cu. ft.	
	=	681.33 cu. Yds.	
Heat Required = 681.33 CY x 100 BTUs per hour	=	68,133 BTUs/hour	
Heater Rental Cost: = 1 Heater x $75.00/week x 4 weeks	=		300.00
Propane Consumption: 68,133 x 30 days x 24 hours	=	49,056,000 BTUs	
Propane Cost: $\dfrac{49,056,000 \times \$2.05}{90,000}$		=	1,117.39
Total Heating Cost:			1,417.39

Figure 8.6   Temporary Heating Calculation

Figure 8.6 shows the calculation of the temporary heating expenses to provide heat for a period of 30 days to the temporary enclosure detailed in Figure 8.4. This estimate is based on a heat consumption rate of 100 **British Thermal Units (BTUs)** per CY of space heated per hour, which is an estimate of the amount of heat required to maintain acceptable working conditions inside the enclosure. The actual heating rate encountered on the project can vary over a wide range of possibilities depending on the quality of the enclosure, the air temperature outside the enclosure, and wind conditions. These conditions are all difficult to accurately predict. Because of this uncertainty, there is a large amount of risk attached to estimates of heating costs. The estimator and the homebuilder's management should be aware of this risk.

Site Access and Storage Space

Consideration of site access and storage space is grouped together because the two are often interconnected. For instance, a city lot may have excellent accessibility but little or no storage space on site, whereas more than adequate storage space may be available on an out-of-town job but access to the site may be limited to a dirt road.

These items need to receive careful attention on the site visit if an accurate assessment of access and storage problems is to be made.

An access road may be in place for a new home in the country, but the cost of maintaining this road for the duration of construction may need to be estimated. Gravel roads must be graded from time to time, and paved roads may require cleaning or even repairs if they are seriously soiled or damaged by traffic to and from the project.

Municipal bylaws in some localities have imposed liability on builders regarding cleanup and restoration of city pavements affected by construction operations. The estimator should check into the regulations that apply to the project since the cost of cleaning up roads can be expensive after trucks have left long trails of mud from a wet site.

The problem of lack of space on congested sites for trailers, storage, and parking can be difficult to solve. Homebuilders have sometimes been able to utilize space on the streets next to the site with municipal approval and upon payment of the necessary fees. Alternative solutions may include renting space on a vacant lot adjacent to or at least near to the site, but whatever the planned solution, the problem has to be confronted at the time of the estimate so that the price of the plan can be determined and included in the bid price.

Site Security

On renovations and custom homes, the owner's requirements may spell out their security needs for the job. These can vary anywhere from the provision of a site fence to an elaborate continuous surveillance system incorporating personal identification cards for site personnel and security checks on all persons working on the project. Requirements for the average residential project include site fencing, temporary covers to door and window openings, and possible after-hours surveillance by a security organization.

Where there are no owner-specified security requirements, the homebuilder will have to decide what measures are to be taken in this regard for the project. In some cases there is a trade off between the cost of losses and the cost of security. Installation of an inexpensive alarm system on storage trailers may be sufficient, or just common sense precautions taken by supervisors and workers during the course of the work may be all that is necessary on certain projects.

Aside from owner requirements, the homebuilder may have to consider the wishes of its insurance company and the recommendations of local law enforcement agencies when reviewing site security needs. But whatever the situation is, decisions have to be in place before the estimator can put a price on them.

Equipment Rentals

The equipment used on the project is priced in three different sections of the estimate using three different pricing methods:

1. Major items of equipment used in excavation and concreting operations are priced under the recaps of that work. This category of equipment includes items such as excavators, bulldozers, graders, rollers, concrete buggies, and concrete pumps. As we have seen in Chapter 9 and Chapter 10, these items are priced using rental rates or ownership costs against the hours the equipment is used on site.
2. Hand tools and other items of equipment valued at less than $1,000 each are priced at the end of the estimate as a percentage of the total labor price of the estimate. See "Small Tools and Consumables" that follows.

3. Items of equipment kept at the site, sometimes for extended periods, and used only intermittently during its time on site. This category of equipment is priced with the general expenses, and it is this equipment that we will consider here.

Some of the more common items of site equipment include:

■ table saws
■ radial-arm saws
■ plate compactors
■ jumping jack compactors
■ air compressors
■ pneumatic tools and attachments
■ generators
■ concrete vibrators
■ surveying instruments
■ welders
■ hoists and cranes
■ cement and mortar mixers
■ forklifts

From the homebuilder's list of site equipment, the estimator selects the pieces required for the job and prices each item by multiplying its price rate by the length of time the item is to be assigned to the project. An amount may also be added to cover the cost of fuel and maintenance expenses on this equipment.

Dewatering Excavations

Dewatering expenses refer to the costs of removing water from excavations and, in some cases, from the finished basements of projects under construction. The usual method of dewatering makes use of submersible electric pumps and hoses operating from sumps excavated below the bottom of the basement.

The amount of dewatering required for a project is yet another area of uncertainty for the estimator. On projects where a large amount of pumping is expected, very little may ultimately be needed and, where little or no moisture was anticipated on a job, extensive dewatering can turn out to be necessary. These errors in judgment are possible even after diligent investigations of the site, previous conditions encountered in the vicinity of the work, and careful study of all available soils reports.

Green Building Provisions

There can be a number of additional expenses associated with green building. See the discussion of green building in Chapter 1. The cost of collecting and sorting waste materials for recycling is an expense that might be incurred by a contractor as part of its green building program. An estimate of this cost is added to the general expenses.

There could also be fees to pay if the contractor is applying to have the project certified under initiatives such as the NAHBGreen program. Again, such fees would be added to project general expenses. To offset the cost of green building, some municipalities and government agencies offer incentives that might be in the form of tax benefits or reductions in building permit fees; the estimator needs to investigate these possibilities and make reductions to the project price where they apply.

Cleanup

Cleanup on a housing project can be divided into five categories:

1. General site cleanup
2. Final site cleanup
3. Cleaning the furnace
4. Cleaning windows
5. Move-in cleanup

The general daily cleanup expenses include the cost of renting garbage containers for most projects plus the cost of constructing temporary garbage chutes on some renovation projects. Expenses include the labor cost of the ongoing cleanup required in connection with the work of the project. However, the cost of cleaning up after specific trades such as concrete work should be included in the cost of the concrete work and, under the terms of subcontracts, subtrades are usually responsible for the cleanup of any garbage resulting from their own work. That cost is to be included in that subtrade's price for its work.

The final site cleanup includes all the costs involved in readying the worksite for handing over the finished project to the owner. Removal of temporary garbage chutes and garbage containers will be required together with the general tidying up of the work site.

When a forced-air furnace is used on a house, the ductwork attached to the furnace is usually installed quite early on the project. It can subsequently become clogged with dust and construction debris as the rest of the project work is completed. This is especially true when other trades use the ductwork for (unofficial) garbage disposal. Consequently, it is considered to be good customer service to have the ductwork cleaned by a specialist company before move in occurs.

When the job approaches the time when the new owner is about to take possession, the move-in cleanup will occur. This involves cleaning carpets and other flooring, cleaning windows, bathroom and kitchen fixtures, and dealing with any spills or splashes that might spoil the appearance of the new home. The work is often subcontracted to a company that provides this cleaning service on the basis of a price per square feet of gross floor area.

Other General Expenses

Other items of general expense that may be required for housing projects include:

1. Stakeout—This is an allowance for the cost of the survey work involved in setting out the lines and levels of the project.
2. Survey and Plot Plan—A survey of the site, together with a plot plan certified by a land surveyor, may be required on some projects. Include the surveyor's fees for this service here.
3. Snow Removal—Homebuilders operating in northern areas find this can be a significant expense when a project is required to be completed in the winter months.
4. Saw Setup—When there is a large amount of carpentry work to be completed at the site, the homebuilder may build a temporary enclosure for the main table saw or radial-arm saw to protect it from the weather and provide a secure lockup for this expensive piece of equipment. The cost to build and later remove the enclosure, together with any other expense to set up the saw, is priced here.
5. Photographs—In their contract with the homebuilder, the owner of a custom home may call for official progress photographs to be taken over the course

of the project. In cases such as this, it may be necessary to include the price of professional photographers who will perform this work or photographs taken by the project supervisor may be acceptable. If it is not a contract requirement, the builder may still wish to keep a photographic record of construction. Whatever the case, an estimate of the cost of photography may be required.

6. Project Signs—The signs displayed outside the site of a project can vary from a small advertisement of the name of the homebuilder to large, elaborate placards on multi-family developments. The latter often includes an artist's rendering of the finished development along with a list naming all the major participants in the project. Site signs will normally be specified in the contract documents outlining in detailed or general terms the owner's requirements. Sign painters are usually hired to produce the signs that are then installed by the homebuilder.

7. Soils Testing—It is sometimes necessary to test the soil-bearing capacity when poor soil conditions are encountered. An engineering company is usually hired to do this work and make recommendations about how to deal with the problem.

8. **New Home Warranty**—Home building contracts often include warranty provisions that call for the homebuilder—at no extra expense—to perform all remedial or maintenance work required on the project over a defined time period. Homebuilders usually contribute to an insurance agreement that provides the required warranty.

9. Temporary Fire Protection—This is normally the cost of providing fire extinguishers on site during the construction period. The estimator should be careful to note any more elaborate protection requirements (such as the provision of fire hydrants) since this can be an expensive item.

10. Wind Bracing Masonry—In many places, the masonry trade does not include in the scope of their work the temporary wind bracing. Shoring or scaffolding may be necessary to support masonry work while it is under construction. When this is required and not included in the masonry quote, the homebuilder should allow for the cost of this work.

11. Municipal Charges—There are many possible municipal charges that a homebuilder may be liable to pay in connection with any construction project. In addition to building permit fees (discussed later), fees may be payable for the use of city streets for unloading, storage or parking, the construction of city sidewalk crossovers, municipal business taxes, or homebuilder license fees to name but a few. Estimators should be familiar with the requirements of their own city, and on out-of-town jobs they must investigate the municipal charges in vicinities that are new to them.

12. Financing Charges—With an accurate forecast of the cash flow of a project, an estimator would be able to calculate the interest charges on the homebuilder's running overdraft which is required to finance the project. However, precise forecasts of the inflow and outflow of cash on a construction project, even when detailed schedules are available, are extremely difficult to produce. Here again the estimator usually has to fall back on previous job records with necessary adjustments for changes in interest rates to determine the probable financing charges on a future project.

Add-Ons

Add-ons are calculated as a percentage of other amounts in the estimate. For instance, the add-on may be a percentage of material or equipment prices (many taxes are calculated in this way). Some add-ons, such as payroll burden, are calculated as a percentage of the price of labor in the estimate. Still further add-ons are calculated as a percentage of the total estimate.

Taxes

There are as many ways of accounting for taxes in an estimate as there are ways of applying taxes to the price of construction work, and, because the substance of a tax can vary widely from place to place, there is no single formula for dealing with taxes. How a tax is calculated depends upon the particulars of how the tax is assessed. For instance:

a. Where a sales tax applies to materials but varies from one material to another, the tax has to be calculated separately on each individual material price.

b. Where a uniform sales tax applies to all project materials, an add-on can be set up at the end of the estimate to add the necessary percentage to the value of the total material content of the estimate.

c. Payroll taxes that apply to the total labor price of the estimate can also be handled with an add-on. This is the case with the **payroll additive** add-on discussed below.

d. A European and Canadian-style value-added tax, which applies to the prices of all the components of an estimate, can be set up as a percentage add-on calculated on the very last line of the estimate so that the applicable rate applies to the final complete estimate price.

Labor Add-Ons

The following general expense items are proportional to the total labor cost of a project, so the dollar value of these items can only be calculated when the total estimated labor cost has been determined. This point is reached in the estimating process when the homebuilder's own work, the subtrade work, and the other general expenses have been priced. See Figure 12.1 for the location of these add-ons on a typical Estimate Summary Sheet.

Small Tools and Consumables

In this expense item, the cost of hand tools and equipment valued at less than $1,000 per piece is priced together with miscellaneous supplies used in the construction process. These supplies would include such items as replacement drill bits, saw blades, chalk lines, batteries, and all the other incidental components consumed in connection with the work of a construction project.

Payroll Additive

The labor component of the estimate has so far been priced using the base wage rates for the trades involved. Every employer also makes a variety of contributions in relation to its total payroll in response to federal, state, and municipal requirements and company policies. Payroll additive may account for any or all of the following items:

1. Social Security Tax—This is a federal government requirement to provide retirement benefits to persons who become eligible.

2. Unemployment Compensation Tax—This is a state tax that is gathered to provide funds to compensate workers in periods of unemployment.

3. Workers' Compensation and Employer's Liability Insurance—As required by state regulations, contributions are collected to provide a fund from

which workers who are injured while working or as a result of working are compensated.

4. Public Liability and Property Damage Insurance—This insurance protects the employer against claims by the public for injuries or damage sustained as a result of the work activities of an employee.

5. Fringe Benefits—This includes a number of items that could be contained in a contract of employment or a union collective agreement. Examples include health care insurance, pensions, and disability insurance.

These contributions vary from place to place and from company to company. The task of the estimator is to ascertain the rates of contributions that apply to the project labor force and allow for them in the estimate.

Permits

Most municipalities levy a building permit fee on all home building projects whether they are new home developments or merely renovation jobs. Permit fees are usually calculated as a percentage of the total construction cost, but there may be minimum charges for small projects. Check with the local planning department to determine the current fees for building permits and include them in the estimate. There are often additional permits and associated fees for plumbing, mechanical, and electrical work. If the subcontractors do not include these amounts in their prices for the job, the estimator needs to investigate what is required and include the applicable amounts in the estimate.

Insurance

Sometimes the owner will provide insurance coverage for the builder they hire but more often, and especially on the larger projects, the builder will be required to obtain insurance. The basic insurance policies for a building project include:

- Property insurance including fire and vandalism
- Liability insurance to cover liability claims arising from the project
- Vehicle and equipment insurance for units used in connection with the work

Sophisticated owners will usually have specific insurance requirements. The estimator is advised to consult their company insurance agent to be sure that the standard policies they use meet these requirements. There may be additional premiums for special coverage, which the estimator needs to include in the price of the job.

Markup and Fee

The **markup and fee** on a job is often calculated as a percentage of the total estimate price. Markup and fee considerations are discussed in Chapter 12.

Example of General Expenses Pricing

Figure 8.7 shows the general expenses for the sample house project.

Comments on General Expenses Shown in Figure 8.7

1. This project is expected to take 4 months to complete based on the duration of previous projects of a similar type and size. There is no need to prepare a detailed schedule for this project.

PROJECT: House Example
LOCATION: Townsville
DATE:
ESTIMATOR: ABF

GENERAL EXPENSES

	Quantity Unit	Unit Rate Labor	Labor $	Unit Rate Material	Material $	Unit Rate Equip.	Equip. $	Subtrade $	Total $
Design	— —	—	—	—	—	—	—	—	0
Schedule	— —	—	—	—	—	—	—	—	0
Project Superintendent	4 Mo.	1,250.00	5000	—	—	—	—	—	5000
Assistant Superintendent	— —	—	—	—	—	—	—	—	0
Stakeout	1 No.	300.00	300	—	—	—	—	—	300
Survey and Plot Plan	1 No.	—	—	—	—	—	—	200	200
Safety and First Aid	1 No.	—	—	35.00	35	—	—	—	35
Rentals: – office trailer	— —	—	—	—	—	—	—	—	0
– office equipment	— —	—	—	—	—	—	—	—	0
– office supplies	— —	—	—	—	—	—	—	—	0
– storage	— —	—	—	—	—	—	—	—	0
– tool lockups	4 Mo.	—	—	—	—	75.00	300	—	300
– toilets	4 Mo.	—	—	—	—	150.00	600	—	600
Permanent Connections:	— —	—	—	—	—	—	—	—	0
– Sewer Connection	1 No.	—	—	—	—	—	—	960	960
– Underground Wiring	1 No.	—	—	—	—	—	—	165	165
Municipal Charges	— —	—	—	—	—	—	—	—	0
Temporary Site Services:	— —	—	—	—	—	—	—	—	0
– Water services	— —	—	—	—	—	—	—	—	0
– Electrical services	1 No.	—	—	—	—	—	—	500	500
– Telephone	4 Mo.	—	—	75.00	300	—	—	—	300
Scaffolds	— —	—	—	—	—	—	—	—	0
Hoardings	— —	—	—	—	—	—	—	—	0
Temporary Heating	— —	—	—	—	—	—	—	—	0
Temporary Fire Protection	— —	—	—	—	—	—	—	—	0
Site Access	— —	—	—	—	—	—	—	—	0
Site Security	— —	—	—	—	—	—	—	—	0
Equipment Rentals	1 No.	See Calculations	—	—	—	—	750	—	750
Truck Rentals	4 Mo.	—	—	—	—	150.00	600	—	600
Dewatering Excavations	— —	—	—	—	—	—	—	—	0
Collecting/Sorting Recyclables	— —	—	—	—	—	—	—	—	0
Cleanup:	— —	—	—	—	—	—	—	—	0
– General site cleanup	4 Mo.	—	—	—	—	125.00	500	—	500
– Final site cleanup	1166 sf	—	—	—	—	—	—	292	292
– Cleaning the furnace	1 No.	—	—	—	—	—	—	185	185
– Cleaning windows	1166 sf	—	—	—	—	—	—	82	82
– Move-in cleanup	1 No.	—	—	—	—	—	—	100	100
Snow Removal	— —	—	—	—	—	—	—	—	0
Photographs	— —	—	—	—	—	—	—	—	0
Project Signs	— —	—	—	—	—	—	—	—	0
Soils Testing	1 No.	—	—	—	—	—	—	110	110
Green Building Certification	— —	—	—	—	—	—	—	—	0
New Home Warranty	1 No.	—	—	—	—	—	—	300	300
TOTALS TO SUMMARY:			5300	—	335	—	2750	2893	11278

Figure 8.7   Priced General Expenses Sheet

2. Because this is such a small project, the cost of site supervision is shared with other projects, so only a portion of the total cost of this item is included in this estimate.

3. The price of labor to stake out the lot is based on the cost of a helper to assist the superintendent with the surveying and setting out work.

4. The survey amount is for a land surveyor to provide a reference point for the superintendent to use for setting out the project.

5. The amount shown against Safety and First Aid is an allowance for first aid supplies consumed during the course of the work. This is based on the average amount spent on this item on previous projects.

6. There is no office trailer required for this project, but there is an allowance for the rental of a lockup for tools and for a chemical toilet facility at the site.

7. Sewer Connection and Underground Wiring Connection amounts are for the municipal charges to make these permanent connections.

8. Charges for the Temporary Electrical services are based on the historical record of the cost of electricity consumed on similar projects in the past.

9. The superintendent's cell phone expenses are also based on historic costs. These charges are shared by a number of projects, so only a portion of the total amount is shown here.

10. Equipment to be used on this project would be listed and priced on the back of the General Expenses Sheet. The note: "SEE CALCULATIONS" is directing the reviewer of this estimate to consult a supplementary calculation sheet usually located on the back of the General Expenses Sheet. The equipment list for this project is as shown in the following example:

EXAMPLE 8.1

1 Compressor	-2 months × $165.00	=	330.00
2 Concrete Vibrators	-2 days × $25.00	=	100.00
1 Crane (hoisting steps)	-1 day × $170.00	=	170.00
			600.00
	Allowance for fuel and maintenance:		150.00
	Total:		750.00

11. The fuel and maintenance allowance above is calculated as 25 percent of the rental rate for these items.

12. The Trucking amount allowed is the portion of the cost of the superintendent's pickup truck on this project.

13. The General site cleanup amount is a contribution to the cost of renting a garbage container for the site.

14. The amounts for Final site cleanup, Cleaning the furnace, Cleaning windows and Move-in cleanup are the prices charged by a subtrade to complete this work to ready the house for handing over to the owner.

15. A sum is included for testing the soil-bearing capacity under the footings.

16. A new home warranty is provided to the purchaser of this home, so the premium for this service is included here.

17. Add-on amounts are not included on this General Expense Sheet since add-ons cannot be calculated until the Summary is completed. Add-ons that apply to this job can be found at the end of the Summary Sheet shown in Figure 12.9 in Chapter 12.

SUMMARY

- General expenses or project overheads comprise the indirect costs expended on a construction project.
- It is usually the builder's responsibility to assess and price the general expenses on a construction project.
- Because there are so many possible general expense items, estimators usually make use of a checklist when general expenses are priced.
- General expense requirements are assessed by examining the plans and specifications and considering the amount of work involved in the project as measured in the takeoff process. Job superintendents and project managers can help the estimator define general expense requirements for a project.
- Some general expense items are calculated as a percentage of the estimate or as a percentage of some part of the estimate. These items are included as add-ons at the end of the estimate.
- Many general expense items are calculated as a function of the duration of the project so at least a rough schedule is required before these items can be evaluated.
- Major items to consider when pricing general expenses include:
  - Site personnel
  - Safety and first aid
  - Travel and accommodations
  - Temporary site offices
  - Temporary site services
  - Hoardings and temporary enclosures
  - Temporary heating
  - Site access and storage
  - Site security
  - Site equipment
  - Trucking
  - Dewatering
  - Site cleanup
- General expenses calculated as add-ons include:
  - Taxes
  - Small tools and consumables
  - Payroll additive
  - Building permits
  - Insurance
  - Markup and fee
- Pricing the general expenses for the house example is demonstrated.

RECOMMENDED RESOURCES

Information	Web Page Address
■ Occupational Health and Safety Administration	http://www.osha.gov/
■ Scaffolding safety	http://www.osha.gov/SLTC/scaffolding/
■ Masonry scaffolding information	http://www.masonrymagazine.com/8-02/scaffolding.html
■ Example of a new home warranty program in Colorado	http://www.blueribbonhomewarranty.com/

REVIEW QUESTIONS

1. Explain what the general expenses of a project are.
2. How can the total duration of the construction work of a project be determined in the short time available in the estimating process?
3. List the three temporary site services that are usually required on a residential project, and describe how each service is priced.
4. List three possible ways of dealing with the problem of limited storage space on a downtown site.
5. List the five categories of site cleanup that have to be considered in a housing estimate.
6. What are the homebuilder's financing charges on a project, and how is the cost of financing charges estimated?
7. Describe how "add-ons" are calculated.
8. Describe two basic insurance policies that a builder is often required to obtain for a project.

PRACTICE PROBLEMS

1. Estimate the cost of temporary heating of a house size 45 feet × 30 feet × 20 feet high. The space is to be heated at the rate of 75 BTUs per CY per hour for a period of two weeks (14 days), 24 hours per day. 75,000 BTU propane heaters can be rented for $25.00 per week. The cost of propane is $0.90 per gallon and assumes 1 gallon provides 90,000 BTUs of heat.
2. Estimate the cost of heating a temporary enclosure for a house size 50 feet × 40 feet × 12 feet high. The space is to be heated at the rate of 100 BTUs per CY per hour for a period of one week (7 days), 24 hours per day. 90,000 BTU propane heaters can be rented for $35.00 per week. The cost of propane is $1.20 per gallon and assumes 1 gallon provides 90,000 BTUs of heat.
3. Use the prices given in Figure 8.7 to estimate the general expense charges for a new 2,500 square feet home that is expected to take 5 months to construct. This home is to be built to NAHBGreen standards, so there are two additional expenses to consider:
   a. Collecting and sorting waste materials for recycling is expected to cost $250 per month.
   b. **Green building certification** fees will be $500.

9

PRICING THE WORK

OBJECTIVES

After reading this chapter and completing the review questions, you should be able to:

■ Describe the pricing process using the homebuilder's standard trade breakdown.
■ Explain estimating risk and why project costs may exceed estimated prices for the work.
■ Explain the influence of risk, job factors, and labor and management factors on pricing.
■ Describe how materials are priced in an estimate.
■ Describe how bills of material are compiled.
■ Explain how a subcontractor's work is priced.
■ Complete the pricing of a residential project including a bill of materials.

KEY TERMS

bill of materials	recap
compaction factor	royalty
cost-plus contract	swell factor
order units and order quantities	takeoff units and takeoff quantities

Introduction

With non-computer estimating (in other words, preparing an estimate by hand rather than by computer), the process of pricing an estimate can be divided into two stages. The first stage consists of processing the takeoff for pricing by sorting and listing all takeoff items, usually into a trade-by-trade breakdown. This sorted list is referred to as the **recap**. The second stage consists of pricing the items on this list. The reason for preparing the recap is to bring together all work of a similar nature so that the estimator can concentrate on the needs of one trade at a time. Also, when the recap is complete, prices of each trade will be presented in summary form that facilitates easier evaluation of the estimate. If items were not recapped into trades but, instead, the takeoff priced directly, it would be far more difficult to answer questions such as what is the total amount in the estimate for rough carpentry materials?

The format of the recap breakdown varies from one homebuilder to another depending on what information they wish to obtain from the priced estimate. Frequently the breakdown reflects a company's coding system for cost accounting where items of work are assigned cost codes often in no particular order. Estimators prefer a more orderly breakdown, usually divided into trades, listed in the order of construction. This makes it easier for the estimator to see if there are any gaps in the estimate, whereas a random list of items is far more difficult to evaluate for deficiencies.

A standard trade breakdown consists of the following items:

1. Excavation and backfill work
2. Concrete work
3. Formwork
4. Rough carpentry work
5. Finish carpentry work
6. Exterior finishes
7. Interior finishes
8. Plumbing
9. HVAC
10. Electrical

In Chapter 2 we recommended measuring assemblies. These often contain a mixture of different trades. For instance, a foundation assembly may include excavation, concrete work, and carpentry. Some estimators takeoff each of the recap trades separately which makes the preparation of the recap much easier, but this approach has few devotees for the reasons discussed in Chapter 2. So, where the work is not taken off trade by trade, the estimator has to prepare a recap. This means the entire takeoff has to be scanned for each of the trades listed on the breakdown. The estimator will first scan for all the items of Excavating and Fill and list these in the recap, then she/he will scan for all the Concrete items, etc. The process will be repeated until the takeoff is fully recapped. This procedure has to be performed with great care, and a system of checking off items as they are carried to the recap is essential to confirm that all items that were in the takeoff have indeed been processed and are now included in the recap.

In computer estimating systems, the process of recapping takeoff items is handled automatically. In most systems, a numerical code is applied to takeoff items which enables the computer to classify items and list them under the appropriate trade section for pricing. Depending on the system used, takeoff items can also be priced automatically as part of this same process. This demonstrates a significant advantage of using a computer in the estimating process because not only does the computerized system eliminate the time an estimator otherwise would have to spend recapping items, but it also ensures all takeoff items will be accounted for. There is very little chance of a takeoff item being overlooked and not priced once it has been entered into the computer.

Returning to the manual process, once the recap is prepared, the pricing procedure can commence. In pricing a construction estimate, there are five price categories that need to be considered:

1. Labor
2. Equipment
3. Materials

4. Subcontractors
5. Job overheads

We will consider the particular factors that affect the level of prices in each of these categories, but there is one factor that impacts upon all the above categories: risk.

Homebuilder's Risk

In Chapter 12, we suggest that the homebuilder, when considering the amount of markup to add to an estimate for a new home, should assess the risk the builder will be assuming if they are awarded a contract to construct the home. In essence, the risk we are contemplating here is the risk of the builder losing money on the project. This usually results from situations where the actual cost of construction exceeds the estimated cost. The precise nature and value of this risk on any given project is not simple to evaluate. First we have to consider the nature of the contract that the homebuilder enters into with the owner, for it is from the terms of this contract that much of the builder's risk emerges.

For example, the contract could state "The builder will be reimbursed all project costs by the owner plus an added amount for profit." This arrangement is known as a cost-plus contract. (See Chapter 1 for a discussion of various contract types.) With a **cost-plus contract** the builder's risk will be reduced to a minimum. A negligent homebuilder who fails to accurately keep track of all costs could still lose money under these terms, but the majority of builders will find cost-plus contracts profitable since most of the risk of price increases is borne by the owner rather than the builder.

If the contract is an agreement to build a house for a fixed price, the type of arrangement used on most homebuilding projects, the builder agrees to build the house for an agreed price. In this situation, if the cost of the work is more than the agreed price, the builder will lose money. Only where the owner changes the contract will the contract price change, so, if the work costs more than the amount of the builder's price, the extra costs will have to be paid by the builder.

There are three general reasons why actual costs may exceed estimates on a project:

1. **Takeoff quantities** are too low
2. Actual productivity does not meet anticipated productivity
3. Subcontractors or material suppliers fail to meet obligations

Inaccurate Takeoffs

The possibility of inaccuracy in the measurement of takeoff quantities is a risk that is fundamental to the estimating process. The principal objective of following the systematic methodology of preparing a quantity takeoff outlined in preceding chapters is to minimize the possibility of the error in the results. But even the most careful estimator can make a mistake in the takeoff—a fact that every experienced estimator will admit—and although computer systems can help reduce this risk, they cannot eliminate the risk. Consequently, senior estimators should review their colleagues' takeoffs to try to identify anything that may have been done in error.

Time and cost restraints usually make it impossible for a second estimator to check every single step that was taken in the preparation of a takeoff, but veteran estimators develop an ability to make broad assessments of takeoffs that can often enable them to detect deficiencies. This assessment, nevertheless, is not perfect. Inevitably some errors will still go undetected. Consequently, there is always some takeoff risk that needs to be addressed when the project fee is considered and the bid price is finalized.

Pricing Labor and Equipment

Most homebuilders hire subcontractors to perform the work of the project. This means the subtrade will assume the risk of any cost over-runs resulting from failure to meet productivity targets. For instance, the subtrade may assume a productivity rate of 50 cubic yards per hour for excavating a basement. So, if the size of the basement is 500 cubic yards, the estimated time for excavating will be 10 hours and the price of the work will be based on this duration. If, for any reason, it takes longer to complete the work, it will cost more than the estimated price and the subtrade will have to bare the loss.

However, some builders like to perform at least some of the work using their own labor and also, sometimes, their own equipment. Each of the categories needs to be assessed and priced in an estimate. The most common method of pricing labor and equipment makes use of unit prices that are applied to listed takeoff quantities to calculate the total labor or equipment price for the work involved.

Apart from the risk of not meeting anticipated productivity rates, there are also risks involved with the wage rates and equipment rates used in pricing the builder's work.

Wage Rates

The unit prices for labor are derived from the base wage rates of the workers involved. Wage rates can vary from state to state, and even from place to place within a state. Thus, before the labor costs of an estimate can be priced, the estimator has to ascertain the craft wage rates that apply at the project location.

In addition to the base wage rates, employers also have to consider such items as social security tax, unemployment tax, workers' compensation insurance, liability insurance, and fringe benefits. The amount of these additional contributions, referred to as payroll additive below, is calculated on the Estimate Summary Sheet as a percentage of the total labor content of the bid price.

While a single wage rate can be used for each trade over the course of a project of short duration, the estimator has to predict how wage rates will change over the term of larger multi-unit projects that may stretch over longer time periods. This assessment can never be done with total certainty because wage rates or benefits may be subject to sudden changes that are difficult to foresee.

Faced with the prospect of a long duration project, historical trends can be studied to help predict the level of future wages. The usual procedure is to examine the shape of the graph of actual wage rates plotted over a number of years and, from this graph, determine the average rate of wage increment. The analyst can proceed to extrapolate along the graph line to arrive at the projected wage rates for the term of the project. However, no matter how scientifically this process is accomplished, estimating future wage rates from previous trends amounts to little more than guesswork.

With small, single dwelling projects of relatively short duration, these complications should not apply.

Equipment Rental Rates

Accurate predictions of future hourly costs of construction equipment can be just as difficult to evaluate as future wage rates. Ownership costs of a builder's own equipment can be analyzed with some certainty (see Chapter 10 for examples), but even these calculations are based on assumptions about such things as life expectancy, maintenance costs, and salvage values, all of which can vary from expectations.

Rental rates of equipment can also be difficult to tie down because they fluctuate in response to a number of variables including the level of economic activity in the construction industry, interest rates, taxes, import duties, and even construction labor wage rates. (Steeply rising wage rates cause managers to substitute equipment for labor wherever possible; this increases the demand for equipment and can raise rental rates.)

If the estimate is prepared some time before the work begins, rental rates paid for the work can be quite different from rates forecast at the time of the estimate. The risk of rate increases can be managed to some extent by procuring guaranteed price quotations from rental companies for major equipment on some projects, but these firm price agreements are not always obtainable.

In addition to the risk of rate increases, there are a number of other questions the estimator should consider regarding equipment rentals, including:

1. What is the age and condition of the equipment being offered?
2. Is the cost of an operator included in the rates quoted?
3. (On longer rentals) Do the rates include the cost of necessary maintenance of the equipment?
4. Is the cost of fuel included in the rental?
5. Is there a minimum rental fee or a minimum rental duration that translates into a minimum rental fee?
6. Will there be additional charges for delivery?
7. Are there any accessories needed to operate the equipment? If so, at what price are they available?
8. What taxes and/or license fees over and above the rental fee are applicable?
9. What insurance coverage is included in the rental agreement?
10. Are there any special qualifications needed before the builder can legally operate the equipment? In some jurisdictions, some proof of operator training is required before certain equipment can be legally used.

Productivity of Labor and Equipment

The productivity of labor and equipment is influenced by a large number of factors that can differ according to the time and place of a project. These factors can be classified into two main groups: job factors and labor and management factors.

Examples of *job factors* would include:

■ Weather conditions expected at the site
■ Access to and around the site
■ Site storage space
■ Project character, its size, and complexity
■ Distance from materials and equipment sources
■ Wage and price levels at the job location

Examples of *labor and management factors* would include:

■ Quality of job supervision
■ Quality of subtrades
■ Motivation and morale of workers
■ Type and quality of tools and equipment
■ Experience and records of similar projects in the past

Job Factors

The builder has very little influence, if any, over the job factors that prevail on an individual project, but the nature of these factors should be carefully investigated since they can have a major impact on the rates of productivity and thus the cost of work done. Each of the job factors listed can vary over a wide range of possible conditions. The estimator has to learn to determine the implications of the particular conditions expected on the project with regard to their impact upon job factors. Below are some observations on the method of investigating job factors and examples of some situations that may be encountered on different projects.

If the estimator is unfamiliar with the prevailing weather conditions at the location of the job, weather records may be obtainable from a weather-monitoring agency in the vicinity of the site. Particular attention should be paid to data about the extreme and average summer and winter temperatures, together with information regarding expected rainfall and snowfall amounts at the site location. Poor weather conditions at a site can make it impossible to attain the productivity levels obtained on previous projects at other locations. On projects away from the home-builder's home territory, a little investigative work about the prevailing weather conditions there can help to avoid over optimistic production forecasts.

The quality of site access can affect productivity in many ways. Poor access to some sites may restrict the size of equipment that can be used for the work so that smaller, less efficient machines have to be used rather than those of a more appropriate size. Because large trucks are not able to get close to a site due to restricted access, they may have to unload some distance away and then have the material moved some time later to the work area. This double handling of materials can significantly raise the cost of work.

Cramped storage space at a site can also necessitate double handling of materials where materials have to be temporarily stored at one location then moved on site some time later. "Just-in-time" delivery of materials in accordance with a precise schedule can be used to avoid this kind of double handling, but working to this kind of schedule requires extensive and careful management of operations that is not always available.

On large multi-unit projects where there is a significant amount of repetitive work, the builder may obtain increased productivity from work improvements attained because operations are repeated. Studies have shown that when a relatively complex work activity is repeated, there is an opportunity to improve production rates because those involved in the work learn to perform it more efficiently. However, situations where a certain operation is performed just once or only a few times exclude the builder from the benefits of a learning rate achieved in repetitious operations. Even on large projects the work may be too diverse to enable the builder's crews to achieve a significant learning rate.

Labor and Management Factors

Like job factors, some of the labor and management factors are also beyond the control of the builder. Over the long term, a builder may be able to develop effective supervisors and workers inside the company but because of the cyclical nature of the construction business, homebuilders often find it difficult to maintain enough employment to retain these good workers. Consequently, the staffing of larger projects requires the builder to hire many workers from outside the company, consequently the quality of supervision and the labor available for the project often depends most upon the economic climate at the time of the work. In a strong economy where there is plenty of construction work underway and builders are busy, the better supervisors and workers will already be fully employed, and so the builder may have to make do with lower quality personnel for the next job. In a weak economy, good workers are far easier to find.

Some labor and management factors can, however, be influenced by the builder. Efforts can be made to cultivate high morale and motivate workers, and a builder can try to ensure that projects are equipped with high-quality, well maintained tools and equipment to pursue the work. The concepts of "total quality management" have been embraced by a number of homebuilders who endeavor, through a process of constant improvement, to develop their personnel into a highly productive work force. These builders are convinced that these efforts will improve their company's competitive edge, which will ensure their future success in the industry.

Use of Cost Records

If an estimator had a range of accurate historic prices for the labor or equipment requirements of each work item on a project, then by accounting for all the factors and variables associated with that project, he would be able to determine the price of each item with some confidence. But how does an estimator come by such a range of prices? What is required is a database of information on the precise cost of work items over a variety of past jobs that were completed under carefully monitored conditions. From this painstakingly gathered historic data, a range of realistic prices for each work item of an estimate can be established. This then allows the estimator to assess the cost of any particular work item on the range and, thus, arrive at a reasonable prediction of the cost of that item.

In order to be effective, an historic cost database used for estimating has to be assembled systematically; a haphazard collection of previous project costs gathered using different systems of measurement and costing will be more of a hindrance to good estimating than a help. A satisfactory cost database can only be developed from an accurate and consistent cost reporting system that is pursued in four key processes:

1. The priced takeoff items on the estimate are coded to produce the project budget. Some of the estimate takeoff items may be combined to provide a simpler breakdown that is easier to use than a detailed estimate in the cost recording system.
2. As the work is performed on the project, labor hours and equipment hours expended are coded to the applicable work items of the budget. It is advisable that project supervisors perform this coding function since they are in the best position to determine exactly which work item their work crews are working on.
3. The quantity of work done in each of the work items underway is measured in accordance with the same rules of measurement followed by the estimators.
4. Company accountants apply wage rates and equipment prices to the hours coded. This determines the labor cost and equipment cost for each work item. These costs are then combined with the work quantities to generate the actual unit costs of the project work items.

Figure 9.1 shows an example of a selection of items that may be found in a cost report prepared in this way. The cost record shows the budget amounts and budget unit prices in the form of an estimate summary. Next, the amounts that have been spent on each of the items in the current month are shown together with the unit costs calculated from the quantity of work done on each item. The amount spent to date and the unit costs to date are then shown. Finally, based on the rates of productivity realized to date, projections of the final costs are indicated.

COST REPORT

ITEM		ESTIMATE SUMMARY						CURRENT MONTH								TOTAL TO DATE								
	QUANT.	LABOR	EQUIP.	MATL.	SUB.	TOTAL		QUANT.	LABOR	EQUIP.	MATL.	SUB.	TOTAL	UNDER/(OVER)		QUANT.	LABOR	EQUIP.	MATL.	SUB.	TOTAL	UNDER/(OVER)	PROJECTED TOTAL	UNDER/(OVER)
		$	$	$	$	$			$	$	$	$	$	$			$	$	$	$	$	$		
Stakeout Building	1 No.	250.00	—	—	—	250.00		—	—	—	—	—	—	—		***1 No.	230.00	—	—	—	230.00	$20	$230	$20
Rentals: Tool Lockups	4 Mo.	—	200.00	—	—	200.00		1 Mo.	—	50.00	—	—	50.00	$0		2 Mo.	—	100.00	—	—	100.00	$0	$200	$0
unit price			50.00			50.00				50.00			50.00	$0				50.00			50.00	$0.00		
Excavation	1 No.	—	—	—	4520.00	4520.00		0 No.	—	—	0.00	—	0.00	$0		***1 No.	—	—	—	4520.00	4520.00	$0	$4,520	$0
Concrete Footings	10 CY	200.00	50.00	1100.00	—	1350.00		0 CY	0.00	0.00	0.00	—	0.00	$0		***10 CY	0.00	0.00	0.00	—	0.00	$0	$1,329	$22
unit price		20.00	5.00	110.00		135.00											18.00	4.85	110.00		132.85	$2.15		
Concrete Walls	30 CY	750.00	225.00	3300.00	—	4275.00		10 CY	280.00	79.50	1100.00	—	1459.50	($35)		20 CY	580.00	161.00	2200.00	—	2941.00	($91)	$4,412	($137)
unit price		25.00	7.50	110.00		142.50			28.00	7.95	110.00		145.95	($3.45)			29.00	8.05	110.00		147.05	($4.55)		
Concrete Slabs	20 CY	700.00	170.00	2100.00	—	2970.00		10 CY	320.00	82.50	1020.00	—	1422.50	$63		10 CY	320.00	82.50	1020.00	—	1422.50	$63	$2,845	$125
unit price		35.00	8.50	105.00		148.50			32.00	8.25	102.00		142.25	$6.25			32.00	8.25	102.00		142.25	$6.25		

** Completed

Figure 9.1 Cost Report

Project Cost Information from Cost Reports

A carefully and systematically prepared cost report is necessary to obtain accurate data for estimating purposes, but it is not sufficient. Before the data from a cost report can be used effectively, two basic questions must be answered:

1. Why did the estimator choose the original unit price for an item?
2. Why did the actual unit price for the item vary from the estimated unit price?

While it is not the objective of the estimating process to predetermine the exact cost of every item of the work, it is useful to learn why the actual cost of an item of work did vary from the estimated cost of that item. If the quality of estimating is to improve, some feedback of information about actual results is necessary. The unit prices used in the estimate could have been inappropriate because of an estimating problem, because of a production problem, or simply due to unfortunate circumstances. We need to know which.

EXAMPLE 9.1

Figure 9.1 shows the actual costs of what we will refer to as Job #1. On this report, the unit cost of labor for *Concrete Walls* in the TOTAL TO DATE column is $29.00. An estimator may conclude that the original estimate unit price ($25.00/CY) was an error and decide to use a unit price of $29.00 per cubic yard in the estimate for the future Job #2. However, the actual price of $29.00 may have been attained because of some special circumstances encountered as the work was performed. This would make it inappropriate to use $29.00 on Job #2 unless these same special circumstances are anticipated for this job.

If the estimator of Job #1 is the same person as the estimator now considering Job #2 and Job #1 was constructed quite recently, then this estimator will be aware of the factors determining the unit prices of Job #1. He will probably have a good idea of the reasons why the final unit costs varied from the unit prices, in which case good use can be made of the cost information contained in the cost record for Job #1. But if the estimator of Job #1 is no longer around, or if Job #1 took place some time ago, the information necessary to properly evaluate the Job #1 cost record may be lacking. Also, if an estimator works on more than thirty estimates per year and numerous projects are underway all generating cost records, it can be difficult to keep track of all the information involved with the result that, when the cost record of a particular project is studied, the estimator may be uncertain of the actual circumstances surrounding that project.

To ensure that the needed project information is available to future estimators, the final cost report on a project should be accompanied by two commentaries, one from the project estimator describing the basis of the unit prices (at least for the substantial items of the estimate). A second record should come from the project supervisor or the estimator if he has been closely observing the progress of the job. This record should describe the actual conditions that precipitated the unit costs realized. Armed with this information, the estimator is far more likely to obtain appropriate unit prices when estimating subsequent projects.

We can add that there are additional benefits that may be obtained from having estimators write a commentary on the unit prices they chose for a project. By performing this task, the estimator is disciplined to seriously think about the factors affecting unit prices rather than just picking a price for vague (if any) reasons. The mere process of writing out their justifications often contributes to clearer thinking in estimators. Certainly, an estimator who has prepared detailed notes can provide a far more coherent explanation when questioned about unit prices at the management review of his estimate.

Bid Pricing Strategies

When a builder uses its own forces, labor, and equipment on a project, they need a strategy in order to produce a competitive bid for the work. One pricing strategy that could be used for a custom home or a multi-unit project consists of adjusting unit prices for job factors and management factors that are relatively certain and then being optimistic with all the other factors. For example, the estimator may know that access to the site is good, the job is uncomplicated, and supervision and labor at the site will be good quality. In a case such as this, the item prices are adjusted to reflect the effect of these factors. However, for weather conditions, worker morale, and other factors where there is uncertainty, the estimator makes optimistic assumptions that keep the prices inexpensive. The total dollar value of these "optimistic" assessments is calculated and, when the fee for the job is determined, the appraised risk factor is accounted for.

There is certainly some danger involved in this strategy, but whether or not the strategy is adopted, estimators should always be encouraged to take time to try to identify the estimating risks because this information is needed to properly complete a bid. Quantifying and subsequently dealing with the project risk is studied later when we consider the process of completing the bid in Chapter 15.

Preparing Bill of Materials

A **bill of materials** is a list of all the materials required for the construction of a project sorted into trades. See Figure 9.2 for an example of the bill of materials format.

In order to prepare a bill of materials from the completed takeoff, list all the materials required for excavation operations, then the materials associated with the concrete work, and so on for each of the following categories:

- Excavation materials—gravels and suchlike
- Concrete materials
- Masonry materials
- Rough carpentry materials
  - Floors systems
  - Wall systems
  - Roofing systems
- Finish carpentry materials
  - Doors
  - Windows
- Other miscellaneous materials

The quantities listed on the bill of materials are expanded to include waste factors, and the units of measurement are converted from **takeoff units** to **order units**, which often differ from takeoff units. For example, floor sheathing is measured in square feet in the takeoff, but the order units are sheets of plywood. Standard sheets of plywood measure 4 feet by 8 feet, so the area of a sheet is 32 square feet.

BILL OF MATERIALS

Page No. [　　　]

JOB .... DATE ....

ESTIMATED ....

ITEM NO.	DESCRIPTION	TAKEOFF QUANTITY	WASTE FACTOR	ORDER QUANTITY	UNIT PRICE	TOTAL

Figure 9.2 Bill of Materials

Here is the calculation for the floor sheathing for the house example taken off in Chapter 5:

EXAMPLE 9.2

¾" T & G floor sheathing		1166.0 sq. ft. (as measured in the takeoff)
Add 10% waste	=	116.6
	=	1282.6 sq. ft.
Divide by 32 sq. ft.	=	40 sheets

The full bill of materials for the house example is included at the end of this chapter.

Pricing Materials Generally

There should be less risk involved in pricing materials than there is with labor and equipment pricing. Once the work item quantities are established in the takeoff process, the estimator should be able to obtain firm prices from material suppliers for the supply and delivery of materials, which will minimize the homebuilder's risk.

Simply using a catalog price for materials may be acceptable for pricing minor items, but where significant quantities are involved, the estimator should ask suppliers for firm prices for the supply of materials that meet the project requirements. Volume discounts may be obtained for large orders, and obtaining firm quotes will reduce the risk of having to pay more if prices rise between the time of the estimate and the time of construction. Homebuilders who complete many projects over a year may be able to negotiate with suppliers to provide all the materials of a certain type at fixed prices for the year. This simplifies the pricing process for the estimator since the same price can then be used for these materials on all projects over the term of the agreement.

There are a number of questions that should be answered before material prices are used in an estimate:

- Do the materials offered by the supplier comply with the specifications? If an owner has selected a certain type of kitchen cabinet, for instance, the estimator should ensure that the supplier is quoting a price for cabinets that meet the owner's requirements.
- Do the prices quoted include for the delivery of the materials to the site or are delivery charges extra?
- Can the homebuilder rely on the supplier's prices to remain firm until the time when the materials are purchased for the job?
- Does the supplier's price include state and/or city sales taxes? Materials may be quoted exclusive of taxes then, where applicable, state and city taxes are added to the item prices for materials used in the estimate.
- Will there be any storage or warehousing requirements for the materials? On-site storage is sometimes required for construction materials before they are incorporated into the structure. The estimator should ensure that storage space will be available at the site. If not included, temporary storage of materials off the site may be necessary which can have significant costs associated with it.
- What are the supplier's terms of offer? Almost all price quotations received from material suppliers will have terms and conditions attached to them. Later in this chapter we will examine a concrete materials quote and the terms and conditions attached to this quote.

Pricing Backfill and Gravel Materials

Whenever it is possible, the builder will use the material excavated at the site for backfill requirements and, thus, eliminate the need to pay for buying materials for this purpose. For instance, if a basement is to be excavated and backfilled, the excavator will leave sufficient excavated material next to the basement so that when the basement is constructed this material can be used to backfill around the finished basement as required. In this case, there will be just two items to price: *Excavation of the basement* (including the cost of removing surplus material) and *Backfill around the basement using the excavated material*. However, where it is impossible to use the excavated material for backfill because it is not suitable for some reason, a third item has to be priced on the estimate: *Supply of basement backfill material*. This item may be priced in two ways:

1. Obtain a quote from a supplier to deliver material to the site at a price per cubic yard; or, especially if large quantities of fill are required,
2. Calculate the cost of obtaining suitable material from a pit or quarry. (The full price of this will include the cost of loading and transporting the material to the site plus the payment of a **royalty** to the owner of the pit.)

An example of pricing the supply of gravel by the second method is shown below.

Swell and Compaction Factors

Before we can calculate a price for supplying gravel, we must consider the effect of material compaction because, no matter which method of pricing the supply of fill material is used, the estimator has to account for the swell or **compaction factors** applicable to that material. Swell and compaction factors are discussed in Chapter 3 where the measurement of excavation work was addressed. However, these factors are not applied at the time of the takeoff since bank measure excavation and backfill quantities are not modified when they are measured. Therefore, the necessary adjustments now have to be made for swell and compaction.

Recall that when soil or gravel is excavated, the resulting material is less dense than it was in the ground before it was disturbed, thus the same material occupies more volume. The increase in volume is known as the **swell factor**. Similarly, backfill material that is compacted will be denser than before it is compacted; so, to fill a certain volume with compacted material, a larger volume of loose material will be required. The **compaction factor** is an allowance for the extra material required to fill volumes with compacted material rather than loose material.

EXAMPLE 9.3

When a supplier quotes a price of $16.00 per cubic yard (loose) for wet sand, the bank measure takeoff quantity has to be increased by a compaction factor. In addition to the extra material required for compaction, there will also be some wastage of material in the operations. It is, therefore, convenient to also add a wastage factor to the compaction factor. In the example below, the compaction factor for wet sand is 13 percent and wastage is a further 10 percent. This gives a combined total of 23 percent to be added for wastage and compaction of the wet sand. Where a quantity of 1000 cubic yard (bank measure) of *Sand Bedding* had been taken off, the price of materials is:

1000 cu. yds. + 23% = 1230 cu. yds. (loose) × $16.00
                      = $19,680

EXAMPLE 9.4

Where the supply of fill material is quoted as a price per ton, the estimator has to use the unit weight of the compacted material to determine the price per cubic yard bank measure. For instance, a price of $15.00 per ton may be offered for the supply of pit run gravel. If the weight of compacted pit run gravel is 3,650 lbs per cubic yard, the price per cubic yard can be calculated as:

$$\text{Price per cubic yard} = \frac{\text{Price per ton} \times 8 \text{ compacted cu. yds. (lbs)}}{2000 \text{ lbs per ton}}$$

So, in this case:

$$\text{Price per cu. yd. (bank measure)} = \frac{\$15.00 \times 3650}{2000}$$
$$= \underline{\$27.38}$$

Calculating Trucking Requirements

Sometimes the estimator has to determine the number of trucks required to transport excavated materials and gravels. When there is a large quantity of material to be transported, the following formula can be used to calculate the optimum number of trucks to use. The formula is based on the premise that it is desirable to have sufficient trucking capacity to ensure that the excavation equipment is able to operate continuously and not have to waste time waiting for trucks. Clearly, three trucks will be required if it takes 10 minutes to load a truck and 20 minutes for that truck to unload and return for another load because while the first truck is away, two further trucks can be loaded. From this intuitive analysis, the following expression is obtained:

$$\textbf{Number of Trucks Required} = \left(\frac{\textbf{Unloading Time}}{\textbf{Loading Time}}\right) + 1$$

Where: Unloading Time = **Round-trip Travel Time + Time to Off-load the Truck**

and:

$$\textbf{Loading Time} = \frac{\textbf{Truck Capacity}}{\textbf{Loader Output}}$$

(Both the truck capacity and the loader output have to be measured in the same units, usually compacted cubic yards.)

The number of trucks obtained from this calculation should always be rounded up no matter how small the decimal. This is because rounding down even a small fraction may result in a shortage of truck capacity. Most estimators consider it better to have over rather than under capacity so that the excavator is kept occupied.

Example of Gravel Supply Price Calculation

Calculate a price of obtaining gravel from a pit located 6 miles from the site where there is a pit royalty of $2.50 per cubic yard using a track loader at the pit to load the gravel at a rate of 50 cubic yards (bank measure) per hour and 10 cubic yard (bank measure) dump trucks to transport the gravel to the site. The rental rate of the loader is $450.00 per day (8 hours), and dump trucks rent for $275.00 per day.

The labor crew for this operation consists of one equipment operator at $24.00 per hour and truck drivers at $18.00 per hour. In this case, the dump trucks travel at an average speed of 20 miles per hour and take 5 minutes to unload the material at the site.

Given this data, the first step in the calculation is to determine how many trucks are required for the operation. So, making use of the formula previously described:

EXAMPLE 9.5

$$\text{Number of Trucks Required} = \left(\frac{\text{Unloading Time}}{\text{Loading Time}}\right) + 1$$

Where, in this case:

$$\text{Unloading Time} = \frac{\text{Round Trip}}{\text{Speed}} + 5 \text{ mins}$$

$$= \frac{2 \times 6 \text{ miles}}{20 \text{ miles per hour} \times 60 \text{ mins}} + 5 \text{ mins}$$

$$= 36 + 5$$

$$= 41 \text{ minutes}$$

$$\text{Loading Time} = \frac{\text{Truck Capacity}}{\text{Loader Output}}$$

$$= \frac{10 \text{ cu. yds.}}{50 \text{ cu. yds. per hour} \times 60 \text{ mins}}$$

$$= 12 \text{ minutes}$$

$$\text{Number of Trucks} = \left(\frac{\text{Unloading Time}}{\text{Loading Time}}\right) + 1$$

$$= \frac{41}{12} + 1$$

$$= 3.42 + 1$$

$$= \underline{4.42}$$

Therefore, five trucks are required.

	Equipment	Labor
Gravel Supply Price:	$	$
Track Loader ($520.00/8 hours)	65.00	
Operator		32.00
Trucks 5 × ($300.00/8 hours)	187.50	
Drivers 5 × $23.00		115.00
	252.50/hour	147.00/hour

Price per cubic yard: (divide by output 50 cu. yds./hour)

	= 5.05/cu. yd.	2.94/cu. yd.

Pricing Concrete Materials

Prices for ready-mixed concrete are obtained from local suppliers who are interested in quoting the supply of all concrete required for a project. Most suppliers maintain a current price list of concrete, but it is always advisable to secure a price quotation specifically for the project being estimated for several reasons: discounts may be available, prices for concrete that meets the specifications are required and, perhaps most importantly, firm prices are required. The prices on a current price list are exactly that, (i.e., prices in effect at this time), but the concrete for the project being estimated may not be required until some time in the future. So what the estimator needs is the price that will be charged for the concrete when it is used on the project, in other words, a price offer that when it is accepted by the builder will bind the supplier to the prices quoted for the full duration of the project. This is what is referred to as a "firm price."

Price quotations obtained from ready-mix concrete suppliers usually have a large number of conditions and extra charges attached to them. Some of the issues that the estimator has to carefully consider when pricing concrete materials include:

1. Does the concrete described in the quote meet the specifications? The estimator should be aware of the strength of concrete required for the project.
2. What are the extra charges for supplying special cements like sulfate-resisting or high-early cements? When using cements other than ordinary Portland Type I cement, concrete prices are invariably increased by a stipulated sum per cubic yard of material supplied.
3. What are the extra charges for air entrainment, calcium chloride, or any other concrete additives required to meet specifications?
4. Are there additional charges for cooling concrete in hot weather or heating concrete in cold weather to account for? In some areas, the cost of heating and/or cooling will be charged automatically for all concrete delivered in certain months of the year.
5. If small quantities of concrete are required on a project, what are the premiums charged on small loads of concrete?
6. If delays are anticipated in unloading the concrete at the site, what will the charges be for keeping the delivery trucks waiting?
7. What are the additional charges when concrete is delivered outside of normal working hours?

Waste Factors

Recall that the quantities of concrete taken off are the unadjusted net amounts shown on the drawings. Allowance for waste and spillage of this material can be made by increasing the takeoff quantities or by raising the price by the percentage factor considered necessary. In the examples that follow, the concrete material quantities have been increased as noted to account for wastage.

The value of waste factors usually lies between 1 percent and 5 percent for concrete placed in formwork and can be as much as 10 percent for concrete placed directly against soil.

Example of Pricing Concrete Materials

The example of a quotation for the supply of ready-mix concrete shown in Figure 9.3 is used to price the following concrete material requirements for work to be completed on a project between December and February:

XYZ CONCRETE PRODUCTS INC.

PROJECT: New Home Building, Townville Subdivision

We are pleased to quote you as follows for the supply of ready mix delivered to the above project:

Mix	Strength (psi)	Aggregate Size	Cement Type	Delivered Price per cu. yd.	Additional for Type III or Type V	Additional for 4-6% Air Entrainment
1.	2000	3/4"	I	$97.00	$5.00	$3.00
2.	2500	3/4"	I	$99.00	$5.25	$4.00
3.	3000	3/4"	I	$102.00	$5.50	$5.00
4.	3500	3/4"	I	$105.00	$6.00	$6.00

The above prices are based on ABC Concrete's standard mix designs.

All products are subject to a municipal sales tax of 7%.

For 1/2" aggregates add $6.50 per cu. yd. to the above prices.
For polypropylene fibers add $18.00 for 2 lbs. per cu. yd.
For pigments (red, black tan, or brown) add $5.50 per cu. yd.

EXTRA CHARGES: Calcium Chloride (1%): $3.00   Per cu. yd.
Calcium Chloride (2%): $4.00   Per cu. yd.
Winter heat between November 25 and March 15: $9.00   Per cu. yd.
Environmental Fee: $3.00   Per cu. yd.

Figure 9.3   Concrete Material Quote

EXAMPLE 9.6

1. 2000 psi concrete, type I cement:

Basic concrete price	$ 97.00
Extra for winter heat	9.00
	106.00
Sales tax 7%	7.42
Total Price	113.42

2. 2500 psi concrete, type V cement, air entrained:

Basic concrete price	99.00
Extra for winter heat	9.00
Extra for type V	5.25
Extra for air entrainment	4.00
	117.25
Sales tax 7%	8.21
Total Price	125.46

(continued)

EXAMPLE 9.6 (continued)

3. 3000 psi concrete, type V cement, air entrained, fiber reinforced:

Basic concrete price		102.00
Extra for winter heat		9.00
Extra for type V		5.50
Extra for air entrainment		5.00
Extra for fiber reinforcing		18.00
		139.50
Sales tax	7%	9.77
Total Price		149.27

Pricing Carpentry Materials

Building code requirements for lumber used in carpentry operations often define the species of lumber required and its use classification within that species. The grade of lumber required can also be specified. This is important where the lumber is to be used structurally.

Softwood lumber rather than hardwood is more commonly used for rough carpentry in most of North America, and species used in the construction industry include Douglas fir, balsam fir, Pacific Coast hemlock, Eastern hemlock, Sitka spruce, white spruce, southern pine, white pine, lodge pole pine, western larch, and various cedars and redwoods. Softwood lumber is classified by use into three groups:

1. Yard Lumber—Lumber used for general building purposes
2. Structural Lumber—Lumber of at least 2" nominal width that will be exposed to structural stresses
3. Factory and Shop Lumber—Lumber selected for remanufacturing use

Softwood lumber is also classified according to the extent of processing in its manufacture:

1. Rough Lumber—Sawn lumber that has not been surfaced
2. Surfaced (Dressed) Lumber—Lumber that has been surfaced by a planer to provide a smooth finish on one or more sides. It is referred to as S1S if one side is dressed, S2S if two sides are dressed, etc.

Worked Lumber—In addition to being surfaced, worked lumber has been further shaped to produce such products as tongue and grooved boards.

The size of lumber is stated in accordance with its nominal size, which is the dimension of the cross section of the piece before it has been surfaced. Thus, 2 × 4-sized pieces were approximately 2" × 4" in cross section before they were surfaced but now that have a cross section of 1 1/2" × 3 1/2" after the lumber is dressed. Softwood lumber is further classified according to its nominal size:

1. Boards—Lumber less than 2" in thickness and 2" or more in width
2. Dimension—Dimension lumber is from 2" up to, but not including, 5" in thickness and 2" or more in width
3. Timber—Lumber which is 5" or over in its least dimension

Lumber Lengths and Waste Factors

Lumber is generally available in lengths that are multiples of two feet. Studs, however, can be obtained pre-cut to the required length. When taking off joist quantities, the estimator will usually allow for length of the joists used even though the span may not be a multiple of 2 feet. For example, if the span of the joists is 12'-6" the estimator will takeoff 14-foot joists. If this is not done, a waste factor of 12 percent (14-12.5 ÷ 12.5) should be allowed on the material for these joists in the pricing process.

Even where the 2-foot multiple lengths are allowed for in the takeoff, a waste factor is still required because components such as plates, beams, headers, etc., usually require some cutting and there has to be allowance for some poor quality lumber and error. This waste factor is usually between 5 percent and 15 percent.

Engineered joists can be obtained pre-cut to required lengths, so there is no need to round up to the nearest 2 feet in the takeoff.

Lumber Grades

Within lumber's use classification, lumber is graded in order to provide information to users so that they can assess its ability to meet requirements.

Yard lumber is graded as being either Selects or Commons, where Selects are pieces of good appearance and finishing qualities suitable for natural or painted finishes, and Commons are pieces suitable for general construction and utility purposes.

Structural lumber is stress graded in order to specify the safe working stresses that can be applied to it. Tables are published that list the allowable stresses for structural lumber in each of the following use categories, including:

1. Light Framing
2. Structural Light Framing
3. Structural Joists and Planks
4. Appearance Framing
5. Decking
6. Beams and Stringers
7. Posts and Columns

Within each of these categories there are one or more grades. Allowable stresses are specified according to species of lumber and grade within each of the above categories. Light Framing, for example, is divided into Construction Grade, Standard Grade, and Utility Grade for each species of lumber used in this capacity. Structural Light Framing generally has five grades for each species:

1. Select Structural
2. No. 1
3. No. 2
4. No. 3
5. Stud Grade

A full specification for structural lumber would call for a certain category, grade, and species of lumber to be used. For instance, a specification may state: "All load-bearing stud walls shall be framed using Structural Light Framing graded lumber of No. 2 or better Hem-Fir." More often, abbreviated versions of such a specification are found calling merely for "No. 2 or better." The task of the estimator, as usual, is to ensure that the lumber prices used in the estimate reflect the specification requirements for lumber.

An example of a lumber company price list is shown on Figure 9.4. This price list will be used for pricing lumber in the examples that follow.

Pricing Rough Hardware

Rough hardware for lumber mostly consists of nails, but other fasteners and such items as joist hangers may also be required. Many estimators allow for rough carpentry by keeping track of hardware costs of previous jobs in terms of a percentage of total carpentry materials costs, and then by applying this historical rate to the estimated cost of lumber for future jobs.

XYZ LUMBER INC.

DOUGLAS FIR AND LARCH S4S
Dimension and Timbers: 2x4 & 4x4 (Std. & Btr.)
                                    2x6, 4x6 and wider (#2 & Btr.)

SIZES	LENGTH	"GREEN"	"DRY"
2x4	8 - 14'		$650.00
	16 - 20'	$650.00	$700.00
2x6	8 - 14'		$650.00
	16 - 20'		$700.00
2x8	8 - 16'	$590.00	
	18'	$630.00	
	20'	$650.00	
2x10	8 - 16'	$630.00	
	18 - 20'	$750.00	

PLANKS AND TIMBERS

S4S	4x4	$530.00
S4S	4x6	$550.00
ROUGH	6x6	$470.00
ROUGH	8x8	$610.00
ROUGH	3x12	$750.00

"S-DRY SPRUCE"

2x3 - 92⅝" Studs		$510.00
2x4 - 92⅝" Studs		$510.00
2x6 - 92⅝" Studs		$510.00
2x3	8' - 12'	$510.00
	14' - 16'	$530.00
2x4	8' - 12'	$530.00
	14' - 16'	$550.00
	18' - 20'	$580.00
2x6	8' x 12'	$530.00
	14' - 16'	$550.00
	18' - 20'	$570.00
2x8	8' x 12'	$550.00
	14' - 16'	$580.00
	18' - 20'	$650.00
2x10	8' - 20'	$700.00

P.W.F. LUMBER (.50 LBS. CU. FT.)

2x4	8 - 14'	$670.00
	16'	$700.00
2x6	10 - 14'	$670.00
	16'	$700.00
2x8	10 - 14'	$670.00
	16'	$700.00

STRAPPING - BRIDGING   Per LF

1x1	Spruce	$0.10
1x2	Spruce	$0.12
1x3	Spruce	$0.20
1x6	Spruce	$0.32
2x2	D. Fir	$0.30

Bridging 2x2 pre-cut for
2x8, 2x10 at 16" on
center per length $0.40

SHEATHING	STD. FIR FIR PER SHEET	SELECT FIR PER SHEET
3/8"	$17.30	$19.80
1/2"	$23.10	$25.50
5/8"	$27.80	$31.35
3/4"	$33.30	$37.10
5/8" T & G	$29.70	$32.25
3/4" T & G	$35.15	$37.95

Figure 9.4   Lumber Quote

An alternative method of estimating hardware costs, which is perhaps more precise, is to takeoff the quantity of specific items of hardware required and/or estimate the quantity of nails required and then price these quantities. For example, the amount of nails required for *rough carpentry* may be estimated by allowing 7 pounds of nails for every 1000 board feet of lumber and 10 pounds of nails for every 1000 square feet of sheathing.

Pricing Subcontractor's Work

On the face of it, the process of pricing subcontract work would appear to be quite simple. Subtrades calculate prices for the work of their trade on a project and submit quotes to the homebuilder. The homebuilder then evaluates the quotes received from subtrades and customarily selects the subtrade with the lowest price for each trade. Once subtrades are selected, their prices are incorporated into the homebuilder's estimate for the project. By subcontracting the work, the risk of cost over-runs is shifted to the subtrade who agrees to perform the work of their trade for the firm price they quote. However, the process is not quite as straightforward as it appears. Some of the problems associated with pricing subcontractors' work are listed here:

- Not all subtrades offer lump-sum bids; many trades bid unit prices for work. For instance, a subcontractor may quote a price of so much per foot to supply and install eaves trough. In this case, the builder will need to takeoff the length of eaves trough required for the project then apply the unit price to the quantity obtained.
- Some subtrades offer prices on an hourly basis. Excavation trades may present a price per hour for the use of operated excavation equipment. In cases such as this, not only will work quantities have to be taken off, but an assessment of a probable productivity rate of this crew will also be needed before a complete price for the work can be calculated.
- The subtrade's interpretation of what is and what is not part of the work of their trade may differ from the homebuilder's interpretation. This problem usually occurs when the design of a house calls for something new and different. The homebuilder's estimator has to ensure that all the work she/he expects a subtrade to perform is, in fact, covered in the subtrade's bid.
- The subcontractor whose price is used in the estimate is unable to perform the work. The most common reason for failure to perform is the financial collapse of the subtrade. While legal action against the defaulting subtrade is possible, suing a bankrupt is a fruitless endeavor. Regular assessments of the financial health of major subtrades can reduce this risk.
- These points will be examined in more detail in Chapter 11: Subcontracting Work.

Example of Subtrade Price to Supply and Place Concrete

The contractor simply calls a subtrade to get a price for the supply and placing of concrete in walls: This means the subtrade has to cost out this work to determine a price for the contractor:

EXAMPLE 9.7

Concrete material price:

2000 psi concrete		$97.00	
Extra for type V cement		$5.00	
Extra for air entrainment		$3.00	
Winter heat		$9.00	
Environmental fee		$3.00	
		$117.00	
Taxes	7%	$8.19	
		=	$125.19/cu. yd.

Equipment:

Pump rental:	2 hours × $116.00	=	$232.00
2 Vibrators:	2 × 2 hours × $9.00	=	$36.00
Transport equipment			$100.00
			$368.00/25 cu. yds.
		=	$14.72/cu. yd.

(continued)

EXAMPLE 9.7 (continued)

Crew:

1 Labor foreman	2 hours × $26.00	=	$52.00
5 Laborers	5 × 2 hours × $23.00	=	$230.00
1 Cement finisher	2 hours × $27.00	=	$54.00
1 Equipment operator	2 hours × $32.00	=	$64.00
			$400.00
Payroll burden	30%		$120.00
			$520.00/25 cu. yds.
		=	$20.80/cu. yd.
			$160.71
Overheads and profit	40%	=	$64.28
			$224.99/cu. yd.

Example of Recap and Pricing for House Example

Figure 9.5a, Figure 9.5b, Figure 9.5c, and Figure 9.5d show the recap and pricing for the sample house project.

Comments on Recap and Pricing Shown in Figure 9.5a, Figure 9.5b, Figure 9.5c, and Figure 9.5d

1. We have modified the pricing sheet because, in this example, materials are priced separately on a bill of materials and subtrades will perform all work. Therefore, there is no need for labor, materials, or equipment columns on this recap.
2. All takeoff items associated with a particular trade are listed below that trade; any work not included for in the subtrade quote will be priced separately. For example, the excavation subtrade includes excavation and backfill of the basement, but gravels and drilled holes are not included. Gravels are priced on the bill of materials, and drilling holes is an additional charge.
3. Concrete is priced on the bill of materials because the material supplier will supply and place concrete for the prices they quote.
4. The rough carpentry framing subtrade will frame the floor, walls, and roof of the house for $2.65 per square foot based on the floor area; so the total price for this house is 1063 square feet × $2.65 = $2,817.00.
5. It is important to determine exactly what each subtrade includes for in their quote. Here the framing subtrade includes for installing the stairs, but the railing around the stair opening is not included and has to be priced separately.
6. The exterior deck and the canopy over the entrance are also extra charges that have to be added because they are not included in the basic framing package.
7. "Supply and Install" indicates that the subtrade will supply and install their work, so there will be nothing listed on the bill of materials for these items.

PRICING SHEET (Subtrades) Page No. [1 of 4]

JOB......... *HOUSE EXAMPLE* .... DATE...................

ESTIMATED................... *ABF* ....

No.	DESCRIPTION	QUANTITY	UNIT	UNIT PRICE	SUBTRADE $	TOTAL $
1.0	*Excavate and Backfill Basement*	1	Lsum	---	$2,385.00	2,385.00
.1	Excavate Basement	541	CY	---	Incl.	---
.2	Backfill Basement	214	CY	---	Incl.	---
.3	4" Diam. Footing Drain	146	LF	$3.00	$438.00	438.00
.4	Drain Gravel	22	CY	---	See Bill of Materials	
.5	Gravel Under Slab-on-Grade	26	CY	---	See Bill of Materials	
.6	Drill 10" Diam Holes	12	LF	$16.50	$198.00	198.00
	Total:					3,021.00
2.0	*Excavate and Backfill Service Trenches*	1	Lsum	---	$1,485.00	1,485.00
.1	Excavate Trench	42	CY	---	Incl.	---
.2	Backfill Trench	41	CY	---	Incl.	---
.3	Sand Bedding	1	CY	---	See Bill of Materials	
	Total:					1,485.00
3.0	*Concrete*	---		---	See Bill of Materials	
.1	Concrete Footings	5	CY	---	Incl.	---
.2	Concrete Walls	26	CY	---	Incl.	---
.3	Concrete Slab-on-Grade	13	CY	---	Incl.	---
.4	Finish and Curing Slabs	1063	SF	---	Incl.	---
.5	6mil Poly Vapor Barrier	1170	SF	---	Incl.	---
.6	Concrete Piles	1	CY	---	Incl.	---
	Total:					0.00
4.0	*Form Footings*	1	Lsum	---	$350.00	350.00
.1	2 x 8 Footing Forms	318	LF	---	Incl.	---
.2	2 x 4 Keyway	135	LF	---	Incl.	---
	Total:					350.00
5.0	*Form Walls and Columns*	1	Lsum	---	$3,285.00	3,285.00
.1	8' High Wall Forms	270	LF	---	Incl.	---
.2	2 x 8 Blockouts	53	LF	---	Incl.	---
.3	#4 Rebar	397	lbs	---	Incl.	---
.4	10" Diam Columns	16	LF	---	Incl.	---
	Total:					3,285.00
6.0	*Damp Proof Walls*	1101	SF	---	$350.00	350.00
7.0	*Rough Carpentry Framing*	1	Lsum	---	$3,100.00	3,100.00
	- Floor System					
.1	2 x 6 Sill Plate	138	LF	---	Incl.	---
.2	½" x 9" Anchor Bolts	27	No.	---	Incl.	---
.3	3" Diam. Teleposts	3	No.	---	Incl.	---
.4	6 x 10 Paralam Beam	40	LF	---	Incl.	---
.5	2 x 10 Engineered Joists	1147	LF	---	Incl.	---
.6	4 x 4 Wood Posts	21	BM	---	Incl.	---
.7	¾" T & G Floor Sheathing	1166	sf	---	Incl.	---

Figure 9.5a   Recap and Pricing Example

PRICING SHEET (Subtrades) Page No. | 2 of 4

JOB......... *HOUSE EXAMPLE* .... DATE...

ESTIMATED................... *ABF* ....

No.	DESCRIPTION	QUANTITY	UNIT	UNIT PRICE	SUBTRADE $	TOTAL $
7.0	- *Wall Systems*			---	---	---
.7	2 x 6 Plates	452	BM	---	Incl.	---
.8	2 x 6 Studs	172	No.	---	Incl.	---
.9	½" OSB Wall Sheathing	1350	SF	---	Incl.	---
.10	2 x 10 Lintels	143	BM	---	Incl.	---
.11	2 x 4 Plates	642	BM	---	Incl.	---
.12	2 x 4 Studs	333	No.	---	Incl.	---
.13	2 x 4 Lintels	27	BM	---	Incl.	---
	- Roof System					
.14	King Post Trusses 28'-10" Span	19	No.	---	Incl.	---
.15	Gable Ends	2	No.	---	Incl.	---
.16	1 x 3 Ribbons	200	LF	---	Incl.	---
.17	2 x 4 Ridge Blocking	29	BM	---	Incl.	---
.18	2 x 4 Barge Rafters	46	BM	---	Incl.	---
.19	2 x 4 Lookouts	107	BM	---	Incl.	---
.20	2 x 6 Rough Fascia	88	BM	---	Incl.	---
.21	Roof Saddle 2'-3" Long	1	No.	---	Incl.	---
.22	2 x 4 Ceiling Blocking	92	BM	---	Incl.	---
.23	½" OSB Roof Sheathing	1522	SF	---	Incl.	---
	- Stairs					
.24	3'-4" Wide Stair with 11-Risers	1	No.	---	Incl.	---
.25	- Ditto - with 3-Risers	1	No.	---	Incl.	---
.26	Basement Handrail	12	LF	---	Incl.	---
.27	Wood Railings 40" High	9	LF	$9.00	81.00	81.00
	Total:					3,181.00
8.0	*Deck*	1	Lsum	---	125.00	125.00
.1	2 x 8 in Beam	63	BM	---	Incl.	---
.2	2 x 8 Joists	181	BM	---	Incl.	---
.3	2 x 4 Cedar Decking	266	LF	---	Incl.	---
.4	40" High Cedar Railing	28	LF	---	Incl.	---
.5	3'-7" Wide Cedar Steps 2-Treads	1	No.	---	Incl.	---
	Total:					125.00
9.0	*Canopy Over Entrance*	1	Lsum	---	50.00	50.00
.1	6 x 6 Posts	2	No.	---	Incl.	---
.2	2 x 10 Joists	65	BM	---	Incl.	---
.3	Joist Hangers	7	No.	---	Incl.	---
	Total:					50.00

Figure 9.5b Recap and Pricing Example

PRICING SHEET (Subtrades)   Page No.   3 of 4

JOB.........*HOUSE EXAMPLE*...DATE...

ESTIMATED.................*ABF*...

No.	DESCRIPTION	QUANTITY	UNIT	UNIT PRICE	SUBTRADE $	TOTAL $
10.0	*Finish Carpentry*	1	Lsum	---	1,950.00	1,950.00
	- Doors					
.1	3'-0" x 6'-8" Exterior Door with 12" Sidelights	1	No.	---	Incl.	---
.2	5'-0" x 6'-8" Patio Door	1	No.	---	Incl.	---
.3	2'-4" x 6'-8" Interior Door	2	No.	---	Incl.	---
.4	2'-6" x 6'-8" Interior Door	4	No.	---	Incl.	---
.5	2'-10" x 6'-8" Interior Door	1	No.	---	Incl.	---
.6	2'-0" x 6'-8" Bifold	1	No.	---	Incl.	---
.7	4'-0" x 6'-8" Bifold	3	No.	---	Incl.	---
.8	3'-0" x 6'-8" Bifold	1	No.	---	Incl.	---
.9	Attic Access Hatch	1	No.	---	Incl.	---
	- Door Hardware	1	No.	---	Incl.	---
.10	4" Butt Hinges	1½	Pr	---	Incl.	---
.11	3½" Butt Hinges	7	Pr	---	Incl.	---
.12	Key-in-the-Knob Lockset	1	No.	---	Incl.	---
.13	Dead Bolts	2	No.	---	Incl.	---
.14	Passage Sets	4	No.	---	Incl.	---
.15	Privacy Sets	3	No.	---	Incl.	---
	- Windows					
.16	36" x 50" Fixed / 36" x 15" Awning	1	No.	---	Incl.	---
.17	27" x 30" Awning - Fixed	3	No.	---	Incl.	---
.18	36" x 30" Awning - Fixed	1	No.	---	Incl.	---
.19	27" x 22" Awning	1	No.	---	Incl.	---
.20	36" x 24" Awning	4	No.	---	Incl.	---
.21	1" x2" Window Trim	105	LF	---	Incl.	---
.22	1" x 3" Baseboard	446	LF	---	Incl.	---
	Total:					1,950.00
11.0	*Cabinets (Supply and Install)*	1	Lsum	---	6,950.00	6,950.00
.1	2'-0" x 3'-0" Floor Mounted c/w Countertop	24.66	LF	---	Incl.	---
.2	1'-0" x 2'-8" Wall Mounted	15.75	LF	---	Incl.	---
.3	2'-0" x 2'-6" Bathroom Vanity	8.66	LF	---	Incl.	---
.4	1'-3" Wide Closet Shelves	31	LF	---	Incl.	---
.5	Adjustable Closet Rods	4	No.	---	Incl.	---
	Total:					6,950.00
12.0	*Bathroom Accessories (Supply and Install)*	1	Lsum	---	518.00	518.00
.1	Toilet Roll Holder	2	No.	---	Incl.	---
.2	5'-4" x 3'-0" Mirror	1	No.	---	Incl.	---
.3	Medicine Cabinet	2	No.	---	Incl.	---
.4	Shower Rod	1	No.	---	Incl.	---
	Total:					518.00
13.0	*Appliances (Supply and Install)*	1	Lsum	---	2,090.00	2,090.00
.1	Dishwasher	1	No.	---	Incl.	---
.2	Fridge	1	No.	---	Incl.	---
.3	Range	1	No.	---	Incl.	---
	Total:					2,090.00

Figure 9.5c   Recap and Pricing Example

PRICING SHEET (Subtrades)

Page No.

JOB.........*HOUSE EXAMPLE*.. DATE...

ESTIMATED..................*ABF*.................................;..

No.	DESCRIPTION	QUANTITY	UNIT	UNIT PRICE	SUBTRADE $	TOTAL $
14.0	*Soffit and Fascia (Supply and Install)*	1	Lsum	---	1,416.00	1,416.00
.1	Vented Aluminum Soffit	298	SF	---	Incl.	---
.2	Aluminum "J" Mould	149	LF	---	Incl.	---
.3	Aluminum Fascia 6" high	157	LF	---	Incl.	---
	Total:					1,416.00
15.0	*Siding (Supply and Install)*	1	Lsum	---	2,225.00	2,225.00
.1	Vinyl Siding	946	SF	---	Incl.	---
.2	Building Wrap	1223	SF	---	Incl.	---
	Total:					2,225.00
16.0	*½" Parging (Supply and Install)*	296	SF	$3.05	902.80	902.80
17.0	*Paint Exterior Windows and Doors*	1	Lsum	---	547.00	547.00
.1	Paint Exterior Windows	112	SF	---	Incl.	---
.2	Pauint Exterior Doors	71	SF	---	Incl.	---
.3	Stain Cedar Deck	88	SF	---	Incl.	---
.4	Stain Cedar Railings	92	SF	---	Incl.	---
	Total:					547.00
18.0	*4" Eaves Gutter (Supply and Install)*	88	LF	$6.25	550.00	550.00
19.0	*3" Downspouts (Supply and Install)*	27	LF	$4.50	121.50	121.50
20.0	*Roofing (Supply and Install)*	1	Lsum	---	2,320.00	2,320.00
.1	210lbs Asphalt Shingles	1611	SF	---	Incl.	---
.2	Ridge Cap	44	LF	---	Incl.	---
.3	4" Wide Drip-edge-flashing	88	LF	---	Incl.	---
	Total:					2,320.00
21.0	*7'-0" x 3'-6" Precast Concrete Steps*	1	No.	---	965.00	965.00
22.0	*Flooring (Supply and Install)*	1	Lsum	---	3,905.00	3,905.00
.1	Carpet	938	SF	---	Incl.	---
.2	Vinyl Flooring	252	SF	---	Incl.	---
.3	Carpet Edging	11	LF	---	Incl.	---
	Total:					3,905.00
23.0	*Drywall Finishes (Supply and Install)*	1	Lsum	---	11,711.00	11,711.00
.1	½" Drywall Ceiling	1157	SF	---	Incl.	---
.2	Textured Ceiling Finish	1157	SF	---	Incl.	---
.3	R35 Loose Insulation	1157	SF	---	Incl.	---
.4	6Mil Ply Vapor Barrier	3649	SF	---	Incl.	---
.5	Insulation Stops	40	No.	---	Incl.	---
.6	R20 Batt Insulation	1101	SF	---	Incl.	---
.7	R12 Batt Insulation	1059	SF	---	Incl.	---
.8	½" Drywall Walls	4627	SF	---	Incl.	---
	Total:					11,711.00
24.0	*Interior Painting (Supply and Install)*	1	Lsum	---	3,234.00	3,234.00
.1	Paint Walls	4627	SF	---	Incl.	---
.2	Paint Doors	460	SF	---	Incl.	---
.3	Paint Wood Railings	32	SF	---	Incl.	---
.4	Paint Shelves	78	SF	---	Incl.	---
	Total:					3,234.00
25.0	*Plumbing - Rough In*	1	Lsum	---	5,060.00	5,060.00
26.0	*Plumbing -Finish*	1	Lsum	---	2,860.00	2,860.00
27.0	*Heating - Rough In*	1	Lsum	---	3,410.00	3,410.00
28.0	*Heating - Finish*	1	Lsum	---	2,640.00	2,640.00
29.0	*Electrical - Rough In*	1	Lsum	---	2,450.00	2,450.00
30.0	*Electrical - Finish*	1	Lsum	---	1,045.00	1,045.00
31.0	*Light Fixtures*	1	Lsum	---	475.00	475.00

Figure 9.5d   Recap and Pricing Example

Example of Bill of Materials

Figure 9.6a and Figure 9.6b show the bill of materials for the sample house project.

Comments on Bill of Materials Shown in Figure 9.6a and Figure 9.6b

1. Appropriate waste factors are applied to takeoff quantities to determine **order quantities**.
2. In the case of gravels and sand, the waste factor also includes for compaction of this material.
3. On this job, concrete in contact with the soil (which includes footings, walls, and columns) requires sulphate-resisting cement Type V. Normal Type I concrete can be used for the slab-on-grade.
4. The wall forms subtrade includes for materials in their quote, so there is no need for a form materials price here.
5. Note that curing compound is measured in square feet on the takeoff but the material is priced per gallon, so the conversion factor of 400 square feet per gallon is used to calculate the number of gallons required. The resulting quantity is rounded up to the nearest whole gallon.
6. Material for damp proofing also has to be converted to gallons using a conversion factor.
7. The 2×6 material specified for the floor system is Standard or better spruce-pine-fir lumber.
8. The 2×6 material specified for the wall plates is #2 or better spruce-pine-fir lumber.
9. Sheathing material is ½" thick oriented strand board.
10. An allowance is made for rough hardware used in the rough carpentry work. The amount is calculated as 1 percent of the total price of carpentry materials.

BILL OF MATERIALS

JOB........ *House Example* .... DATE...................................

ESTIMATED............ *ABF* ....

ITEM No.	DESCRIPTION	TAKEOFF QUANTITY		WASTE FACTOR	ORDER QUANTITY		UNIT PRICE	TOTAL $
1.0	*Excavation and Gravel Materials*							
.1	4" Dia Footing Drain	146	LF	10%	161	LF	$3.25	$523.25
.2	Drain Gravel (Supply and Place)	22	CY	25%	28	CY	$30.00	$840.00
.3	Sand Bedding (Supply and Place)	1	CY	50%	2	CY	$17.50	$35.00
.4	Gravel Under Slab-on-Grade (Supply and Place)	26	CY	25%	33	CY	$36.00	$1,188.00
	Total:							$2,586.25
2.0	*Concrete Materials*							
.1	2500psi Concrete Type V Cement (Supply and Place)	32	CY	5%	34	CY	$225.00	$7,650.00
.2	2500psi Concrete Type I Cement (Supply and Place)	13	CY	5%	14	CY	$220.00	$3,080.00
.3	Curing Compound (Supply and Place)	1063	SF	10%	1169	SF	---	---
	@ 400 SF per Gallon =	---		---	3	Gal	$8.50	$25.50
.4	2 x 8 Forms	318	LF	10%	350	LF	$2.00	$700.00
.5	2 x 4 Keyway	135	LF	5%	142	LF	$1.25	$177.50
.6	8' High Wall Forms	(Materials included in subtrade price - See Pricing Sheet)						
.7	2 x 8 Blockouts	53	LF	10%	58	LF	$1.80	$104.40
.8	Bituminous Damp Proofing	1101	SF	10%	1211	SF	---	---
	@ 25 SF per Gallon =	---		---	49	Gal	$3.95	$193.55
.9	#4 Rebar	397	lbs	10%	437	lbs	$0.85	$371.45
.10	6mil Vapor Barrier (Incl. VB for Walls & Ceilings)	4819	SF	10%	5301	SF	$0.06	$318.06
.11	10" Diam. Sono Tubes	16	LF	10%	18	LF	$2.00	$36.00
	Total:							$12,656.46
3.0	*Carpentry Materials*							
	- Floor System							
.1	2 x 6 - Std&Btr SPF	138	BM	5%	145	BM	$0.55	$79.75
.2	3" Diam. Teleposts	3	No.	5%	3	No.	$40.00	$120.00
.3	6 x 10 Parallam Beam	40	LF	5%	42	LF	$25.00	$1,050.00
.4	2 x 10 Engineered Joists	1147	LF	5%	1204	LF	$3.15	$3,792.60
.5	4 x 4 - 8' #2&Btr SPF	2	No.	---	2	No.	$10.50	$21.00
.6	¾" T & G Floor Sheathing	1166	SF	10%	1283	SF	---	---
	@ 32SF per Sheet =	---		---	41	Sht	$35.15	$1,441.15
	- Wall Systems							
.7	2 x 6 #2&Btr SPF	452	BM	10%	497	BM	$0.55	$273.35
.8	2 x 6 Studs SPF	172	No.	10%	189	No.	$4.40	$831.60
.9	½" OSB Wall Sheathing	1350	SF	10%	1485	SF	---	---
	@ 32SF per Sheet =	---		---	47	Sht	$16.00	$752.00
.10	2 x 10 #2&Btr SPF	143	BM	10%	157	BM	$0.70	$109.90
.11	2 x 4 #2&Btr SPF	669	BM	10%	736	BM	$0.51	$375.36
.12	2 x 4 Studs SPF	333	No.	10%	366	No.	$2.72	$995.52
	- Roof System							
.13	King Post Trusses 28'-10" Span	19	No.	---	19	No.	$77.50	$1,472.50
.14	Gable Ends	2	No.	---	2	No.	$80.00	$160.00
.15	1 x 3 #2&Btr SPF	200	LF	10%	220	LF	$0.30	$66.00
.16	2 x 4 #2&Btr SPF	274	BM	10%	301	BM	$0.51	$153.51
.17	2 x 6 #2&Btr SPF	88	BM	10%	97	BM	$0.51	$49.47
.18	Roof Saddle 2'-3" Long	1	No.	---	1	No.	$30.00	$30.00
.19	½" OSB Roof Sheathing	1522	SF	10%	1674	SF	$0.50	$837.00

Figure 9.6a Bill of Materials for House Example

BILL OF MATERIALS

Page No. 2 of 2

JOB........*House Example*.. DATE...

ESTIMATED............*ABF*...

ITEM No.	DESCRIPTION	NET QUANTITY		WASTE FACTOR	ORDER QUANTITY		UNIT PRICE	TOTAL
	- Basement Stairs							
.20	3'-4" Wide Stair with 11-Risers	1	No.	---	1	No.	$150.00	$150.00
.21	- Ditto - with 3-Risers	1	No.	---	1	No.	$50.00	$50.00
.22	Basement Handrail	12	LF	10%	13	LF	$2.50	$32.50
.23	Wood Railings 40" High	9	LF	10%	10	LF	$11.00	$110.00
	- Deck							
.24	2 x 8 #2&Btr SPF	244	BM	10%	268	BM	$0.65	$174.20
.25	2 x 4 Std&Btr Cedar	266	LF	10%	293	LF	$0.55	$161.15
.26	40" High Cedar Railing	28	LF	10%	31	LF	$9.50	$294.50
.27	3'-7" Wide Cedar Steps 2-Treads	1	No.	---	1	No.	$75.00	$75.00
	- Canopy Over Entrance							
.28	6 x 6 8' #2&Btr SPF	2	No.	---	2	No.	$43.20	$86.40
.29	2 x 10 #2&Btr SPF	65	BM	10%	72	BM	$0.70	$50.40
.30	Joist Hangers	7	No.	10%	8	No.	$0.75	$6.00
.31	Other Rough Hardware	Item		(1% of Carpentry Materials)			---	$138.01
	Total:							$13,938.87
4.0	*Finish Carpentry*							
	- Doors							
.1	3'-0" x 6'-8" Exterior Door with 12" Sidelights	1	No.	---	1	No.	$600.00	$600.00
.2	5'-0" x 6'-8" Patio Door	1	No.	---	1	No.	$1,400.00	$1,400.00
.3	2'-4" x 6'-8" Interior Door	2	No.	---	2	No.	$175.00	$350.00
.4	2'-6" x 6'-8" Interior Door	6	No.	---	6	No.	$180.00	$1,080.00
.5	2'-10" x 6'-8" Interior Door	1	No.	---	1	No.	$185.00	$185.00
.6	2'-0" x 6'-8" Bifold	1	No.	---	1	No.	$115.00	$115.00
.7	4'-0" x 6'-8" Bifold	3	No.	---	3	No.	$180.00	$540.00
.8	3'-0" x 6'-8" Bifold	1	No.	---	1	No.	$120.00	$120.00
.9	Attic Access Hatch	1	No.	---	1	No.	$50.00	$50.00
	- Door Hardware							
.10	4" Butt Hinges	1½	Pr	---	1.5	Pr	$30.00	$45.00
.11	3½" Butt Hinges	7	Pr	---	7	Pr	$22.00	$154.00
.12	Key-in-the-Knob Lockset	1	No.	---	1	No.	$150.00	$150.00
.13	Dead Bolts	2	No.	---	2	No.	$40.00	$80.00
.14	Passage Sets	4	No.	---	4	No.	$48.00	$192.00
.15	Privacy Sets	3	No.	---	3	No.	$55.00	$165.00
	- Windows							
.16	36" x 50" Fixed / 36" x 15" Awning	1	No.	---	1	No.	$540.00	$540.00
.17	27" x 30" Awning - Fixed	3	No.	---	3	No.	$250.00	$750.00
.18	36" x 30" Awning - Fixed	1	No.	---	1	No.	$270.00	$270.00
.19	27" x 22" Awning	1	No.	---	1	No.	$300.00	$300.00
.20	36" x 24" Awning	4	No.	---	4	No.	$370.00	$1,480.00
.21	1" x2" Window Trim	105	LF	10%	116	LF	$6.65	$771.40
.22	1" x 3" Baseboard	446	LF	10%	491	LF	$1.45	$711.95
	Total:							$10,049.35

Figure 9.6b   Bill of Materials for House Example

SUMMARY

- Pricing an estimate consists of first sorting and listing all takeoff items according to the required trade breakdown (the recap) and second, applying prices to this sorted list of items.
- If computer estimating is used, the process of sorting into a recap is automated using a numerical item coding system in the takeoff.
- There are five price categories considered in the pricing of an estimate:
  - Labor
  - Equipment
  - Materials
  - Subcontractors
  - Job overheads
- There are three general reasons why project costs may exceed estimated prices:
  - Takeoff quantities are too low
  - Actual productivity does not meet anticipated productivity
  - Subcontractors or material suppliers fail to meet obligations
- The estimator's work is usually reviewed in order to reduce the risk of takeoff errors.
- There are two main components to consider when pricing labor and equipment, both of which may be difficult to predict:
  - The hourly wage rate of labor or hourly cost of equipment
  - The productivity of labor or equipment
- In order to price labor, the estimator has to determine the wage rates that apply to the project and predict how these rates will change over the life of the project.
- Equipment rates depend on ownership costs or rental rates, both of which can be difficult to predict for a project.
- The productivity of labor and equipment is governed by two groups of factors:
  - Job factors that have to do with the nature and location of the particular project under consideration
  - Labor and management factors that mainly relate to the quality of supervision and the skills of workers on the project
- The builder will have little, if any, influence over job factors but must carefully examine these factors before pricing the work, including:
  - Probable weather conditions
  - Site access
  - Repetitive work
- A database of accurate historic labor and equipment costs is a useful source of information when pricing an estimate. This is generally obtained from previous project cost reports.
- It is also beneficial to have information about why the final cost of work shown on a cost report varied from the estimated costs.
- There is less risk of cost over-runs with material estimates since firm prices can be obtained from suppliers for the supply of materials.
- Quantities of materials measured in *order units* are often calculated because the units that materials are packaged in often differ from *takeoff units*.
- Add a waste factor and a swell factor to *takeoff quantities* of gravel to determine *order quantities* of this material.
- The optimum number of trucks required for transporting excavation material is calculated from the following variables:
  - Haul distance
  - Speed of trucks

- Capacity of trucks
- Output of loaders
■ The price of concrete depends mainly upon the strength required; it is also subject to a number of additional charges including:
  - Type of cement
  - Air entrainment
  - Summer cooling and winter heating
  - Different aggregate sizes
  - Fiber reinforcement
  - Cement additives
  - Environmental fees
■ Lumber prices vary with species, grade, and the type of lumber processing involved.
■ While pricing subcontract work may appear to be straightforward, the subtrade estimates the price of their work and submits the price to the homebuilder. In practice there can be many problems associated with pricing this work. This includes concern about the subtrade's ability to perform the work and the problem of overlap or gaps in the coverage of subtrade work.

RECOMMENDED RESOURCES

Information	Web Page Address
■ For quick quotes on any type of construction work	http://www.get-a-quote.net/
■ For pricing information, RSMeans is a comprehensive source	http://rsmeans.reedconstructiondata.com/

REVIEW QUESTIONS

1. Describe the two stages involved in pricing an estimate.
2. Describe how computers can help with the first stage included above.
3. List ten items often found on a homebuilder's standard trade breakdown.
4. What is the "builder's risk," and how does it relate to the type of construction contract to be used on a project?
5. List three general reasons why actual costs may exceed estimated costs on a lump-sum contract.
6. What strategy do most homebuilders use to reduce the risk of over-runs on the cost of labor and equipment on a project?
7. What ten questions should be considered regarding equipment rental rates?
8. Why should an estimator be on the lookout for situations where construction operations are repetitive on a project?
9. How does site access affect the price of work?
10. How are cost records used by an estimator?
11. Before an estimator can use the data from a cost record effectively, what two questions have to be answered?
12. Give three reasons why pricing subcontracted work may not be as straightforward as it appears to be.

PRACTICE PROBLEMS

1. Use the conversion factor on the sample bill of materials to calculate the number of gallons of damp proofing material required for an order quantity of 2300 square feet.
2. If the takeoff area of floor sheathing is 5000 square feet and the waste factor is 10 percent, what is the order quantity measured in sheets of material?
3. From the concrete quote in Figure 9.3, calculate the price of 2000 psi concrete with sulphate-resisting cement place in winter months.
4. Develop a subcontractor's price per cubic yard to supply and place 2500 psi concrete in slabs-on-grade based on Example 9.7 and the concrete quote in Figure 9.3. Assume that the crew can place 30 cubic yards of concrete in 2 hours.

10

PRICING EQUIPMENT

OBJECTIVES

After reading this chapter and completing the review questions, you should be able to:

■ Describe how equipment is usually priced in an estimate.
■ Explain under what circumstances it can be financially beneficial to own rather than rent construction equipment.
■ List and describe the types of expenses that should be accounted for when calculating the ownership cost of equipment.
■ Determine the complete hourly ownership cost of an item of equipment.
■ Calculate the rental rate for an item of equipment.

KEY TERMS

company overhead costs	financing expenses	maintenance and repair costs
depreciation	fuel and lubrication costs	straight-line depreciation
equipment operator costs	insurance costs	tire depreciation

Introduction

Pricing Builder-Owned Equipment

Equipment used in the process of constructing a building is usually priced in an estimate by multiplying unit prices by the period of time the equipment is used on the project. This time period may be expressed in terms of hours, weeks, or even months, whichever is appropriate. In the situation where the builder's equipment is rented, the unit prices used will be the rental rates obtained from equipment rental companies. Where the equipment is owned by the builder, unit prices will have to be calculated from ownership costs as detailed below, but this raises the question: should a builder rent or own construction equipment?

Renting or Purchasing Equipment

For the builder who is new in the construction business, the decision whether to rent or purchase equipment is usually quite easy to make because, lacking surplus cash and without a well-established credit rating, the only viable alternative is renting. For the more mature construction business, the decision may be a great deal more difficult. This builder, who is more likely to have funds and credit sources available, has to determine if equipment investments are justified. The builder should buy construction equipment only when the investment offers additional benefits to those obtained from renting equipment.

A builder does not necessarily have to own any construction equipment in order to carry on business. In most parts of the country there are many companies in the construction equipment rental business offering competitive rental rates on a large selection of equipment, and there can be distinct advantages to renting equipment, including:

- The builder does not have to maintain a large inventory of specialized plant and equipment where individual items are used infrequently.
- The builder has continuous access to the newest and most efficient items of equipment available.
- There is little or no need for equipment warehouse and storage facilities.
- There is a reduced need for the builder to employ maintenance staff and operate facilities for their use.
- Accounting for equipment costs can be simpler when equipment is rented.
- There may be significant savings on company insurance premiums when a builder is not maintaining an inventory of plant and equipment.

However, the cost of owning equipment can be considerably less than rental costs when the equipment is heavily used. If an item of equipment is used only occasionally by a builder, it is generally most cost effective to rent the item. However, if this item is used every week on one job after another over a long period, significant savings can be obtained from owning this equipment. Builders may gain further advantage from equipment ownership due to the perception that these builders are more financially stable and committed than those that own no equipment. The builder's company name is often displayed on its equipment; this adds to the view that this is a solid, well-established organization.

Where a comparison of equipment ownership with the rental alternative strictly on the basis of cost is needed, the full cost per unit of time of owning an item of equipment has to be determined. To estimate the full ownership cost, the following aspects of equipment ownership have to be considered:

- **Depreciation** expense
- **Maintenance and repair costs**
- **Financing expenses**
- **Taxes**
- **Insurance costs**
- **Storage costs**
- **Fuel and lubrication costs**

Depreciation

In everyday usage, the term depreciation refers to the decline in market value of an asset. To accountants, the term has a more narrow meaning having to do with allocating the acquisition cost of an asset over its useful life. The amount of depreciation calculated may or may not reflect the true loss of market value of the piece of equipment over its life; more often than not, it does not. Also, the depreciation amounts considered here are not related to tax considerations. For tax purposes, a completely different depreciation schedule may be adopted.

In our appraisal of depreciation, some factors are explicit while other factors have to be estimated. Generally what we know is that the item of equipment costs a certain amount to acquire (the initial cost), it will be used for a number of years (the useful life), and this item will be sold at the end of this period for a sum of money (the salvage value). There is, however, some uncertainty about the exact duration of the useful life of the equipment and about the precise amount of salvage value that will be realized when it is disposed of. Each of these values, therefore, has to be estimated when assessing depreciation.

The process of allocating the cost of an item over its useful life is known as amortization. There are several calculation methods available to determine amortization values but, because it is not difficult to calculate and it offers other advantages, here we will adopt the straight-line method.

Straight-Line Depreciation

The straight-line method is a commonly used method of calculating depreciation generally. Most estimators use this method when calculating equipment rental rates in particular. Depreciation on a straight-line basis is allocated equally per year over the useful life of the asset, thus the annual depreciation amount is constant and is equal to the cost of the asset minus any salvage value divided by the years of life of the asset:

$$\text{Annual Depreciation} = \frac{\text{Initial Cost} - \text{Estimated Salvage Value}}{\text{Estimated Useful Life (years)}}$$

EXAMPLE 10.1

The total initial cost of a hydraulic crane is $105,000; the useful life is expected to be 5 years, and the estimated salvage value at the end of this period is $55,000:

$$\text{Annual Depreciation} = \frac{\$105,000 - \$55,000}{5}$$
$$= \$10,000 \text{ per year}$$

Figure 10.1 shows the depreciation schedule based on **straight-line depreciation** over the life of this compressor.

Although the straight-line method of depreciation does not reflect the fact that depreciation usually occurs at an accelerated rate in the early years of the life of an asset, the method of calculation used here makes some allowance for this as explained below. There are alternative methods of depreciation that try to address the deficiencies of the straight-line method, but they are beyond the scope of this text.

Rubber-Tired Equipment

Because the life expectancy of rubber tires is generally far less than the life of the equipment they are used on, the depreciation rate of tires will be quite different from

Straight-Line Depreciation		
End of Year	Depreciation for the Year	Book Value
	$	$
0	0	105,000
1	10,000	95,000
2	10,000	85,000
3	10,000	75,000
4	10,000	65,000
5	10,000	55,000

Figure 10.1   Depreciation Schedule

the depreciation rate on the rest of the vehicle. The repair and maintenance cost of tires as a percentage of their depreciation will also be different from the percentage associated with the repair and maintenance of the vehicle.

Consequently, when considering the depreciation of a rubber-tired vehicle, the cost of a set of replacement tires should be deducted from the initial cost of the vehicle and the depreciation on the tires and the depreciation on the vehicle, without the tires should each be calculated separately. The repair and maintenance of the vehicle and of the tires can then also be calculated separately.

Maintenance and Repairs

The cost of maintenance and repairs is a significant part of equipment ownership costs. Builders who own equipment agree that good maintenance, including periodic wear measurement, timely attention to recommended service, and daily cleaning when conditions warrant it, can extend the life of equipment and actually reduce the operating costs by minimizing the effects of adverse conditions. All items of equipment used by a builder will require maintenance and probably also some repairs during the course of their useful life. A larger company that owns many items of equipment usually sets up facilities with workers qualified to perform the necessary maintenance operations on equipment. Smaller companies will generally have their equipment serviced and repaired at specialty shops. Whatever the case, the costs that are calculated are added into the total ownership costs of equipment.

Construction operations can subject equipment to considerable wear and tear, but the amount of wear varies enormously between different items of equipment used and between different job conditions. The rates used in the following examples are based on the average costs of maintenance and repair, but since these costs do vary so much, the builder formulating equipment operating prices should adjust the rates for maintenance and repairs according to the conditions the equipment is to work under. As with many estimating situations, good records of previous maintenance costs will much improve the quality of the estimator's assessment of probable future maintenance costs.

Maintenance and repair costs are calculated as a percentage of the annual depreciation costs for each item of equipment. When depreciation is calculated using the straight-line method as in the examples that follow, the result is a constant amount being charged yearly for depreciation and then a second constant amount is allowed for maintenance and repairs. Realistically, depreciation will be high in the early years of ownership, while actual maintenance and repair costs in these years should be low. The relative values of yearly depreciation and maintenance costs will gradually reverse until, in the later years, low depreciation will be accompanied by high maintenance and repair bills. Using a constant amount yearly for these two expenses, therefore, is reasonable because at any point in time one factor will be too high but it will be compensated for by the other factor, which will be too low at that time.

Some examples of maintenance rates expressed as a percentage of the depreciation costs are shown below:

Portable Compressors	83%
Generators	100%
Self-Propelled Roller	120%
Fork Lift Trucks	100%
Hydraulic Lifting Cranes	120%
Hydraulic Backhoes	100%
Telescopic Boom Excavators	100%
Front-End Loaders	110%
Dump Trucks	130%

Financing Expenses

Whether the owner of construction equipment purchases the equipment using cash or whether the purchase is financed by a loan from a lending institution, there is going to be an interest expense involved. The interest expense is the cost of using capital, whether it is your money or borrowed. Where the builder's own cash is used, it is the amount that would have been earned had the money been invested elsewhere. When the purchase is financed by a loan, the interest expense is the interest charged on the loan. In both cases, the interest expense is calculated by applying an interest rate to the owner's *average annual investment* in the unit. The *average annual investment* is the value approximately midway between the total initial cost of the unit and its salvage value. Thus:

$$\text{Average Annual Investment} = \frac{(\text{Total Initial Cost} + \text{Salvage Value})}{2}$$

The interest rate used to calculate the financing expense will vary from time to time, from place to place, and also from one company to another depending mostly on their credit rating and how good a deal they can secure from the lender. In the examples that follow, we will use a rate of 6 percent.

Taxes, Insurance, and Storage Costs

Just as with investment expenses, variations can be expected in the cost of the annual taxes, insurance premiums, and storage costs together with fees for licenses required and other fees expended on an item of equipment. Where these expenses are known, they should be added into the calculation of the annual ownership costs of the equipment. In the case where information on these costs is not available, they may be calculated as a percentage of the average annual investment cost of the piece of equipment.

The interest expense rate and the rate for taxes, insurance, and storage costs are often combined to give a total equipment overhead rate. Below we will use an equipment overhead rate of 11 percent that comprises 6 percent for the investment rate and 5 percent to cover taxes, insurance, and storage costs.

Fuel and Lubrication Costs

Fuel consumption and the consumption of lubrication oil can be closely monitored in the field. Data from these field observations will enable the estimator to quite accurately predict future rates of consumption under similar working conditions. However, if there is no access to this information, likely fuel consumption information may be obtained from equipment manufacturers.

As a guide to fuel consumption, a gasoline engine operating under normal conditions will consume approximately 0.06 gallons of fuel for each horsepower-hour developed. A diesel engine is slightly more efficient at 0.04 gallons of fuel for each horsepower-hour. These rates of fuel consumption apply when the engine is operating at full throttle. To account for times when the engine is idling or there is a break in the work, an operating factor is introduced. This operating factor is an assessment of the load under which the equipment engine is operating. An engine continually producing full-rated horsepower is operating at a factor of 100 percent. Construction equipment never operates at this level for extended periods, so the operating factor used in calculating overall fuel consumption is always a value less than 100 percent.

The operating factor is yet another variable with a wide range of possible values responding to the many different conditions that might be encountered when the equipment under consideration is used. In the examples that follow, the specific operating factors used can be no more than averages reflecting normal work conditions. Here again, there is no good substitute for hard data carefully obtained in the observation of actual operations in progress.

EXAMPLE 10.2

If the hydraulic crane mentioned in Example 1 was equipped with a diesel engine rated at 106 hp operating at a factor of 50%, the fuel consumption can be determined:

$$\text{Fuel Consumption} = \frac{(106 \times 0.04 \times 50)}{100}$$
$$= 2.12 \text{ gallons/hour}$$

The amount of lubricating oil consumed will vary with the size of the engine, the capacity of the crankcase, the condition of the engine components, the frequency of oil changes, and the general level of maintenance. An allowance in the order of 10 percent of the fuel costs is used as an average value in the following examples.

Equipment Operator Costs

Whether a builder decides to rent equipment or own the equipment used on its projects, the cost of labor operating the equipment has to be considered. In some situations, rentals may be available that include an **equipment operator** as part of the rental agreement. This variety of rental agreement is sometimes available for excavation equipment, and it can be a preferred alternative when the rental company offers a high-caliber operator who is familiar with the particular excavation unit and is capable of high productivity.

Some equipment is rented without an operator. So, just as in the case where the builder is using company-owned equipment, the labor costs for operating the equipment have to be calculated and added to the estimate. The usual way to price these costs is to apply the operator's hourly wage alongside the equipment hourly rate, and then use the expected productivity of the equipment to determine a price per measured unit for labor and a price per measured unit for equipment. Example 10.3 on page 211 illustrates this method of pricing equipment and operator's costs.

EXAMPLE 10.3

Where the hourly cost of an excavator is $84.00, the wage of an operator for this equipment is $32.00 per hour, and the expected productivity of the excavator is 50 cubic yards per hour, the unit prices for labor and equipment would be calculated as:

EQUIPMENT	LABOR
$\dfrac{\$84.00}{50}$	$\dfrac{\$32.00}{50}$
= $1.68 per cu. yds.	= $0.64 per cu. yds.

These unit prices can now be applied to the total quantity of excavation that this equipment is expected to perform in accordance with the takeoff.

Company Overhead Costs

Where the equipment ownership costs calculated in accordance with this chapter are to be used as a basis of rental rates charged by the builder to others for the use of the builder's equipment, the full rental rates should include an amount for **company overhead costs** and an amount for profit. Company overhead costs are basically the fixed costs associated with running a business. They may include the cost of maintaining a furnished office, office equipment, and personnel, together with all the other costs of business operation.

Since the rental rate quoted by a builder to another party for the use of the builder's equipment is, in a sense, a kind of bid, the same considerations should be applied to the markup on the rental rate as are applied to markup on any of builder's bids. Markup, comprising an amount for company overhead costs and an amount for profit, is dealt with in some detail in Chapter 13 where the markup included in a bid price is discussed.

Examples

EXAMPLE 10.4

Calculate the ownership cost per hour for a generator powered by a 20-hp gasoline engine based on the following data:

Engine:	20 hp gasoline
Operating Factor:	75%
Purchase Price:	$20,000
Freight Charges:	$600

(continued)

EXAMPLE 10.4 (continued)

Estimated Salvage Value: $4,000
Useful Life: 6 years
Hours Used per Year: 1,000
Maintenance & Repairs: 100% of Annual Depreciation
Equipment Overhead Rate: 11%
Fuel Price: $3.80 per gallon

First, preliminary calculations are made to determine the average annual investment and the fuel consumption rate:

$$\text{Average Annual Investment} = \frac{\text{Total Cost} + \text{Salvage Value}}{2}$$

$$= \frac{(\$20,000 + \$600 + \$4,000)}{2}$$

$$= \$12,300$$

Fuel Consumption $= 20 \times 0.06 \times 75\%$ gallons/hour

$$= 0.90 \text{ gallons/hour}$$

The annual cost of depreciation, maintenance and repairs, and equipment overheads can now be calculated:

Annual Costs:

Depreciation $= \dfrac{\text{Initial Cost} - \text{Estimated Salvage Value}}{\text{Estimated Life (years)}}$

$= \dfrac{\$20,600 - \$4,000}{6}$ $= \$2,767$

Maintenance and Repairs = 100% of Annual Depreciation $= \$2,767$

Equipment Overheads   = 11% × Average Annual Investment

$= 0.11 (\$12,300)$ $= \underline{\$1,353}$

Total Annual Costs: $\underline{\$6,887}$

Now the hourly costs can be calculated, including the cost of fuel and lube oil required:

Hourly Costs:

Vehicle Cost $= \dfrac{\text{Total Annual Cost}}{\text{Hours Used per Year}}$

$= \dfrac{\$6,887}{1,000}$ $= \$6.89$

Fuel Cost   = Fuel Consumption × Cost of Fuel

= 0.90 gallons × $3.80 per gallon $= \$3.42$

Lube Oil   = 10% of Fuel Cost

$= 0.1 \times \$3.42$ $= \underline{\$0.34}$

Generator Cost per Hour: $\underline{\$10.65}$

Whenever this generator is used on a project, it will be charged to that project at a price $10.65 times the number of hours it is used.

The item of equipment considered in the next example is a rubber-tired front-end loader. In this case, the depreciation cost of the tires has to be calculated separately from the depreciation cost of the rest of the vehicle.

EXAMPLE 10.5

Calculate the ownership cost per hour for a front-end loader powered by a 51 hp diesel engine based on the following data:

Engine:	51 hp diesel
Operating Factor:	60%
Purchase Price:	$50,000 (including tires)
Frieght Charges:	$2,000
Estimated Salvage Value:	$9,000
Useful Life:	7 years
Hours Used per Year:	1,600
Maintenance & Repairs:	110% of Annual Depreciation
Tire Cost:	$1,000
Tire Life:	2,000 hours
Maint. & Repairs (Tires):	15% of Tire Depreciation
Equipment Overhead Rate:	11%
Fuel Price:	$4.00 per gallon

First, preliminary calculations are made to determine the average annual investment and the fuel consumption rate:

$$\text{Average Annual Investment} = \frac{\text{Total Cost + Salvage}}{2}$$

$$= \frac{\$50,000 + \$2,000 + \$9,000}{2}$$

$$= \$30,500$$

$$\text{Fuel Consumption} = 51 \times 0.04 \times 60\% \text{ gallons/hour}$$

$$= 1.22 \text{ gallons/hour}$$

The annual cost of depreciation, maintenance and repairs, and equipment overheads can now be calculated:

Annual Costs:

$$\text{Depreciation} = \frac{\text{Initial Cost} - \text{Tire Cost} - \text{Salvage Value}}{\text{Life (years)}}$$

$$= \frac{\$50,000 + \$2,000 - \$1,000 - \$9,000}{7}$$

$$= \$6,000$$

Maintenance and Repairs = 110% of Annual Depreciation = $6,600

Equipment Overheads = 11% × Average Annual Investment

= 0.11 ($30,500)   = $3,355

Total Annual Costs—Vehicle: $15,955

(*continued*)

EXAMPLE 10.5 (continued)

Now the hourly costs can be calculated, including the cost of fuel and lube oil required:

Hourly Costs:

Vehicle Cost $= \dfrac{\text{Total Annual Cost}}{\text{Hours Used per Year}}$

$= \dfrac{\$15,955}{1,600}$ $= \$9.97$

Tire Depreciation $= \dfrac{\text{Cost of Tires}}{\text{Life in Hours}}$

$= \dfrac{\$1,000}{2,000}$ $= \$0.50$

Tire Maintenance $= 15\%$ of Tire Depreciation

$= 0.15 \times \$0.50$ $= \$0.08$

Fuel Cost $= $ Fuel Consumption $\times$ Cost of Fuel

$= 1.22$ gallons $\times \$4.00$ per gallon $= \$4.88$

Lube Oil $= 10\%$ of Fuel Cost

$= 0.1 \times \$4.88$ $= \$0.49$

Loader Cost per Hour: $\underline{\$15.91}$

EXAMPLE 10.6

What would the estimated unit prices for backfilling basements using the loader described in Example 5 be, together with an operator at a wage of $20.00 per hour when the expected productivity of this unit is 15 cu.yds./hr.?

	EQUIPMENT	LABOR
Unit Prices $=$	$\dfrac{\$15.91}{15 \text{ cu. yd.}}$	$\dfrac{\$32.00}{15 \text{ cu. yd.}}$
$=$	$\$1.06$ per cu. yd.	$\$2.13$ per cu. yd.

EXAMPLE 10.7

What would the hourly charge-out rates (rental rates) for the above loader and operator be based on the following overheads and profit requirements?

The company overhead on equipment is 20%

The company overhead on labor is 50%

The required profit margin is 10%

(continued)

EXAMPLE 10.7 (continued)

Charge-Out Rate for Excavator

Excavator Cost per Hour		= $15.91
Company Overhead:	20%	= $ 3.18
Subtotal:		$19.09
Profit Margin:	10%	= $ 1.91
Charge-Out Rate—Loader:		$21.00

Charge-Out Rate for Operator

Operator Cost per Hour		= $32.00
Company Overhead:	50%	= $16.00
Subtotal:		$48.00
Profit Margin:	10%	= $ 4.80
Charge-Out Rate—Operator:		$52.80

So the total charge-out rate for this loader and operator would be $73.80 per hour.

Use of Spreadsheets

Because of the repetitive nature of the calculations involved, a computer spreadsheet program is very useful for calculating equipment ownership costs. Use of computer spreadsheet applications is particularly beneficial where it is necessary to calculate the ownership costs of large numbers of equipment items. Also, the process of updating the data on which the ownership cost calculations are based can be accomplished far more conveniently when the original calculations are stored in the form of spreadsheet calculations.

A spreadsheet template is set up to provide the basic format of the calculation process. The data applicable to the specific item of equipment is then inserted to generate the ownership cost of that item. This process can be repeated for any number of equipment units, thereby enabling the estimator to calculate their ownership costs in a matter of minutes. Figure 10.2 shows the format of a spreadsheet template for this use. Figure 10.3 and Figure 10.4 show the computer calculations of the ownership costs of the items of equipment considered in Example 10.4 and Example 10.5 above.

EQUIPMENT OWNERSHIP COSTS

Equipment Item:
Operating Factor:
Purchase Price:
Freight:
Salvage Value:
Life Expectancy:
Hours per Year:
Maintenance and Repairs:
Tire Cost:
Tire Life:
Maintenance & Repairs on Tires:
Equipment Overhead:
Fuel Cost:

Average Annual Investment: =

Fuel Consumption: =

ANNUAL COSTS:

    Depreciation on Equipment: =

    Maintenance and Repairs: =

    Equipment Overhead: =

    Total Annual Cost   = $ _____

HOURLY COST:

    Equipment Cost: =

    Tire Depreciation: =

    Maintenance & Repairs on Tires: =

    Fuel Cost: =

    Lube Oil Cost: =

TOTAL COST PER HOUR   = $ _____

Figure 10.2 Equipment Ownership Template

EQUIPMENT OWNERSHIP COSTS

Equipment Item:	Generator with 20 hp Gasoline Engine
Operating Factor:	75.00%
Purchase Price:	$20,000
Freight:	$600
Salvage Value:	$4,000
Life Expectancy:	6 Years
Hours per Year:	1,000
Maintenance and Repairs:	100% of Annual Depreciation
Tire Cost:	No Tires
Tire Life:	N/A
Maintenance & Repairs on Tires:	N/A
Equipment Overhead:	11% of Average Annual Investment
Fuel Cost:	$3.80 per gallon

Average Annual Investment:	($26,600 + $6,000) / 2	=	$12,300
Fuel Consumption:	20 x 75% x 0.06	=	0.9 Gals. per Hour

ANNUAL COSTS:

Depreciation on Equipment:	($20,600 - $6,000) / 6	=	$2,767
Maintenance and Repairs:	100% of Depreciation	=	$2,767
Equipment Overhead:	11% of $12,300	=	$1,353
Total Annual Cost		=	$6,886

HOURLY COST:

Equipment Cost:	$6,886 / 1,000	=	$6.89
Tire Depreciation:	No Tires	=	N/A
Maintenance & Repairs on Tires:	No Tires	=	N/A
Fuel Cost:	0.90 x $3.80	=	$3.42
Lube Oil Cost		=	$0.34
TOTAL COST PER HOUR		=	**$10.65**

Figure 10.3   Generator Ownership Cost Calculation

EQUIPMENT OWNERSHIP COSTS

Equipment Item:	Front-End Loader with 51 hp Diesel Engine
Operating Factor:	60.00%
Purchase Price:	$50,000
Freight:	$2,000
Salvage Value:	$9,000
Life Expectancy:	7 Years
Hours per Year:	1,600
Maintenance and Repairs:	110% of Annual Depreciation
Tire Cost:	$1,000
Tire Life:	2,000 Hours
Maintenance & Repairs on Tires:	15% of Tire Depreciation
Equipment Overhead:	11% of Average Annual Investment
Fuel Cost:	$4.00 per gallon

Average Annual Investment:	($52,000 + $9,000) / 2	=	$30,500
Fuel Consumption:	51 x 60% x 0.04	=	1.22 Gals. per Hour

ANNUAL COSTS:

Depreciation on Equipment:	($52,000 - $1,000 - $9,000) / 7	=	$6,000
Maintenance and Repairs:	110% of Depreciation	=	$6,600
Equipment Overhead:	11% of $30,500	=	$3,355
Total Annual Cost		=	$15,955

HOURLY COST:

Equipment Cost:	$15,955 / 1,600	=	$9.97
Tire Depreciation:	$1,000 / 2,000	=	$0.50
Maintenance & Repairs on Tires:	15% of $0.50	=	$0.08
Fuel Cost:	1.22 x $4.00	=	$4.88
Lube Oil Cost		=	$0.49
TOTAL COST PER HOUR		=	**$15.91**

Figure 10.4   Front-End Loader Ownership Cost Calculation

SUMMARY

- Construction equipment is usually priced in an estimate by multiplying unit prices by the time the equipment is used on the project.
- Advantages of renting equipment:
  - The builder does not have to maintain a large inventory of specialized plant and equipment where individual items are used infrequently.
  - The builder has continuous access to the newest and most efficient items of equipment available.
  - There is little or no need for equipment warehouse and storage facilities.
  - There is a reduced need for the builder to employ maintenance staff and operate facilities for their use.
- Accounting for equipment costs can be simpler when equipment is rented.
- There may be significant savings on company insurance premiums when a builder is not maintaining an inventory of plant and equipment.
- There are financial advantages to owning equipment when a builder can provide steady use of the equipment.
- The following expenses should be considered when determining the full ownership cost of material:
  - Depreciation
  - Maintenance and repair
  - Financing
  - Taxes
  - Insurance
  - Storage
  - Fuel and lubrication
- When straight-line depreciation is used, the amount of depreciation from initial cost to salvage value is distributed equally each year over the life of the asset.
- With rubber-tired vehicles, the **tire depreciation** is calculated separately from the vehicle depreciation.
- Maintenance and repair costs on construction equipment vary considerably depending on the type of equipment and the job conditions encountered.
- Whether an equipment purchase is financed by loans or by use of the builder's cash, interest charges will apply. These are calculated as a percentage of the average annual investment amount.
- Taxes, insurance charges, and storage costs vary over a wide range of values depending on particular circumstances. An allowance for these costs is also calculated as a percentage of the average annual investment amount.
- Fuel costs depend upon the type of engine and are proportional to the engine's horsepower rating. An operating factor that is always less than 100 percent is also introduced to account for the fact that the engine does not operate continuously at full throttle.
- Examples of calculating the hourly ownership costs of various items of equipment taking into account all of the above factors are provided.
- An allowance for company overheads and profit should be added to hourly ownership costs of an item of equipment when quoting a price for the rental of the item.

RECOMMENDED RESOURCES

Information	Web Page Address
■ General information and equipment prices (Engineering News Record web page)	http://enr.construction.com
■ Use a Web search engine and key words: "equipment prices" to obtain vendors' prices of new and used equipment. Example: Iron Planet of Pleasanton, California	http://www.ironplanet.com

REVIEW QUESTIONS

1. Describe how equipment is usually priced in an estimate.
2. When deciding whether to rent or buy construction equipment, what factors should a builder consider?
3. What are the advantages and disadvantages to the builder of renting rather than buying equipment?
4. Why is the "straight-line method" an unsatisfactory way to calculate depreciation on construction equipment?
5. Why is tire depreciation calculated separately from vehicle depreciation?
6. How is the amount for maintenance and repairs on equipment usually calculated?
7. Explain the term *operating factor* in connection with fuel cost.
8. What factors should be considered when estimating the ownership costs of equipment?
9. What is included in *equipment overheads?*

PRACTICE PROBLEMS

1. Calculate the ownership cost per hour for a crawler-type hydraulic crane powered by a 350 hp diesel engine based on the following data:

Engine:	350 hp diesel
Operating Factor:	60%
Purchase Price:	$570,000
Freight Charges:	$2,500
Estimated Salvage Value:	$350,000
Useful Life:	5 years
Hours Used per Year:	2,000
Maintenance & Repairs:	120% of annual depreciation
Equipment Overhead Rate:	11%
Diesel Fuel Price:	$1.80 per gallon

2. What would the hourly charge-out rates (rental rates) for the above crane and operator be based on the following overheads and profit requirements?
   • The operator's wage is $20.00 per hour
   • The company overhead on equipment is 20%
   • The company overhead on labor is 50%
   • The required profit margin is 10%

3. Calculate the ownership cost per hour for a mobile compressor unit powered by a 60 hp gasoline engine based on the following data:

Engine:	60 hp gasoline
Operating Factor:	45%
Purchase Price:	$25,000 (including tires)

Freight Charges: $1,500
Estimated Salvage Value: $7,500
Useful Life: 8 years
Hours Used per Year: 1,700
Maintenance & Repairs: 83% of annual depreciation
Tire Cost: $1,000
Tire Life: 5,000 hours
Maint. & Repairs (Tires):   15% of tire depreciation
Equipment Overhead Rate: 11%
Gasoline Fuel Price: $2.50 per gallon

4. A homebuilder is trying to decide whether it is cheaper to lease or buy a pickup truck. With all the miles she figures, she is going to put on the truck each month the lease works out to $4.25 per hour. How does the ownership cost of the same vehicle compare based on the following data? You can ignore maintenance, tire, and fuel costs as these will be the same in both cases.

Engine: 248 hp diesel
Purchase Price: $30,440
Freight Charges: $1,000
Estimated Salvage Value: $15,000
Useful Life: 5 Years
Hours Used per Year: 2,000
Equipment Overhead Rate: 11%

11

SUBCONTRACTOR WORK

OBJECTIVES

After reading this chapter and completing the review questions, you should be able to:

- Describe how subtrade work is estimated, and explain the role of the builder's estimator in this process.
- Identify potential problems with subcontractors on a project, and describe what can be done at the time of the estimate to minimize these problems.
- Explain how the competency of subtrades can be evaluated, and list what should be considered in this evaluation.
- Explain how surety bonds may be used to help manage subtrades.
- Describe subtrade pre-bid proposals, and explain their use with subcontractor quotes.
- Describe how subtrade quotes are evaluated, and identify scope-of-work problems associated with subcontractor quotes.

KEY TERMS

bid closing time	pre-bid subtrade proposals	unknown subtrades
bonding of subtrades		unsolicited subtrade bids
hoardings	scope of work	

Introduction

Subcontractors often perform 80 percent or more of the work on residential projects. Often all of the on-site work is put in place by subtrades, which leaves the builder to merely coordinate the activities of the subtrades as construction manager. Outwardly, this would appear to simplify the work of the estimator, since each of the subtrades quoting the job will normally be responsible for estimating their own portion of the work. In theory, all that is left for the builder's estimator to do is find the subtrades with the best prices for each trade and then gather the trade prices to determine the total price of the work. However, in reality it may not be quite that simple. There can be serious problems with the practice of subcontracting for a number of reasons, including:

1. Unreliability of some subcontractors
2. Errors in subtrade quotes
3. Overlap or gaps in the work subcontracted
4. Conditional quotes from subcontractors
5. Congestion at the time of bid closings

Unreliable Subcontractors

When an estimator is preparing a price for an project, she or he places a great deal of trust in subtrades to deliver the quotes they have promised so that they can be incorporated into the overall bid price. Further trust will be placed in the subtrade whose quote is accepted to go on to complete their work on time and in an acceptable manner. While most subtrades can be relied upon to provide the quotes they have promised to submit and at least try to meet their contractual obligations, there are still too many occasions when subcontractors break their agreements leaving the builder in much difficulty.

Subcontractor defaults can occur in numerous ways and at various stages in the construction process. When a builder is preparing a price for a custom home, a subtrade may be late with their quote or completely fail to prepare the price they had indicated they would be submitting. After a bid has been submitted to the owner, a subtrade may try to withdraw their quote that the builder used in the overall price for the project. A subtrade may not show up when their work is required to begin, or a subtrade may simply fail to perform their work in accordance with their contractual obligations.

There can sometimes be legal recourse against a defaulting subtrade, but this often provides little compensation to the builder who has suffered the consequences of the default. An estimator may be called upon to assess the reliability of a subcontractor and the risk of their default. This assessment is considered later under "Evaluating Subcontractors."

Errors in Subtrade Quotes

One of the reasons for the deficient performance of subtrades is the occurrence of errors in the estimates prepared by the subtrades. Accepting a subtrade bid that contains a mistake can lead to endless trouble for a builder, so the estimator has to make every effort to ferret out subtrade bids that contain errors if the consequential problems are to be avoided. When a quote is received from a subtrade and is obviously out of line with expectations, it is advisable for the estimator to contact the subtrade before the builder uses the price. There is usually no need to discuss details with this subtrade; merely indicating that they should check their quote for errors should be sufficient warning to them.

Overlap and Gaps in the Subcontracted Work

Ensuring that subtrades properly cover all the work of the project can be difficult since it is easy for a section of work to be excluded from the subtrade's scope and be overlooked in the estimator's takeoff. Similarly, discovering that some work has been covered more than once in the estimate is also possible. These cases where there are gaps or overlap in coverage occur when work is not clearly defined as being part of the scope of a single subcontractor's work. Avoiding these problems calls for a great deal of investigative work by the estimator. This topic is further discussed under "Scope of Work."

Conditional Subtrade Bids

The estimator's task becomes far more complicated when subtrades attach conditions to their bids. Where all of the subtrades quoting a certain trade base their bids on the same scope of work and comply with the same set of specifications, selecting a subtrade price should not be difficult. But when each subtrades has a different set of bid inclusions and exclusions and each expects the builder to supply a variety of different items, or the subtrades bid according to various alternative specifications, selecting a subtrade can become a horrendous task for the builder's estimator.

Congestion at Bid Closing

Where a builder is in a bid competition to obtain a contract for a custom home or a multi-unit project, the owner usually requires bids be submitted by a certain time on a certain date—the **bid closing time**. The builder notifies subtrades of this and asks for subtrade quotes be submitted in time for the builder to compile its bid for the owner. However, instead of sending in their quotes one or two days before the time of the owner's bid closing, subtrades tend to hold onto them until just before the bid closing. This happens mostly because subtrades are afraid that their prices will be disclosed which would enable a competitor to underprice them. Consequently, most of the subtrade bids arrive at the builder's office, usually by fax, within an hour of the bid closing time. This puts the estimator under great pressure: first sorting the faxes into trades, then interpreting the quotes and any conditions attached, and finally, trying to decide which quote or combination of quotes provide the best price for the work.

Managing Subcontract Pricing Procedures

The problems we have described can be controlled only where the builder applies effective management practices throughout the estimating process. The underlying objective of the subcontract pricing procedures detailed below is to provide a well-organized system of dealing with subtrades in the estimating process. Following this system allows the builder first to clearly identify the subtrade needs for the project being estimated and second, to ensure that the subtrades selected to perform the work are capable of meeting their contractual obligations.

List of Subtrades for a Bid

As soon as drawings and specifications are obtained for a competitive bid project and the estimator has reviewed the documents, a list of the subtrades required for the project should be prepared. This list is compiled and subtrades are notified early in the estimating process to give the subtrade bidders as much time as possible to put together their estimates. Many of the design questions raised in the estimating process relate to subtrade work, so subtrades have to be encouraged to make an early start on their takeoffs so that the problems they encounter can be properly dealt with in the time period when estimates are being prepared.

The list of subtrades required should not vary too much from one project to another so that a standard list can be maintained by the estimator to use on each project. The estimator then checks off the subtrades required for this project from the standard list of subtrades. Figure 11.1 is an example of the kind of checklist of trades estimators use to indicate the particular trades needed on a project. However, every

SUBTRADE LIST

Project: Estimator:

Bid Closing Date:

	Required	Comments
Demolition		
Earthwork		
Landscaping		
Paving and Surfacing		
Concrete		
Precast Concrete		
Damp proofing		
Parging		
Formwork		
Reinforcing Steel		
Masonry		
Structural Steel		
Metal Decking		
Miscellaneous Metals		
Carpentry Framing		
Engineered Joists		
Glue-Lam Construction		
Finish Carpentry		
Cabinets		
Bathroom Accessories		
Appliances		
Siding		
Stucco		
Insulation		
Roofing		
Soffit and Fascia		
Eaves Trough		
Doors		
Windows		
Flooring		
Drywall		
Tile Work		
Painting		
Plumbing		
HVAC		
Electrical		

Figure 11.1   Subtrade Checklist

project seems to have at least one aspect that is different from others so the estimator needs to carefully examine the project specifications and drawings for any items that are out of the ordinary; especially items that call for a specialized subtrade. Extra trades are then added to the subtrade list to meet specific requirements of a project.

The subtrade list is first used to determine which trades have to be contacted for bid requests on the project. The list can later be incorporated into the estimate summary and used to ensure that all subtrade work necessary for the project is duly priced as required.

Contacting Subtrades

Once the subtrade needs of a project have been defined, the next task is to inform subcontractors that the builder is going to prepare a bid for the project and that bids for their trade will be required. Most builders maintain a directory of the subtrades they prefer to deal with. These trades are usually alerted by e-mail that a price is required for a new project.

One way to organize this process is to have a subtrade checklist like the one shown on Figure 11.1 set up on a computer spreadsheet, and then proceed as follows:

1. The estimator checks off the trades required for the project and adds any required trades that are not on the basic list.
2. A copy of this checklist is e-mailed to all the subtrades on the builder's mailing list.
3. Each subcontractor is asked to return the checklist with their name against all the trades on the checklist that they are prepared to submit prices for.

Subtrades are quickly notified of a project, and the estimator promptly determines if all the subtrade prices necessary for a project are going to be obtained from the subcontractors who normally work for that builder.

When the builder's subtrade directory does not cover all the trades necessary for the job or where prices from additional subtrades are needed to improve competition, soliciting prices by advertising in trade publications or newspapers may have to be considered. This, however, introduces the problem of receiving bids from unknown subcontractors.

Unknown Subtrades

Having to award contracts to companies from outside their select group of preferred subtrades can be a stressful experience for some builders. It may, however, be necessary for a number of reasons. For instance, having to find a subtrade to perform some specialized work may compel a builder to deal with an unknown subcontractor.

Unsolicited Subtrade Bids

More commonly, a builder will be faced with an unfamiliar subtrade when a low price is received for a trade from a company who is not known to the builder. This can put the builder in a difficult position: on one hand, the builder may need to use the low price to maintain a competitive bid and, on the other hand, the builder may be hesitant to accept the risk of using an unknown subcontractor. Legal and ethical considerations also impact upon this situation. One course of action that has been adopted by some builders faced with this decision is to use the low price in the bid but not commit to using the subtrade who submitted the price. This solution, as well as being ethically questionable, may also be contrary to bid conditions. In most major contract bids, the builder will be required to name its subcontractors on the bid document. Failure to properly comply with such bid requirements can put the builder into a vulnerable legal position.

The problem a builder faces when it receives an unsolicited low bid can be further aggravated by time constrains. While there is no sure recipe that can be followed to avoid all the potential problems a builder can have with a subcontractor, there are certain precautions (outlined below) that the prudent builder can take to minimize the risk. But when a builder is confronted with a low price from an unknown subtrade just hours or even minutes before the bid to the owner is due to be submitted, there is little or no time available to make effective inquiries about the subtrade.

This problem of time constraints can be avoided to some extent if the estimator takes some precautions earlier in the estimating process. First, the estimator must try to learn which subtrades are planning to bid on a project. This information can sometimes be obtained from the builder's own subtrades. They frequently hear from their own industry contacts details about which of their competitors intend to bid on a job. If new subtrade names come to the attention of the estimator during the course of these inquiries, investigation and evaluation of the unfamiliar subtrades can begin in a timely fashion.

Evaluating Subcontractors

The builder's two main concerns when considering an unfamiliar subtrade are first, is the subtrade capable of meeting the quality and schedule requirements of its work? And second, does the subtrade have the financial capability required?

The estimator or another member of the builder's team should evaluate potential subcontractors in these two areas so that an assessment of risk can be made before a subtrade is named in a bid. Carrying an unknown subtrade in a bid without carefully considering the consequences is asking for trouble.

An effective evaluation of subtrade companies would involve an investigation of a number of factors including:

1. The general reputation of the company
2. The quality of previous work performed by the company
3. The quality of the company's management
4. The company's disposition in terms of their general cooperativeness and the ease in dealing with company site and office personnel
5. The financial status of the company

This kind of information can be obtainable from several sources. The company itself may be willing to supply general details about the number of years it has been in business, the number of people it employs, and it might offer references, especially if it is eager to secure contract work. Examining a list of the equipment that is owned by the company can sometimes make an assessment of the stature of a subtrade. A company's status may also be gauged from whether and how long it has been a member of a trade association. Construction companies who have previously hired the subtrade may be willing to comment on their experience with the subtrade and other industry contacts, such as material suppliers, who have had contact with the subtrade may offer their opinions about the company.

When investigating the financial condition of a subtrade, a useful source of information can be bonding companies. If the subtrade wishes to bid on larger value projects, it must be able to obtain surety bonds. The builder's own bonding company which is usually concerned about the business health of the builder can be approached to share any financial information they may have about the subtrade. The builder's bank can also be a source of financial data on the subtrade. Lastly, the builder may be able to pick up a financial report on the subtrade, for a fee, from a credit-reporting agency.

The reader should note that subcontracting problems are not exclusive to the subtrades new to the contractor. Even subtrades that a builder has worked with over many years can be the source of problem or two, so some of the inquiries previously suggested for unfamiliar subtrades may also be appropriately applied from time to time in an effort to ensure that the regular subtrades remain reliable.

Bonding of Subtrades

Under the terms of the contract it has with the owner, the builder is always held responsible for the performance of the work including the performance of the work done by subcontractors. If a subtrade fails to perform as required, the prime builder will have to cope with the problem within the project budget. Certainly there will be no additional funds available from the owner to rectify the defaults of a subtrade.

Just as the owner, in an effort to manage the risk of a non-performing builder, may require the builder to obtain surety bonds to guarantee contract performance, so, too, can the builder require major subtrades to provide such bonds to guarantee the performance of their work.

There has been much discussion in construction management texts about the advantages and disadvantages of surety bonds as a means of managing construction risk; suffice it to state here that some builders do consider it worthwhile to require subcontractor surety bonds even where it can add more than 1 percent to subtrade prices. Builders who demand bonded subtrades will impose a bid condition on subtrades specifying that the subtrade shall obtain a performance bond that names the builder as obligee. Then, should the subtrade default on its contractual obligations, the builder will be able to look to the bonding company to finance the due performance of the subtrade's work.

Subtrades may also be required to provide bid bonds. A subtrade bid bond would give the builder the right to damages from the bonding company in the event that a subtrade fails to execute the subcontract after the builder has properly accepted their bid.

When subtrade bonds are requested on a project, subtrades may be instructed to include the cost of bonding in their bid prices, or the bid conditions may call for the price of the bonding to be stated separately from the bid price. The second alternative informs the builder of the cost saving of waiving the bonding requirement, an option that might be considered with a low-risk, well-established subcontractor.

Pre-Bid Subtrade Proposals

Most subcontractors now prefer to submit their bids by e-mail, but there are still some who fax their quotes to the builder. These quotes provide a hard copy of the subtrade's quote before the builder's bid goes to the owner, but conditions attached to the quotes often vary from subtrade to subtrade. This makes bid analysis difficult to pursue especially, as mentioned above, when many of the subtrade quotes are received just before the builder's bid is due to go to the owner.

Subcontractors generally cannot be persuaded to submit their bids earlier, but some time has to be available if any analysis of subtrade bids is to be possible. A practical solution to this dilemma is to have subtrade bidders submit pre-bid proposals so that the builder's estimator has the opportunity to examine subtrade bids and make comparisons where a number of bids are received for the same work.

WWW WINDOWS INC.

PRICE QUOTATION

Date:

To: XYZ Construction Inc.

Re: Custom Home
Townsville

We propose to supply and install the following materials for the above-named project:

1. Aluminum casement windows including glass and glazing

2. Glazing to wood interior partitions: (separate price)

3. Bathroom mirrors: (separate price)

Note: If we are awarded items 1, 2, and 3 above, we offer a discount of 10% on the prices quoted.

EXCLUSIONS:
1. Final cleaning
2. Breakage by others
3. Scaffolding and hoisting
4. Heating and hoarding
5. Flashings
6. Any other glass or glazing not listed above
7. Temporary enclosures

Figure 11.2   Subtrade's Pre-Bid Proposal

A pre-bid proposal is a copy of the subtrade's bid without prices but complete in all other details including the conditions the bidder wishes to attach to the bid. Figure 11.2 shows an example of a pre-bid proposal from a glass and windows subcontractor.

Analyzing Subtrade Bids

The estimator has to ensure that all of the work required from a trade, as defined by the project drawings and specifications, is included in the subtrade bid. One problem is that individual subtrades may have different interpretations about the scope of work of their trade. For example, a subtrade bidding masonry work may include in their quote the insulation under the brickwork, while another may not. Taking into account all the subtrades on a project, the number of possible combinations of trades and part trades is endless.

When a bid proposal has been obtained from a subtrade, it should list inclusions that detail all the work that bidder is offering to do for the price quoted. However, this list may or may not coincide with the inclusions listed by other subcontractors bidding the same trade. Meaningful bid comparisons can only be made where the bidders are offering to perform the same work. Excluded items should also be listed on the subtrade's proposal. Most exclusions are items of work that, by standard practice, are always omitted from the work scope of the trade. Exclusions from one trade are sometimes covered elsewhere in the estimate by other trades or by the builder in its work. The estimator has to be constantly on the lookout for items excluded from the work of a trade and not priced anywhere else in the estimate.

The subtrade in the Figure 11.2 example has listed all the work items it intends to price. Note that flashings are listed on this quote as an exclusion; this is typical of the kind of item that is excluded by one trade but may not be included by any other subtrade on the project. If this is the case here, the estimator will have to make sure an additional price is obtained to cover the cost of these flashings.

Subtrade bids can be further complicated by multiple trade bids and discounts that are sometimes offered. Subcontractors frequently offer prices for a number of different trades on a project. They may state on their quote that they are only interested in a contract for the entire group of trades they have included in their bid. Alternatively, the bidder may allow the builder to pick and choose which trades it wishes to award contracts for, but, as an inducement, the subtrade may offer a discount if they are awarded a contract for all of the trades they submitted bids on. In the example in Figure 11.2, if the builder accepts all the subtrade's prices, a price reduction of 10 percent is offered.

To facilitate subtrade bid analysis, a spreadsheet like the one shown in Figure 11.3 can be set up from the information obtained in pre-bid proposals for each of the trades required for the estimate. Each spreadsheet is compiled sometime before the bid closing day and lists the names of bidders together with the items of work to be included in their bids so that, when prices are received, they can be easily inserted onto the spreadsheet to provide a speedy bid analysis. Note that in this example a combination of prices from the various bidders gives the least total price for this trade.

Scope of Work

In the residential sector, the scope of work covered by a particular trade is usually determined from local trade practices. It is important for an estimator to become familiar with trade definitions that outline the scope of work of each trade. Asking many questions and learning from the experience of dealing with local subtrades is the only way to gain this knowledge.

Trade scope definitions are sometimes available from local bid depositories for commercial projects, but they seldom apply to residential construction. However, these definitions could be useful to a residential estimator as a guideline to a trade's scope of work. For instance, such a scope could be used as a checklist for asking a subtrade what is and is not included in their quote. Adopting this practice can make commercial scope definitions a valuable aid in the process of analyzing subtrade bids.

The following is an example of the trade scope definitions for the masonry trade. The reader is cautioned that this definition is provided for illustrative purposes only; it is not intended to be complete, and specific trade rules may vary from location to location.

SUBTRADE ANALYSIS

Project: Laboratory Building - Townsville   Page No:

TRADES: **8500 Metal Windows**   Date:
                     8800 Glazing

SUBTRADES	Casements & Glzg.	Glzg. Wood Partitions	Bathroom Mirrors	Flashings	Discounts	BASE BID TOTAL
WWW WINDOWS INC.	$20,000	$900	$180	NIC	($2,108)	$19,372
				$400	(SMITH)	(With
						discount)
XXX WINDOWS INC.	$19,500	inc.	inc.	$250	no	$19,750
YYY WINDOWS INC.	$18,900	$500	$220	$375	no	$19,995
ZZZ WINDOWS INC.	$18,400	$850	$180	$290	no	$19,720
Combination:	ZZZ	YYY	ZZZ	XXX	=	$19,330

Figure 11.3 Subtrade Bid Analysis

Example—Masonry Scope of Work

The following items shall be *included* in the work of the masonry trade:

- Brickwork
- Concrete blocks, glazed blocks, and glass blocks
- Building in of miscellaneous metals (supplied by others) in masonry
- Built-in metal flashing—supplied by others
- Caulking (other than firestopping), masonry to masonry
- Cavity wall flashings behind masonry
- Wall insulation complete with adhesives, anchors, etc., behind masonry where masonry is the last component to be installed
- Clay flue lining
- Clay tiles
- Cleaning masonry
- Clear waterproofing of exterior masonry
- Concrete and grout fill in masonry only
- Masonry anchors and ties
- Damp-proof course in masonry
- Expansion and control joints in masonry or masonry to other materials, including caulking where it forms part of the expansion or control joint

- Guarantee
- Hoisting
- Inspection
- Installation only of precast concrete band courses, copings, sills, and heads which fall within masonry cladding, including supply and installation of insulation and vapor barrier behind these items
- Loose fill or foam insulation in masonry
- Mortar
- Non-rigid and fabric flashing when built into masonry
- Parging of cavity masonry walls only when specified
- Protection of other trades' work from damage by this trade
- Reinforcing steel placed only in masonry—supplied by others to the job site, clearly tagged, and identified as to end use
- Scaffolding
- Shop drawings
- Stone (natural and artificial), granite, terra cotta facing over 2" thick, including anchorages
- Testing of material of this trade where specifically called for in the masonry specifications
- Trade cleanup
- Weep hole and venting devices
- Wire reinforcing to mortar joints

The following items shall be *excluded* from the work of the masonry trade:

- Anchor slot
- Demolition
- Fabrication and tying of reinforcing steel
- Fire stopping
- General and final cleanup
- **Hoardings**
- Installation of frames
- Lateral support devices on top of masonry walls
- Temporary heat (including mortar shack), light, power, sanitation, and water
- Temporary support and shoring for lintels
- Welding
- Wind-bracing

SUMMARY

- Subcontractors often perform 80 percent or more of the work on residential projects.
- Subtrades will normally be responsible for estimating their own portion of the work which would appear to make it easy for the general contractor to price subtrade work, but in reality it may not be quite that simple.
- Problems with subtrade bids can result from:
  - Unreliable subcontractors
  - Errors in subtrade bids
  - Overlap or gaps in the scope of work subcontracted
  - Conditional subtrade bids
  - Bid closing congestion

- Effective management practices are required to deal with subcontractors in the estimating process. This includes:
  - Listing required subtrades
  - Contacting prospective subtrades
  - Ensuring subtrade requirements are covered
- Problems can also develop when the subtrade with the low price is unknown to the general contractor.
- Contractors need to evaluate subtrades by asking:
  - Is the subtrade capable of meeting the quality and schedule requirements of its work?
  - Does the subtrade have the financial capability required?
- The evaluation of a subcontractor needs to consider:
  - The general reputation of the company
  - The quality of previous work performed by the company
  - The quality of the company's management
  - The company's disposition in terms of their general cooperativeness and the ease of dealing with company site and office personnel
  - The financial status of the company
- One possible way to manage the risk of non-performing subtrades is to have subtrades provide performance bonds.
- Another tool in subtrade management is the **pre-bid subtrade proposal**. This outlines for the contractor's estimator the terms of the subtrade bid in good time before the subtrade releases its price at the last minute.
- The builder's estimator needs to carefully analyze subtrade bids to ensure that all of the work required from a trade, as defined by the project drawings and specifications, is included in the subtrade bid price.
- Estimators need to be familiar with the scope of work covered by local residential sector subtrades.
- Scope of work definitions that are available for subtrades on commercial projects can help in analyzing residential subtrade quotes.

RECOMMENDED RESOURCES

Information	Web Page Address
■ Free copies of subcontractor forms and many other contract documents are available at the web site: *all about forms.com*	http://www.allaboutforms.com

REVIEW QUESTIONS

1. What should an estimator do when a bid is received from a subcontractor with a very low price that suggests that the subtrade has made an estimating error?
2. Discuss overlap and gaps in subcontracted work and the problems associated with each condition.
3. Why are conditional bids from subtrades undesirable?
4. How can the problem of bid closing congestion associated with receiving subtrade bids be reduced?
5. What are the general contractor's two main concerns when a low bid has been received from a subcontractor who is new to the general contractor?

6. What are the advantages and disadvantages to a general contractor of having subtrades provide surety bonds?
7. What is a pre-bid subtrade proposal, and of what benefit is it to the contractor's estimator?
8. Why should a residential estimator take an interest in local building trade practices?

PRACTICE PROBLEMS

1. Use a spreadsheet to calculate which of the following three bids (or combination of bids) for masonry work on a house offers the best price to the contractor:

Bid from Subtrade A
Included: Brick facings $13,900; Brick chimney $6,750; Brick fireplace $4,000
Excluded: Weatherproofing; Flashings to openings; Scaffolding

Bid from Subtrade B
Included: Brick facings $15,808; Flashings to openings $1,795; Brick chimney $4,900; Brick fireplace $2,500; Weatherproofing as specified $1,200
Note: We offer a discount of $3,000 if we are awarded all of this work.
Excluded: Scaffolding

Bid from Subtrade C
Included: Brick facings $15,900; Brick chimney $5,120
Note: If we are awarded both the facings and the chimney, we can offer a 10% discount.
Excluded: Brick fireplace; weatherproofing; flashings to openings; scaffolding

2. Use the spreadsheet shown in Figure 11.4 to determine the best combination of prices for structural and miscellaneous metals on this project.

SUBTRADE ANALYSIS

Project: **Multi-Unit Condo - Lennoxville**   Page No:

TRADES: **Structural & Misc Metals**   Date:

SUBTRADES	Anchor Bolts	Misc Metals Supply	Structural Steel Fabricate	All Steel Install	Painting Steel	Metal Deck S & I	Discount	TOTAL PRICE
AAA Bolts	$950	NIC	NIC	NIC	NIC	NIC	no	---
BBB Steel	$1,050	$19,450	NIC	NIC	NIC	NIC	10% Discount (If awarded both)	---
CCC Steel	$1,200	$18,900	$240,220	NIC	NIC	NIC	($5,000) (If awarded all)	---
DDD Painting	NIC	NIC	NIC	NIC	$15,272	NIC	no	---
EEE Steel	$1,165	$24,300	$182,650	$175,000	$20,000	NIC	($4,000) (If awarded all)	---
FFF Steel	NIC	$25,000	$192,620	$159,000	$18,500	NIC	8% Discount (If awarded all)	---
GGG Steel	NIC	$27,500	$195,900	$174,900	$20,350	$87,300	20% Discount (If awarded all)	---
HHH Decking	NIC	NIC	NIC	NIC	NIC	$65,475	no	---
III Decking	NIC	$26,000	NIC	NIC	NIC	$72,000	10% Discount (If awarded both)	---
Combination 1								
Combination 2								
Combination 3								
Combination 4								
Best Combination:								

Figure 11.4   Subtrade Price Evaluation

12

ESTIMATE SUMMARIES AND BIDS

OBJECTIVES

After reading this chapter and completing the review questions, you should be able to:

- Describe the last steps in the estimating process including summarizing prices, markup considerations, and determining the total price.
- Explain the use of different summary formats, and distinguish between a summary for the sales price of a home and a bid summary.
- Describe the items that are addressed in the advance stage of the bid process.
- Explain why bid clarifications are sometimes required.
- Explain how bid markup is calculated including an assessment of risk.
- Outline the tasks that should be addressed after bids are submitted.
- Complete an estimate to determine the sales price of a home and a bid estimate using manual methods.

KEY TERMS

allowances	bid closing	consent of surety
alternative prices (alternates)	bid markup	instructions to bidders
	bid runner	post-bid review
bid bond	bid summary	profit margin
bid clarifications	company overheads	qualified bids

Introduction

In this chapter we consider the last stage of the estimate process when all of the component prices are gathered together in a summary and the builder considers the amount of markup to add to the price. The format of the summary depends on what the estimate is to be used for; an estimate used for calculating the sales price of a home will be summarized differently from an estimate used to prepare a bid.

Builders use the summary to show all the constituent prices of an estimate including the price of the builder's own work, subtrade work, general expenses, overheads, and the fee. This is usually displayed on standard summary sheets. When the estimating process is complete, the summary will present total amounts for each part of the estimate, but the summary sheet can also be used at the start of an estimate to identify the items that will later have to be priced in the estimate. Estimators often begin the estimating process by checking off on a standard summary sheet the trades that will be completed by "own forces" and those that will be subcontracted. Later, this document becomes a gathering place for subtrade prices as prices are obtained for each trade.

Summary Formats

Figure 12.1 shows an example of a summary sheet format suitable for use in calculating the sales price of a new home.

A builder may use a number of different summary formats on a project for different users of the information. For example, there could be a summary format for the job superintendent showing the budget in a way that is useful in the construction process. Another summary could be set up for company accountants showing project prices in a format that is more useful for accounting purposes. Many different formats can quite easily be generated using computer estimating programs.

The purchaser of the new home is another possible user of the estimate summary. The builder may wish to show the homebuyer a summary in the form of a price breakdown. This can be useful to the builder, especially with semi-custom homes where there are a large number of price changes for modifications to a standard design. This summary would not need to be as detailed as the estimator or superintendent's summary, and the builder would probably not want to list the add-ons on this document. Instead, add-on amounts can be distributed to all the other amounts in the summary. This would increase these other prices slightly, but the overall price of the estimate will remain the same. This process, which is sometime referred to as *allocating add-ons,* can be done automatically and far more quickly when using some of the more advanced computer estimating software.

Figure 12.2 shows an example of a summary format suitable for the purchaser on a semi-custom home.

Bid Summary

On the **bid summary** shown on Figure 12.3, prices of the builder's own work are entered at the top of the sheet so that subtotal 1 provides an estimate of direct cost of the builder's work. It is useful to know this amount because some add-ons may be calculated from it. Subtrade prices and the price of general expenses are recorded on the summary as they become available until the bid total at the foot of the summary is finally established. This total can then be transferred to the bid documents.

To successfully prepare a bid that is free from errors and complete in all essentials, the estimating process has to be carefully planned and organized from the point when the bid documents are first reviewed up to the last act in the process, which is the bid submission. The busiest period, and the period when the process is most error prone, is the last few hours before the final price is determined. Every task that can possibly be accomplished before **bid closing** day has to be out of the way so that these last few hours can be devoted to those activities that can only occur at this time.

No.	DESCRIPTION	LABOR $	MATL. $	EQUIP. $	SUBTRADE $	OTHER $	TOTAL $
	PROJECT: GROSS FLOOR AREA: LOCATION: DATE: ESTIMATOR:			**ESTIMATE SUMMARY**			
	SUBCONTRACTORS						
1.0	Excavate and Backfill Basement						
2.0	Excavate & Backfill Service Trenches						
3.0	Concrete						
4.0	Form Footings						
5.0	Form Walls and Columns						
6.0	Damp Proof Walls						
7.0	Rough Carpentry Framing						
8.0	Deck						
9.0	Canopy Over Entrance						
10.0	Finish Carpentry						
11.0	Cabinets (Supply and Install)						
12.0	Bathroom Accessories (Supply and Install)						
13.0	Appliances (Supply and Install)						
14.0	Soffit and Fascia (Supply and Install)						
15.0	Siding (Supply and Install)						
16.0	$1/2$" Parging (Supply and Install)						
17.0	Paint Exterior Windows and Doors						
18.0	4" Eaves Gutter (Supply and Install)						
19.0	3" Downspouts (Supply and Install)						
20.0	Roofing (Supply and Install)						
21.0	7'-0" x 3'-6" Precast Concrete Steps						
22.0	Flooring (Supply and Install)						
23.0	Drywall Finishes (Supply and Install)						
24.0	Interior Painting (Supply and Install)						
25.0	Plumbing - Rough In						
26.0	Plumbing - Finish						
27.0	Heating - Rough In						
28.0	Heating - Finish						
29.0	Electrical - Rough In						
30.0	Electrical - Finish						
31.0	Light Fixtures						
	Subtotal 1:						
	MATERIALS						
32.0	Excavation & Gravel Materials						
33.0	Concrete Materials						
34.0	Carpentry Materials						
35.0	Finish Carpentry						
	Subtotal 2:						
36.0	**GENERAL EXPENSES**						
	Subtotal 3:						
	ADD-ONS						
37.0	SMALL TOOLS						
38.0	PAYROLL ADDITIVE						
39.0	BUILDING PERMIT						
40.0	INSURANCE						
41.0	FEE						
42.0	ADJUSTMENT						
	Price per S.F. =				**TOTAL PRICE**		

Figure 12.1   Sales Price Estimate Summary

No.	DESCRIPTION	BASE PRICE $	ADJUSTMENT $	TOTAL $
	SUBCONTRACTORS			
	Excavation and Backfill			
	Concrete Work			
	Rough Carpentry Framing			
	Finish Carpentry			
	Cabinets			
	Bathroom Accessories			
	Appliances			
	Siding and Soffit			
	Paint Exterior Windows and Doors			
	Roofing			
	Flooring			
	Drywall Finishes			
	Interior Painting			
	Plumbing			
	Heating			
	Electrical			
	Light Fixtures			
	Subtotal 1:			
	BILL OF MATERIALS			
	Excavation and Gravel Materials			
	Concrete Materials			
	Carpentry Materials			
	Finish Carpentry			
	Subtotal 2:			
	GENERAL EXPENSES			
	TOTAL PRICE:			

HOUSE PLAN
LOCATION:
PURCHASER:
DATE:

PRICE BREAKDOWN

Figure 12.2   Sales Price Breakdown for Purchaser

DESCRIPTION	LABOR $	MATL. $	EQUIP. $	SUBS. $	OTHER $	TOTAL $
PROJECT: GROSS FLOOR AREA: LOCATION: DATE: ESTIMATOR:			**ESTIMATE SUMMARY**			
OWN WORK:						
Excavation & Fill						
Concrete Work						
Formwork						
Miscellaneous						
Masonry						
Rough Carpentry						
Finish Carpentry						
Exterior Finishes						
Interior Finishes						
Subtotal 1:						
SUBCONTRACTORS						
Demolition						
Landscaping						
Piling						
Reinforcing Steel						
Precast						
Masonry						
Structural and Misc. Steel						
Carpentry						
Millwork						
Roofing						
Caulking and Damp Proofing						
Doors and Frames						
Windows						
Resilient Flooring						
Carpet						
Drywall						
Acoustic Ceiling						
Painting						
Specialties						
Plumbing						
Heating and Ventilating						
Electrical						
Cash Allowances						
Subtotal 2:						
GENERAL EXPENSES						
Subtotal 3:						
ADD-ONS						
Small Tools						
Payroll Additive						
Building Permit						
Performance Bond						
Insurance						
Fee						
Adjustment						
BID TOTAL:						

Figure 12.3   Bid Summary

In Advance of Bid Closing

A residential construction bid is compiled in three stages: the advance stage, the review stage, and the closing stage. The following tasks are performed in the advance stage, which should be concluded at least two workdays before the day of the bid closing:

1. Complete the quantity takeoff, the recaps, the general expenses, and the summary sheet, except for the subcontractor prices.
2. Calculate all required unit prices and **alternative prices** relating to the builder's work.
3. Obtain **bid bonds**, **consent of surety** forms, and any other documents required to accompany the bid.
4. As far as possible at this stage, fill out the bid forms and have them signed by company officers, witnessed, and sealed.
5. Make duplicate copies of all bid documents for retention.
6. Prepare an envelope for delivery of bid documents.
7. Have a senior estimator hold on to all bid documents and the bid envelope until the day of the bid closing.
8. Prepare a plan for the delivery of the bid. Ensure that the **bid runner** (the person who delivers the bid) is carefully briefed about their role and how they are to enter the final items onto the bid forms if this is required of them.
9. Brief the office staff regarding their roles on bid closing day.
10. Check all fax machines, computers, and telephones to ensure that they will be operational for closing day. Backup fax machines and computers may be required for large multi-unit project bids.

Many estimators prepare a checklist of the items that have to be completed during the entire estimating process to ensure that all needs are taken care of in good time. See Figure 12.4.

There are few things more frustrating and more liable to trigger panic amongst the estimators closing the bid than learning moments before the bid is to be submitted that an item required for the bid, such as a bid bond or a signature on a key document, is missing. Those of us who have been involved in numerous closings can attest to the truth of Murphy's Law: "Whatever can go wrong, will go wrong" in these situations.

The Bid Form

The bid, which may sometimes be called a tender or a proposal, can generally be presented in any form the builder chooses. Figure 12.5 shows an example of a simplified bid form that could be used by a homebuilder on smaller projects. The builder may wish to attach terms and conditions to this document, in which case it may be wise to seek the advice of a lawyer when drafting the bid form.

Standard bid forms, obtainable from many construction associations, can also be used for this purpose. Sometimes, especially when the owner is a public agency, the **instructions to bidders** may specify that bidders have to use the bid forms provided by the owner; these are usually found in the project specifications package.

In all cases, the bid form should contain certain basic information:

- The name and address of the project
- The identity of the owner
- The identity of the bidding builder
- A description of the work to be done
- A list of the bid drawings and specifications
- The bid price

CHECKLIST	
PROJECT: LOCATION: DATE: ESTIMATOR:	

TASK	COMPLETED
Obtain Bid Documents	
Notify Subtrades	
Prepare Summary Sheet	
Takeoff Own Work	
Visit Site	
Price Own Work	
Price General Expenses	
Price Subtrades	
Calculate Unit Prices	
Alternate Prices	
Cash Allowances	
Obtain Bid Bond	
Fill Out Bid Forms	
Prepare Bid Envelope	
Assign Bid Runner	
Review Estimate	
Bid Closing	

Figure 12.4   Estimator's Checklist

- The duration or completion date of the project
- The signatures of the bidders where they are not incorporated, or the corporate seal where the bidder is a corporation, all duly witnessed
- The time, date, and place of the bid

These items are generally considered to be the essential ingredients of a bid, but many bids have been written in an informal style omitting one or more of the items listed above. The lack of some detail in a bid may have no effect on the course of the contract, but owners and builders alike have sometimes wished the bid had been more explicit so that a problem might have been avoided.

In addition to the basic bid information, the bid form may also have a number of other items attached to it, including:

- A list of unit prices for pricing changes in the work
- Alternative prices
- Separate prices
- Bid security
- Verification that the contractor is licensed (required in some states)
- Certification that the owner owns the property
- Payment terms
- Interest rate on late payments

BID FORM

Bid Number:_____   Date: _____

Project Name: _____

Project Address: _____

City: _____   State:_____   Zip: _____

Project Work:_____

_____

Drawings:_____   Dated:_____

Specifications: _____   Dated:_____

Design Consultant: _____

Submitted To:

Owner:_____

Address: _____

City: _____   State:_____   Zip: _____

Telephone:_____   Fax: _____

Work to Begin: _____

Work Completed By: _____

To provide all materials, equipment, and labor required to complete the project as described in the drawings and specifications noted above for the sum of:

_____ *dollars.*   ($ _____)

Builder's Name:_____

By: _____

Witness:_____

Builder's License Number: _____

Address: _____

City: _____   State:_____   Zip: _____

Telephone:_____   Fax: _____

Figure 12.5   Sample Bid Form

- A note that the price is based on certain soil conditions. If conditions vary, the contract price may be adjusted
- Other bid conditions

If they are not mentioned in the bid, many of these issues will be dealt with under the terms of the contract between the owner and the contractor. A copy of the proposed contract with all its provisions should be included in the specification package for the project.

Alternative Prices

Alternative prices are quoted for proposed changes in the specifications. The builder normally bids a price for the project as specified, then quotes alternative prices to perform the work in accordance with the different requirements called for in the alternates. For instance, if the specifications indicate that 26-ounce carpet is required, this type of carpet will be priced in the base bid. Then the owner may request an alternative price to substitute 32-ounce carpet; the extra cost of supplying the heavier carpet will be calculated and quoted separately in the bid. Alternative prices are usually expressed as additions to or deletions from the base bid and are often referred to as alternates. In this example, the alternate submitted with the bid may state: "To substitute 32-ounce carpet for 26-ounce carpet, add $45,000."

Bid documents call for alternative prices where owners or their design consultants wish to provide some price flexibility in the bid. This allows them to substitute lower-priced alternatives where the bid price exceeds the owner's budget and vice versa. A few alternative prices can be reasonably expected on a bid, but in some cases several dozen prices have been required. This puts the estimator under added pressure on bid closing day since many of the alternatives involve subtrades so alternates cannot be priced until information is received with the subtrade bids that arrive in the last few hours of the estimating period.

Alternates can be extremely complex, indeed alternative prices have been requested for the deletion of entire sections of a building comprising several floors in some cases. This kind of alternate will impact upon the prices of most of the subtrades on the project. When an alternate involves a significant number of trades, the estimator is advised to set up a spreadsheet in advance to streamline the process of pricing the alternate when subtrade bids are received.

Figure 12.6 shows a template for pricing some example alternates on a custom home. This kind of spreadsheet can be set up early in the estimating process with the spaces for prices left blank until the pricing information is obtained for the subtrades.

Alternative prices may also be introduced into the bid on the builder's own initiative. These alternatives are invariably offers to perform the work for a reduced price on the condition that certain changes are made to the design or specifications. They can be in the form of a simple substitution of a product of equal quality but lower cost for the one specified, or the alternative may consist of offering a completely different design for the project.

Allowances

Builders can be instructed to include in their bid a stipulated sum of money that is to be expended on a specific work item on the project. These sums, which are usually referred to as cash **allowances**, are not normally required to be listed on the bid documents like alternate prices, but they are to be included in the bidder's price.

The instructions to bidders could say, for example, "Allow $2,000 for the finish hardware as specified." Contract conditions will normally go on to describe how the cash allowances are to be disbursed. Instructions usually state that the cash allowance shall include the actual supply cost of the goods required for the work but shall not include for the builder's markup. The builder is required to include the profit that applies to the cash allowance work elsewhere in its bid price.

A second type of allowance, often called a "contingency sum," differs from that previously described in that the funds are to be made available for a broader purpose. A bidder may be instructed to include a contingency sum of $10,000 in its bid

BID ALTERNATES

PROJECT: _Custom Home 0619_
LOCATION: _____
DATE: _____
ESTIMATOR: _ABF_ _____

NUMBER	ALTERNATIVE PRICES	ADD $	DEDUCT $
ALT. 1	To substitute cedar shingles for asphalt system as specified	10,000	—
ALT. 2	To reduce attic insulation as specfied	—	5,000
ALT. 3	To add air conditioning unit as specified	15,674	—
ALT. 4	To delete basement development:		
	— Partitions and doors	—	4,500
	— Floor finishes	—	3,005
	— Drywall	—	2,256
	— Painting	—	982
	— Washroom accessories	—	350
	— Plumbing and heating	—	3,850
	— Electrical	—	1,900
	TOTAL DEDUCTION	—	16,843

Figure 12.6   Bid Alternates Spreadsheet

price for extra work authorized on the project, in which case the cost of extra work will be paid out of this sum.

When the estimator first reviews the bid documents at the start of the estimating process, any allowances specified should be highlighted. Many estimators set up the summary sheet for the estimate at this time and enter details of the bid cash allowances on the sheet at once so that they are not later forgotten.

Bid Security

Sometimes, especially on larger projects, the builders are required to include bid security with their bids. This can be in the form of a bid bond or could consist of a certified check or other form of negotiable instrument. The purpose of the security is to guarantee that if the bid is accepted within the time specified for acceptance of bids, the builder would enter into a contract and furnish satisfactory contract bonds to the owner, otherwise the bid security is forfeited.

Bid bonds are issued by surety companies and are favored by builders over other forms of bid security since their use does not require large sums of money to be tied up for extended time periods. Certain owners, especially government bodies, require surety bonds be drafted in a particular form, in which case the estimator has to ensure that the bonds obtained do comply with these specifications. Requests for bonds should be placed as soon as the bonding provisions of a project are determined.

The face value of a bid bond is usually set at 10 percent of the bid price, but a 5 percent value has been specified for some projects. A second form of bid security, called a Consent of Surety, is also frequently required to accompany bids. A Consent of Surety is a written statement made by a bonding company indicating that they will provide surety bonds as required by the contract to the builder named in the statement. The estimator should be able to collect this document at the same time and from the same source as the bid bond.

Owner's Bid Conditions

The owner often lists on the bid forms certain conditions that are imposed on bidders. The list of owner's conditions has been known to run to many pages on some bid forms, but some of the more usual examples included declarations that:

1. The bidder has examined all of the contract documents and addenda as listed by the bidder on the bid
2. The bidder has inspected the site of the work and is familiar with all of the conditions relating to the construction of the project
3. The bidder agrees to provide all labor, materials, equipment, and any other requirements necessary to construct the complete project in strict accordance with the contract documents
4. The bidder agrees to give the owner 30 days to consider the bid (the acceptance period)
5. If, within 12 days after the bid is accepted, the bidder whose bid is accepted within the specified acceptance period refuses or fails, to either (1) enter into a contract with the owner or (2) provide contract performance security or both, the bidder shall be liable to the owner for the difference in money between the amount of its bid and the amount for which the contract is entered into with some other person

(Note that the bid security is in place to guarantee this last condition.)

Many of the conditions found in the bid documents are placed there to shift a risk factor from the owner to the builder. The estimator should highlight these conditions and bring them to the attention of their company managers so that the consequence of assuming the risk involved can be assessed. It is important for the estimator to ensure that all risks imposed by both bid and contract conditions have all been addressed in the estimate.

The Pre-Bid Review

The objectives of the pre-bid review are threefold: to ensure that there are no obvious omissions from the bid, to consider ways of improving the competitiveness of the bid price, and to consider an appropriate fee for the project. While the estimators involved in preparing the bid will have previously considered all three of these elements during the estimating process, the additional experience and different viewpoints contributed by other people at the review meeting can often be of great benefit.

Ideally, a pre-bid review meeting will take place at least two workdays before the bid closing day. This gives the estimator time to take action on the decisions that arise from the meeting and allows sufficient time for a brief follow-up meeting when it is necessary.

At least the project estimator and his supervisor, who is usually the chief estimator, will attend the pre-bid meeting. Whether other participants are present depends upon both the size of the organization and the size of the project to be bid. In a small company, the estimator and supervisor may be the only people involved with bids. In contrast, when a large project bid is closed at the offices of a major builder, a company vice president or even the president may attend, together with a regional general manager, an operations manager, a project manager, the project estimating team, and the company's chief estimator.

Before the review, the takeoff of the builder's work should be complete and priced, the general expenses should have been priced as far as possible, and pre-bid subtrade proposals should have been obtained from the major subcontractors on the project. By the time of the meeting, the chief estimator should have looked over the bid documents and discussed the estimate generally and the prices specifically with the estimator. It is also advisable for everybody else attending the review to examine the drawings and, at least, glance over the other bid documents before the meeting commences, otherwise much of the meeting time will be taken up "enlightening" these attendees about the basic characteristics of the project.

The principal topics discussed at the pre-bid review, together with typical questions raised about these subjects, are listed below:

1. The nature and scope of the project work: Is it clear what is involved in the project? Does the estimate address all of the constituent parts that it should?
2. The construction methods proposed for the project: Can more effective or more efficient methods be used? If so, what will the effect be on the price of the work?
3. The supervisory needs of the project: What key positions will need to be filled on the project? Will there be company personnel available when these positions have to be filled?
4. The labor resources required for the project: What will be the project labor demands? Are any problems anticipated in meeting these demands?
5. The equipment needs of the project and availability of key items: What are the main items of equipment required? Will there be any problems meeting these requirements?
6. The prices used in the estimate: Is there agreement regarding the prices of key labor and equipment items used in the estimate?
7. The project risks: Which of the takeoff is considered to be risky, and how much has been allowed to cover the risk involved?
8. The general expense prices: What has been allowed for in the key items of general expenses? Can anything be done to improve the competitiveness of the prices used?
9. The response from subcontractors: Will prices be forthcoming for all the trades involved in the work? What problems have been identified by subcontractors?
10. The markup fee to be added to the estimate: What is the exposure to risk contained in this bid? What is an appropriate fee to include in the bid price? (See Bid Markup)
11. Clarifications to be submitted with the bid: Is it necessary to include a statement about the basis of the builder's price with the bid? (See Bid Clarifications)

All of these items—and any others that affect the bid—should be discussed in an open and mutually supportive climate at the bid review meeting. It should be made clear that the object of the meeting is not to criticize the work of the estimators, but rather to discuss what can be done to maximize the quality of the bid. Good bid reviews are infused with team spirit that can restore the confidence of an estimator who is suffering from doubts and the feeling of aloneness that can torment even the best of the profession.

Important decisions can be reached in many of the areas discussed at the meeting, but two topics especially (the **bid markup** and the need for **bid clarifications**) require careful consideration.

Bid Markup

A markup or fee is added to the estimated price of a project. This has two components: **company overheads** and **profit margin**.

Company Overheads

Company overheads encompass all the expenses, apart from the project costs, incurred in operating a building company. Some expenses are difficult to distinguish as either company overheads or project expenses, but any company expenditure that is not the direct cost of project work or a part of the general expenses of a project is normally considered to be company overhead. Company overheads usually include the following expenses:

1. The cost of providing a single or multiple company office premises including mortgage or rent costs
2. Office utilities
3. Office furnishings and equipment
4. Office maintenance and cleaning
5. Executive and office personnel salaries and benefits
6. Company travel and entertainment costs
7. Accounting and legal consultant's fees
8. Advertising
9. Business taxes and licenses
10. Interest and bank charges

The amount of company overhead establishes the minimum fee a builder needs to obtain from each project.

> **EXAMPLE 12.1**
>
> In the case where a company has an annual overhead of $1,000,000 and the revenue obtained from completed projects per year is estimated to be $10 million, the minimum fee is calculated as:
>
> $$\frac{\$1,000,000}{\$10,000,000} = 10\%$$

Profit Margin

Profit margin, defined as the amount of money remaining after all project expenses and company expenses have been paid, can be assessed in two ways in a building business: as a return on the investment, or as compensation for the risk assumed by the company owners. Owners who have extensive investments in their company, usually in the form of housing lots, may wish to analyze their profits in terms of a percentage of the amount they have invested in the business. An option these owners always have is to sell all the assets of the company and invest the funds elsewhere. Many building company owners, however, have relatively small amounts invested in the operation. For these owners, the risk assessment is far more relevant.

Spec Housing Risk

For the speculative builder, the risk lies in the time it takes to sell its homes. When determining the markup to include in a house price, the question these builders have to consider is: What level of markup is sufficient to compensate for this risk involved? Keep in mind that much of the company overhead will be incurred whether houses are being sold or not. Returning to the figures previously quoted, if the builder has a bad year and annual sales slip from $10 million down to $5 million, company overheads will probably still remain at $1 million; this would mean that 20 percent of revenue would now have to go to pay overheads. Unless sales prices can be raised, profit margins will be squeezed. In fact, when sales are slow, some builders are tempted to reduce prices in order to improve their competitive position. This strategy will only serve to reduce profit margins even further.

Risk in Bid Work

In the case of builder obtaining work from bid competitions, the question of risk is considered from the beginning of the pricing process, and possibly before this in the takeoff section of the estimate. For example, a certain item of work may possibly require additional materials. Here the risk-averse estimator will include the extra material in the takeoff *just in case it is needed*.

It is difficult to deny the claim that every price in the estimate for a bid has an element of risk attached to it, but some items in the estimate are more risky than others. As outlined in the discussion on risk in Chapter 9, the items that hold the most risk for the builder are those where the price depends upon the productivity of the builder's workforce or upon the duration of the project. The value of these items (the risk component) is equal to the sum of the labor and equipment prices of the estimate plus the price of time-dependent general expense items. This amount is considered to be the risk component of the estimate.

Once the value of the risk component is established, the amount of fee to be included in a bid price can be calculated as a percentage of this risk component. The percentage amount used in the fee calculation is determined by a large number of factors, many of which where touched upon in items discussed in the pre-bid review listed above, but the factors affecting the fee markup can be summarized into three main categories:

1. The residential construction market and the level of competition expected
2. The builder's desire for and capacity to handle more work
3. The desirability of the project under consideration

However, calculating the project fee as a percentage of the value of risk in a bid has a certain danger attached to it. The reasons for applying a low fee to a bid (high competition, the builder's need for work, or a desirable project) can also influence the estimator into pricing the estimate optimistically. This can then produce a compounding effect since the low prices used in the estimate will lead to a low-risk component that will have a low percentage fee applied to it. The complete reverse of this is also possible: high prices, or a high risk factor resulting in an excessively high bid price. To avoid these extremes, the estimators need to be impartial in the pricing process and try to avoid undue optimism or pessimism.

Bid Clarifications

The instructions to bidders, especially on public projects, will frequently state that **qualified bids**, or bids that have conditions attached to them, will be considered invalid and these bids may be rejected. One of the reasons why it is important to obtain clear answers to the questions raised in the estimating process is that it is not advisable to state on the bid that it is based on this or that interpretation since the presence of such a statement can cause the bid to be rejected.

However, a situation occasionally arises where some aspect of the bid is still not perfectly clear even after explanations are received from the designers and the builder wishes to avoid the consequences of conflicting interpretations of the contract documents. In such a case, the group gathered together at the pre-bid review has to weigh the risk of these consequences against the risk of having the bid rejected. Sometimes the anticipated dangers are large enough to justify attaching the condition to the bid and risking rejection.

The Bid Closing

The process of closing a bid is a time of great excitement in the builder's office. From the first day the project drawings and specifications arrived in the office, the estimator has been keenly aware of the bid deadline; first the days are seen to slip by, then the hours are noted, and finally the minutes tick away to bid closing time. Those estimators who enjoy the thrill of a bid closing may approach the final day with eager anticipation, but there are some estimators who approach it with dread.

After the bid has been reviewed, the amount of fee set, and the estimator has made the changes to the bid that were decided upon in the review process, all that remains to be done is add in the subtrade prices and complete the bid documents. This is what some estimators refer to as the crazy part of the process.

On a project with more than thirty subtrades when, over a period of just a few hours, as many as ten bids can be received for each separate subtrade, the activity level in the prime builder's office can reach frenzied proportions. Under these conditions where each subtrade price can range up to $1 million or more and a single mistake can cost hundreds of thousands of dollars, the stress on estimators can be enormous. Cool heads and total control is essential if panic is to be avoided.

The computer can be a useful tool in the bid closing process and, if used effectively, can reduce the strain on the estimators considerably. However, in the last stage of the bid, successful bidding, more than ever, depends upon good organization.

Staffing the Bid Closing

On small building projects with few subtrade prices involved, the bid closing may be handled comfortably by no more than the estimator and a single assistant. But when many subtrade prices are expected for a project, additional people will be required to assist in the process. The number of estimators and assistants engaged in a bid closing has to be sufficient to deal with all telephone calls and gather all fax sheets promptly, leaving the project estimator and the chief estimator free to analyze the incoming bids and complete the pricing of the estimate. A team of six to eight people is usually sufficient to cope with the task, but on very large projects or in situations where several projects close at the same time, this number may need to be further increased.

Where the bid is to be hand delivered, which is the case with many bids, an additional person has to be added to the bid closing team for any size of project. This is the bid runner, the person who is to physically deliver the bid. Whether the place of the bid closing is just across town or in another city many miles away, the bid runner will usually have to write on the bid documents the final bid information. This may consist of merely the final bid price, but many other items of information can be undetermined when the bid runner leaves the office with the bid documents. Subtrade names, alternative, separate, and breakdown prices are just a few of the possible items that may need to be completed on the documents by the bid runner.

The bid runner will be equipped with a cellular phone so that, once he or she reaches the location of the bid closing and finds a place nearby to complete the documents, they can establish contact with the office and receive the final bid details. Certain subtrades, being aware of this arrangement, will deliberately withhold their price until the last moment to avoid the possibility of a builder disclosing their bid to other subtrades. Sometimes, therefore, the builder can still receive bids in the last few minutes before the closing. If care is not taken at this point, a disastrous situation can develop where the bid is delayed until just after the closing time and is disqualified for being too late.

Some builders have found it necessary to set a policy and inform all subtrades that no bid will be accepted after a stipulated time. This can be set at one hour or even half an hour before the owner's bid closing time. The builder then has time to handle the last arriving bids and the bid runner has time to complete the bid and hand it in before the time for accepting bids expires.

Receiving Subtrade Bids

As we have previously stated, the subcontractor's preferred method of submitting bids is usually by means of e-mail, but some bid prices are sent by fax or even over the telephone. Apart from the price for the work, the person in the builder's office receiving the bid has to ensure that all required information is obtained from the bidder, including:

1. The name of the subtrade bidding
2. The trade being bid
3. The name of the person submitting the bid
4. The telephone number where this person can be contacted
5. The base bid price
6. Whether or not the bid is based on plans and specifications and whether everything required of this trade is covered by the bid price
7. Alternative, unit, and separate prices if required
8. The taxes applicable to this item and whether they have or have not been included in the price
9. Any conditions attached to the bid
10. The time and date the bid was received
11. The name of the person receiving the bid

Figure 12.7 shows a standard form that is used for logging this information when telephone bids are received. This form provides a checklist that the person receiving the bid can follow to ensure that no details are lacking.

Although it is unusual for a telephone to malfunction, computers are far more prone to problems. It is a good idea to have a contingency plan in the office to deal with a situation where computer problems occur just as subtrade bids are

TELEPHONE BID

Trade Classification:

Date:

Project:

Bidder: By:

Address: Phone:

BASE BID:

TAXES: — State   ☐ Included   ☐ Excluded

              — Federal   ☐ Included   ☐ Excluded

REMARKS:

ALTERNATES:

Taken By:

Figure 12.7   Telephone Bid Form

due to be received. This may require you to inform all bidders that they direct their bids to an alternative e-mail address or submit their bids, for now, just by telephone.

Subtrade Conditions

In Chapter 11 we discussed the problems that result from subcontractors submitting conditional bids; the estimator closing a bid should be reminded to examine the terms of subtrade bids carefully. Where a subcontractor's bid has conditions attached, the estimator needs to evaluate the consequences of accepting those conditions. For example, if the subtrade excludes some work that is normally performed by that trade, the estimator will have to calculate the cost of the exclusion and add it to the subtrade's bid price.

Sometimes the condition imposed by the subtrade is something the builder needs to discuss with the owner. For instance, the subtrade may offer a price on the condition that they have the use of a certain sized storage space at the site. If the builder is aware of the condition early enough, the question can be raised with the owner's design consultant and confirmation of whether or not the required space is available obtained before the bid closing. But when the builder is made aware of this condition only an hour or so before the bid closing time, there is no time to discuss the matter. This is the kind of situation where the benefit of pre-bid subtrade proposals is most appreciated.

Naming Subcontractors

Compiling a list of proposed subcontractors can be yet another difficult task for the bidder. The owner and/or the consultant may ask the bidder to list for their approval all the subtrades the builder is going to use on the project. Because of time constraints, because of subtrade bid conditions, and because of the overwhelming need to put together the most competitive price, the final list of subcontractors may not be available until moments before the bid is to be handed in. Often having just a few more minutes after the bid is submitted would be useful to the builder for sorting out exactly whose prices were used in the last-minute scramble to finish off the bid.

While a number of owners and their design consultants are sympathetic about the builder's difficulties and allow builders to submit subtrade lists after the bid closing, others point out, with some justification, that to allow even a short time after the bid for builders to make their list of subtrades invites bad practices. It makes it easier for the unscrupulous builder to engage in bid shopping of subtrade prices. The required list of proposed subcontractors, therefore, can be expected to remain as a pre-bid condition on most contracts.

Bid Breakdown

To further complicate the task of the estimator who is trying to close out a bid, some owners call for a cost breakdown to be attached to the bid documents. This breakdown is often required in a form similar to the estimate summary where a price is indicated for each of the trades involved in the work. The information is generally requested to provide a basis for evaluating monthly progress payments on the project. More reasonable contract requirements usually state that the bidder awarded the contract shall submit this breakdown to the owner within a stated time of the contract award date.

Preparation of this information at the time of the bid is a major chore for the estimator because the prices stated have to include profit and overhead amounts, and many of these prices will not be available until the day of the bid closing, sometimes only minutes before the closing. In some cases, bidders just "rough out" the breakdown and submit the approximate values with the bid. In a few cases, builders have risked having their bids disqualified for failing to include the breakdown with their bid. These builders may state on their bid that the breakdown is not available at this time but will be forwarded to the owner within 24 hours of the bid closing time.

Computerized Bid Closing

Computer programs are available to perform all the calculations and price compilations that are necessary in the closing stages of an estimate. Merely having the bid summary set up on a computer spreadsheet provides a great advantage over manual methods. When a subtrade price is changed on the summary, the computer can be preprogrammed to automatically recalculate all consequential amounts and immediately display the revised bid price. There are software packages especially designed for bid closing situations that offer a number of useful additional features.

Estimating staff may still be required to man the computers and telephones to collect the subtrade prices but, with bid-closing software operating on a computer network, prices can be constantly inputted into the system from any of the computers on the network. The estimator compiling the bid is then able to continually analyze the incoming prices and make selections from the combined input from all the estimators situated at their own terminals.

Bid-closing software is normally compatible with other estimating software so that prices estimated on one system can also be used as input into the bid-closing system and changes made in material, labor, or equipment prices in one part of the estimate will automatically be reflected in the bid price.

After the Bid

Once the bid is submitted and the bid-closing time passes, those who were involved in the estimate, after perhaps sharing a moment of euphoria in the knowledge that the process is finally at an end, turn their attention to the bid results. A different excitement now grips the team from the one they felt in the final stages of the bid. The office scene is now calm after the chaos of the bid closing, but there is an underlying tension, firstly about the position of the bid: Was it low? Was it high? And, secondly, about the bid itself: Was everything covered? Were there any mistakes?

Many owners these days open the bids in public shortly after the closing, in which case the anxieties of the estimators will soon be relieved or, in the case of a very low bid, they may be multiplied. Whatever the results, a review of the bid should be undertaken and all the loose ends left by the estimating process need to be tidied up.

After the bid, the items which need to be attended to include:

1. Tabulation of the bid results—Figure 12.8 shows a typical bid results table. Copies of the bid results table are usually distributed to senior management personnel and to all of the people who were present at the bid review meeting.
2. Compile the bid documents—Drawings, specifications, and other documents provided by the design consultants for the bid are gathered together.

BID RESULTS

PROJECT: _The Parks Home_

FLOOR AREA: _2750 SF_

LOCATION: _Townsville_

DATE: _____

ESTIMATOR: _ABF_

	Bid Price $	Per SF $	
1. Builder ABC	375,694	136.62	1.00
2. Builder DEF	388,279	141.19	1.03
3. Builder GHI	397,239	144.45	1.06
4. Builder JKL	426,754	155.18	1.14
5. Builder MNO	572,156	208.06	1.52
6. Builder PQR	582,000	211.64	1.55

Figure 12.8 Tabulation of Bid Results

3. Return bid documents—If the bid is not one of the three low prices, the bid documents are returned to the design consultants and the deposit left for these documents is claimed.

4. Store bid documents—If the bid results have not been made public or if the bid is one of the three low bids, the bid documents are retained pending the award of contract. Should a contract be awarded to this builder, these documents will provide evidence of the basis of the bid so the builder should hold them in safe storage. In no circumstances should these documents be sent back either to the designer or the owner.

5. Compile and store the estimate file—All the price quotations from material suppliers and subcontractors should be collected together with all other estimate documents in an estimate file. This will be the principal resource used in the review process described below. Afterwards, whatever the outcome of the bid, this file should be kept in safe storage for future reference.

6. Contact subcontractors—How builders deal with subcontractors at this stage is a matter of individual company policy. Immediately after the bid closing, some builders are willing to disclose whether a subtrade's price was or was not used by the builder in its bid. Some builders will disclose this information to subtrades at a later stage when the bid results are made public, while other builders refuse to even provide this information. These last builders, if they are awarded the contract, will notify only those subtrades that they intend to award contracts to, otherwise subcontractors will not be contacted.

Post-Bid Review

If possible, the group of people who attended the pre-bid review should return for the **post-bid review**. Depending upon the bid results, there will often be many questions that arise after the bid is submitted. In general terms, if the bid was not successful, the prime consideration will be: What needs to be done in the estimating process to improve the success of our bids? Clearly, if mistakes were made, action needs to be taken to try to avoid repeating the same mistakes again. If problems were encountered in the estimating process, the group needs to consider what can be done on future bids to prevent these problems from recurring. If a bid was extremely high or low, the major component prices of the bid should be examined to try to determine why the overall price was so extreme.

Sometimes, after all the prices used in the bid have been poured over and all kinds of alternative approaches and possible lower subtrade prices have been considered, the builder finds that the price of low bid can be reached only if the estimate was to be priced below cost with no fee included. In this situation, the most feasible explanation seems to be that the low bid contains an error and this builder is consoled only by the notion that the successful builder will lose so much money on the project that it will not be around too much longer to cause any further problems.

Summary Example of House Project

Figure 12.9 shows the summary of prices for the sample house project.

Notes on Summary Shown in Figure 12.9

1. This is a summary to determine the sales price of a spec home.
2. All of the work of this project is to be performed by subtrades, so the *own work* section at the top of the summary has been removed.
3. The price for supplying and placing concrete is included in the bill of materials, so there is no price against concrete in the *subcontractor's* section.
4. The prices for plumbing, heating, and electrical work are divided into rough in and finish prices to reflect the way these trades are paid.
5. The price of general expenses is entered from the general expense sheet.
6. The add-on for payroll additive is calculated as 25 percent of the total value of labor:

   $25\% \times \$5,300 = \$1,325.00$

7. The remaining add-ons, except for the fee that is a lump sum, are calculated as a percentage of the bid total. A trial-and-error method is used to estimate the total price. First an approximate value is used for the bid total and the summary completed. The resulting bid total is compared with the value used to calculate the add-ons. If there is a large error, the process is repeated until the margin of error is within acceptable limits. These are the final calculations:
   a. Building Permit $0.7\% \times \$150,854 = \$1,056.00$
   b. Insurance $2.5\% \times \$150,854 = \$3,771.00$
8. The cost per square foot is calculated at the end of the summary sheet by dividing the bid total by the gross floor area of the project that is listed at the top of the page. This provides a check on the accuracy of the bid total—an SF cost, that appears too high or too low would sound a warning to the estimator.

	PROJECT:	House Example					
	GROSS FLOOR AREA:	1157 SF		ESTIMATE SUMMARY			
	LOCATION:	Townsville					
	DATE:						
	ESTIMATOR:	ABF					

No.	DESCRIPTION	LABOR $	MATL. $	EQIUP. $	SUBTRADE $	OTHER $	TOTAL $
	SUBCONTRACTORS						
1.0	Excavate and Backfill Basement	---	---	---	3,021.00	---	3,021.00
2.0	Excavate & Backfill Service Trenches	---	---	---	1,485.00	---	1,485.00
3.0	Concrete	---	---	---	See Bill of Materials		0.00
4.0	Form Footings	---	---	---	350.00	---	350.00
5.0	Form Walls and Columns	---	---	---	3,285.00	---	3,285.00
6.0	Damp Proof Walls	---	---	---	350.00	---	350.00
7.0	Rough Carpentry Framing	---	---	---	3,181.00	---	3,181.00
8.0	Deck	---	---	---	125.00	---	125.00
9.0	Canopy Over Entrance	---	---	---	50.00	---	50.00
10.0	Finish Carpentry	---	---	---	1,950.00	---	1,950.00
11.0	Cabinets (Supply and Install)	---	---	---	6,950.00	---	6,950.00
12.0	Bathroom Accessories (Supply and Install)	---	---	---	518.00	---	518.00
13.0	Appliances (Supply and Install)	---	---	---	2,090.00	---	2,090.00
14.0	Soffit and Fascia (Supply and Install)	---	---	---	1,416.00	---	1,416.00
15.0	Siding (Supply and Install)	---	---	---	2,225.00	---	2,225.00
16.0	½" Parging (Supply and Install)	---	---	---	902.80	---	902.80
17.0	Paint Exterior Windows and Doors	---	---	---	547.00	---	547.00
18.0	4" Eaves Gutter (Supply and Install)	---	---	---	550.00	---	550.00
19.0	3" Downspouts (Supply and Install)	---	---	---	121.50	---	121.50
20.0	Roofing (Supply and Install)	---	---	---	2,320.00	---	2,320.00
21.0	7'-0" x 3'-6" Precast Concrete Steps	---	---	---	965.00	---	965.00
22.0	Flooring (Supply and Install)	---	---	---	3,905.00	---	3,905.00
23.0	Drywall Finishes (Supply and Install)	---	---	---	11,711.00	---	11,711.00
24.0	Interior Painting (Supply and Install)	---	---	---	3,234.00	---	3,234.00
25.0	Plumbing - Rough In	---	---	---	5,060.00	---	5,060.00
26.0	Plumbing -Finish	---	---	---	2,860.00	---	2,860.00
27.0	Heating - Rough In	---	---	---	3,410.00	---	3,410.00
28.0	Heating - Finish	---	---	---	2,640.00	---	2,640.00
29.0	Electrical - Rough In	---	---	---	2,450.00	---	2,450.00
30.0	Electrical - Finish	---	---	---	1,045.00	---	1,045.00
31.0	Light Fixtures	---	---	---	475.00	---	475.00
	Subtotal 1:	0.00	0.00	0.00	69,192.30	0.00	69,192.30
	MATERIALS						
32.0	Excavation and Gravel Materials	---	2,586.25	---	---	---	2,586.25
33.0	Concrete Materials	---	12,656.46	---	---	---	12,656.46
34.0	Carpentry Materials	---	13,938.87	---	---	---	13,938.87
35.0	Finish Carpentry	---	10,049.35	---	---	---	10,049.35
	Subtotal 2:	0.00	39,230.93	0.00	69,192.30	0.00	108,423.23
36.0	**GENERAL EXPENSES**	5,300.00	335.00	2,750.00	2,893.12	---	11,278.12
	Subtotal 3:	5,300.00	39,565.93	2,750.00	72,085.42	0.00	119,701.35
	ADD-ONS						
37.0	SMALL TOOLS	---	---	---	---	---	---
38.0	PAYROLL ADDITIVE	---	---	---	---	1,325.00	1,325.00
39.0	BUILDING PERMIT	---	---	---	---	1,056.00	1,056.00
40.0	INSURANCE	---	---	---	---	3,771.00	3,771.00
41.0	FEE	---	---	---	---	25,000.00	25,000.00
42.0	ADJUSTMENT	---	---	---	---	---	0.00
						TOTAL PRICE:	150,853.35

Price per S.F. = $130.38

Figure 12.9 House Example Summary

SUMMARY

- This chapter considered the last stages in summarizing and completing an estimate.
- Various summary formats may be required for different purposes:
  - To determine the sales price of house
  - To use in supervising the construction work
  - To use in accounting
  - To provide a price breakdown for a purchaser
  - To determine a bid price
- A residential construction bid is assembled in three stages:
  - Advance stage
  - Review stage
  - Closing stage
- To be concluded successfully, the bid process has to be carefully planned and organized.
- As much of the estimate as possible is finalized in the advance stage; also, a plan for the delivery of the bid is formulated here.
- In addition, the builder has to address the following items in the advance stage:
  - Completion of bid forms
  - Calculation and listing of unit prices
  - **Alternative prices (alternates)**
  - Separate prices
  - Allowances
  - Bid security—bid bonds and deposits
  - Bid conditions
- Standard bid forms that address essential requirements are available from construction associations.
- Builders are sometimes instructed to use owner-supplied bid forms.
- A pre-bid review meeting will usually take place at least two workdays before the bid-closing day. This review is conducted for three main reasons:
  - To ensure that there are no obvious omissions from the bid
  - To consider ways of improving the competitiveness of the bid price
  - To consider an appropriate fee (bid markup) for the project
- The bid markup has two components: a contribution to cover company overheads and a profit margin that is generally assessed as compensation for the risk taken by the builder.
- Company overheads are all the expenses incurred in operating the company.
- Builders sometimes need to include a clause with the bid to explain the basis of the bid. These bid *clarifications* can be dangerous since they may give the owner grounds to disqualify a formal bid.
- A bid closing can be a period of hectic activity in a builder's office with a large number of competing subcontractors submitting bids for often more than thirty different trades.
- The bid closing needs to be properly staffed and organized on large project bids to deal effectively with all the fax and telephone bids received over a relatively short time period.
- Having to name subcontractors and provide price breakdowns can complicate a bid.
- A number of tasks follow the bid submission, including:
  - Tabulate bid results
  - Gather together all documents

- Return bid drawings and specifications
- Store bid documents
- Compile an estimate file
- Handle subcontractor inquiries
- Conduct a post-bid review

■ Manually summarizing and concluding an estimate to determine the selling price of a house is demonstrated.

RECOMMENDED RESOURCES

Information	Web Page Address
■ Bid forms can be purchased from The Contractor's Group	http://www.thecontractorsgroup.com/form-bid-proposal.htm
■ United States Small Business Administration explains surety bonds	http://www.sba.gov/category/navigation-structure/loans-grants/bonds/surety-bonds
■ Information about bidding software can be obtained by entering the key words: **Bidding Software** into your Internet search engine	

REVIEW QUESTIONS

1. Describe four different uses for an estimate summary.
2. Explain why custom home builders sometimes provide price breakdowns to the home purchaser.
3. At which point in the bid processes are errors most likely to occur? Explain why.
4. Why is the planning of the delivery of a bid an important part of the advance stage of the bidding process?
5. Why do owners sometimes call for alternative prices to be quoted with bids for custom homes?
6. If a "cash allowance" for floor finishes was specified in the contract documents of a project, what action would the estimator who is preparing a bid on the project have to take?
7. What is the most common form of bid security for large project bids?
8. What is the purpose of bid security?
9. What are the objectives of and who should attend the pre-bid review?
10. Describe the financial risks taken by a spec homebuilder.
11. Why do some builders include "bid clarifications" with their bid even when they risk having their bid rejected by doing this?
12. Why is a list of subtrades sometimes required, and why do bidders sometimes find it difficult to compile the list?
13. How can computer-estimating software help with bid closings?
14. What is the purpose of a post-bid review?

PRACTICE PROBLEMS

1. Complete the Bid Alternatives Chart in Figure 12.10 to determine the *Total Addition* to or *Total Deduction* from the contract price based on the following changes:
   a. To substitute Menalta facing bricks for the specified Glengary bricks, deduct $7,250

BID ALTERNATES

PROJECT: _____

LOCATION: _____

DATE: _____

ESTIMATOR: _____

NUMBER	ALTERNATIVE PRICES	ADD $	DEDUCT $

Figure 12.10   Practice Problem #1

    b. To substitute aluminum frame windows for wood frames, deduct $1,222

    c. To reduce the size of the living room and master bedroom as discussed, deduct $7,190

    d. To finish the basement, add the following amounts:

i.	Replace stairs	$4,800
ii.	Partition walls	$1,400
iii.	Passage doors	$560
iv.	Hardware	$450
v.	Floor finishes	$2,725
vi.	Wall finishes	$1,082
vii.	Increase furnace capacity	$1,500
viii.	Electrical	$2,100

2. A homebuilder anticipates that he will build 50 homes next year and sell them at an average price of $385,000 each. What is the minimum markup he can add to the price of these homes if the company overheads for next year are as listed in Figure 12.11?

Company Overheads for Next Year:

Office Rental	$24,000.00
Office Utilities	$2,800.00
Office Furnishings and Equipment	$3,000.00
Office Maintenance and Cleaning	$26,400.00
Executive and Office Personnel Salaries	$710,000.00
Company Travel and Entertainment	$5,000.00
Accounting and Legal Fees	$12,000.00
Advertizing	$25,000.00
Business Taxes	$130,000.00
Interest and Bank Charges	$216,800.00

Figure 12.11   Company Overheads

13

ESTIMATES FOR REMODELING WORK

OBJECTIVES

After reading this chapter and completing the review questions, you should be able to:

- Describe how the remodeling business differs from the housing construction business.
- Explain how an accurate remodeling estimate is compiled.
- Describe how demolition and other remodeling work is measured.
- Explain the importance of making sketches, both in scope definition and in the takeoff process.
- Describe how an on-site takeoff can be prepared in a systematic manner.
- Describe how remodeling work is priced.
- Prepare a complete bid estimate including general expenses and summary for a remodeling project.

KEY TERMS

bill of materials	combination contracts	on-site takeoff
cost plus contract	lump sum contract	unit price contract

Introduction

The remodeling business is different from home building in a number of ways. The remodeling contractor deals far more closely with the customer in a way similar to the direct-sales retail business. There are generally no architects or designers between the homeowner and the remodeler and, consequently, no drawings to inform the builder what is required. The remodeler is mostly dealing with structures that are in place rather than ideas that are expressed in plans and specifications. This has positive and negative aspects: positive in that much of what is required can be seen and physically measured, but negative in that some items are concealed and, therefore, cannot be properly assessed until the work begins. There is often more uncertainty on remodeling projects; problems can easily arise because plans do not turn out to be quite what was expected.

For the remodeler to succeed, they have to sell their expertise, show that they can solve the project problems, and that they are going to provide the quality and service the home renovator is looking for. In this process, price may not be as important as it is to the regular homebuilder, but the remodeler still has to make a profit in order to stay in business. To achieve this, accurate estimating remains a priority.

Contracts for Remodeling Work

Lump Sum Contracts

The type and extent of estimating required for a project depends mostly upon the contract terms for the project. The most commonly used contract for remodeling work is the **lump sum contract** with a firm price. The builder is usually selected on the basis of low price following a bid competition. If the scope of work is well defined and the owner has no wish to become involved in the construction process, this is probably their best option. Their only concern might be that the builder meets their obligation to perform all work according to requirements, particularly with regard to quality and schedule. The form of contract also provides the builder with the advantage of independence free from the owner's interference.

An accurate estimate is required for a builder to make a profit on a lump sum contract because the risk of extra costs is mostly borne by the builder. But while the owner enjoys the benefit of having a firm price for the job, a lump sum contract lacks flexibility. If there are many changes to the design during construction, the work will be disrupted and the builder will usually charge high prices for the changes. This will result in a project that neither meets the budget nor the schedule. When design changes are anticipated or when there is insufficient time to finalize a design before work begins, it may be preferable to use a cost plus form of contract.

Cost Plus and Management Contracts

Under the terms of a **cost plus contract**, the builder will be reimbursed for the costs incurred while doing the work and will also receive a fee for the job. As discussed in Chapter 1, the fee may be a percentage of the costs, a fixed amount, or a variable amount with incentives for efficiency. This arrangement allows the owner to make virtually any changes they wish; they can even design the project as it is built if they choose. Builders can still compete for the project, in this case on the basis of fee rather than total price. Once they have been awarded the contract, however, they will be guaranteed a profit on the job. There is less of a need for an accurate estimate with a cost plus contract, but owners often call for a rough estimate before work begins in order to provide them with a target budget or, possibly, to determine the maximum price that they want the builder to complete the work for.

A variation on the cost plus contract is a construction management contract. Here the builder is hired as a professional construction manager to oversee the project for an agreed fee. The cost of the work including labor, materials, equipment, subtrades, and overheads are paid directly by the owner. Budget estimating is commonly used with management contracts in an attempt to control costs, but the flexibility of a cost plus and management contracts does, however, come at a price for the homeowner. Homeowners will undoubtedly pay more for the job under these terms than they would have done had they taken the time to set up a lump sum agreement.

Unit Price Contracts

Unit price contracts are used when the amount of work required on a project is uncertain for some reason. This situation occurs from time to time with remodeling contracts. For instance, when there are underground features to be removed, such as old cables, pipes, footings, and so on, no one can be altogether certain how much there is or how deep you will have to excavate to reach these items. If a unit price contract is adopted, the builder will quote a series of unit prices for the type of work involved in the job and be paid for the actual quantities of work completed multiplied by the applicable unit prices.

Estimating unit prices can be complicated, especially where the prices have to include for the builder's overheads and profit, and they may be further complicated if the owner wishes to use the same price for both added and deleted work on the project. The way to proceed is to work out what it would cost to complete a certain amount of work including all required overheads and profit. Then determine the unit price by dividing this total price by quantity of the work.

Combination Contracts

On larger projects, the contract arrangements may consist of a combination of contract types. On one job, for example, the builder was to complete the main part of the work for a fixed lump sum amount, all underground work was paid for at the unit prices in the contract, and any extra work was paid for at cost plus at an agreed percentage. In this way, the owner sets out to take advantage of the strengths of each form of contract on a single project.

The Remodeling Estimate

Budget estimating and preliminary estimating techniques are discussed in Chapter 1. These techniques are often used on remodeling jobs in the initial stages just as they are on housing projects. But what if a more accurate estimate is required in order to quote a firm price bid for a remodeling job? There are three key steps in preparing an accurate estimate for this kind of project when a lump sum price is called for:

1. Clearly identify all the work to be done
2. Measure the work to be done
3. Price the work

The most important and usually the most difficult to achieve of these steps is the first. Good estimating procedures, as described in the other chapters of this text, will take care of steps 2 and 3, but if the work has not been properly defined, no amount of good estimating practice will improve a bad estimate. The estimator has to be able to ascertain what work is required to achieve the results the homeowner is looking for. This first calls for good communication with the owner; the estimator has to ask many questions to the homeowner, show the owner, by means of sketches and descriptions, what they plan to do and, perhaps most importantly, obtain the owner's approval of these plans. This process also calls for a certain discipline. The estimator has to keep in mind that they are there to build the project the owner wants, not what the estimator or the remodeling contractor wants.

Once the project goal is confirmed, the estimator next has to establish what exactly is in place before work begins, then go on to decide what has to be done to the existing structure to achieve the desired end results. This, above all, relies upon the estimator's knowledge and experience. Perhaps even more than on housing jobs, the remodeling estimator needs to be very familiar with construction practices and have a wealth of experience in this work in order to deal with the many problems that arise on remodeling projects. Making sketches showing the existing building and the size and shape of features to be changed will help. The estimator will often also need to consult with designers, engineers, and trades such as plumbers and electricians to receive their input when putting an estimate together.

At the end of this first step in the estimate, the estimator should have:

1. A clear outline of what the owner wants in the finished project (this is usually detailed on sketches and in notes that are approved by the homeowner)
2. A detailed analysis of the existing building with sketches and notes of what is now in place
3. A detailed plan of what has to be done and how it is to be done to build the finished project

Measuring the Work

Remodeling work may include any or all of the trades required for homebuilding:

- Excavation and backfill
- Concrete work
- Masonry work
- Metal work
- Rough carpentry
- Finish carpentry
- Interior finishes
- Exterior finishes
- Plumbing work
- HVAC work
- Electrical work

We have discussed the methods of measurement for these trades in previous chapters with regard to homebuilding estimates. The same methods are followed in a remodeling takeoff. There is, however, one further trade that is often involved with remodeling work that has not been considered previously: demolition work.

Measuring Demolition Work

Once again, for consistency and to enable a takeoff prepared by one estimator to be easily understood by another, it is advisable to follow some generally accepted rules of measurement:

1. Measure demolition work in the same way and using the same units as new work is measured.
2. Where information is lacking, the estimator shall specify any assumptions made.
3. Describe dust curtains, temporary partitions, and suchlike, and measure in square feet.
4. Include loading and disposal of garbage in the price of demolition materials.
5. Measure in square feet finishes to floors, walls, ceilings, roofs, etc., if they are to be removed separately from what they are attached.
6. Enumerate doors and windows if they are to be separately removed from walls; otherwise, they are included in the price of wall demolition.
7. Describe cutting openings in slabs and walls, and measure in square feet.
8. Describe stating size of concrete strip footings to be removed and measure in linear feet.

9. Describe concrete pad footings and irregular-sized footings to be removed and measure in cubic feet.
10. Describe stating thickness of concrete walls to be removed and measure in square feet.
11. Describe stating thickness of concrete slabs to be removed and measure in square feet.
12. Describe other concrete items to be removed and measure in cubic feet.
13. Describe stating thickness of masonry walls to be removed and measure in square feet.
14. Describe other masonry items to be removed and measure in cubic feet.
15. Describe and enumerate structural steel items to be removed.
16. Describe and enumerate miscellaneous steel items to be removed.
17. Describe carpentry floor systems to be removed and measure in square feet.
18. Describe carpentry wall systems to be removed and measure in square feet.
19. Describe carpentry roof systems to be removed and measure in square feet.
20. Describe other items of carpentry to be removed and measure in the same units as the installation of these items.
21. Describe items of finish carpentry and millwork to be removed and measure in the same units as the installation of these items.
22. Describe items of plumbing to be removed and measure in the same units as the installation of these items.
23. Describe HVAC items to be removed and measure in the same units as the installation of these items.
24. Describe electrical items to be removed and measure in the same units as the installation of these items.
25. Describe items of equipment to be removed and measure in the same units as the installation of these items.
26. Necessary cleanup is deemed to be included in the price of demolition items.

The Remodeling Takeoff

All the comments regarding takeoffs found in Chapter 2 apply as much to remodeling takeoffs as they do to regular house takeoffs but, because detailed plans and specifications are seldom available on remodeling projects, certain elements of the takeoff become more important than they were.

We have already referred to the use of sketches in defining what exists before work begins and also what is to be done on the job. In the takeoff process, these sketches show where the work dimensions come from and also clarify what the estimator is measuring in the takeoff.

Specification notes are also useful to indicate what has been allowed for in the takeoff. Items such as the strength of concrete, the grade of lumber, the type of floor finish, the type of paint, and many more should all be defined so that anyone who reviews the estimate can see what has been allowed for in the price. It is important to inform the homeowner about the basis of the price that you quote for the work: spell out precisely what you are going to do for the price you quote and what materials you are going to use. It might even be worthwhile to describe how you propose to do the work, especially if you anticipate some temporary inconvenience or disruption to the owner and their property.

The takeoff will generally follow the order of construction, but on larger jobs where there are many rooms or even multiple buildings to be considered, it is best to proceed on a building-by-building than room-by-room basis. All the work that is needed in one room is measured before passing on to the next room. Then, once the

```
┌─────────────────────────────────────┐
│                                     │
│        Takeoff Checklist:           │
│                                     │
│         1  Floors                   │
│         2  Baseboards               │
│         3  Walls                    │
│         4  Partitions               │
│         5  Windows                  │
│         6  Doors                    │
│         7  Bifolds                  │
│         8  Hardware                 │
│         9  Ceilings                 │
│        10  Cabinets                 │
│        11  Plumbing                 │
│        12  HVAC                     │
│        13  Electrical               │
│        14  Other Work               │
│                                     │
└─────────────────────────────────────┘
```

Figure 13.1   Room Takeoff Checklist

work in one building is all measured, you can move on to the next building. Estimators sometimes use a simple checklist to ensure all aspects are considered for each room on the project. See Figure 13.1 for an example of such a checklist.

On-Site Takeoff

Much of the information required for a remodeling takeoff will be obtained from the site of property that is to be remodeled. If good notes, measurements, and sketches are made on site, the takeoff will be much easier to complete. In fact, a good portion of the takeoff can be done at the site of the work, leaving only the final calculations and the pricing to be completed back at the office. Electronic tape measures with built-in calculators can be used to speed up the process of measuring on site. Inexpensive versions of these tools are available that use sound waves to measure distances, but if more accuracy is required, especially when distances exceed 40 feet, the more expensive laser models may be required.

To ensure that everything that needs to be measured and noted on site is done, the estimator has to proceed in a systematic fashion such as this:

Exterior Work

- North elevation
- South elevation
- East elevation
- West elevation
- Roofing
- Outbuildings
- Site surfaces
- Fences
- Landscaping

Interior Work

- ■ Main Floor
  Garage
  Entrance hall
  Kitchen
  Living room
  Family room
  Bathroom 1
  Main stairway
- ■ Second floor
  Bathroom 2
  Bedroom 1
  Bedroom 2, etc.
- ■ Basement
  Basement stairway
  Basement room 1
  Basement room 2, etc.

In each room, a checklist such as that shown in Figure 13.1 can be used to make sure that all the work required in that room is measured in the takeoff.

Pricing Remodeling Work

The process of pricing an estimate for a remodeling project is very similar to the process followed in an estimate for a residential project generally that is described in Chapter 9. The estimator will recap the takeoff items ready for pricing; a trade format is most often used. On a small project where there are few items in the takeoff, the estimator may choose to bypass the recap and price the takeoff items directly. Figure 13.2 shows a type of stationery that is designed for this purpose.

Pricing Labor and Equipment

Whereas in the home building industry most of the on-site work is done by sub-trades, much of the work in the remodeling sector is completed by the builder's own workers. Subcontractors are still employed, particularly for plumbing, HVAC, and electrical work, but on the smaller jobs, the builder may use its own personnel even for the work of these specialized trades. In order to price the work completed by its own forces, the builder has to calculate unit prices based on the anticipated productivity of its workers and the equipment they use.

As mentioned in Chapter 9, the productivity of labor and equipment depends upon two main factors: *Job Factors* and *Labor and Management Factors.*

Remodeling Job Factors

The job factors that are particularly important to productivity on remodeling projects include:

- ■ Weather conditions—especially with outside work such as roofing and landscaping
- ■ Access to and around the work area—poor access to the job may prevent the builder from using certain equipment, while a restricted workspace will usually reduce the efficiency of workers
- ■ Size of the project—larger jobs, especially when there is repetition of activities, are generally more productive than smaller jobs

No.	DESCRIPTION	No. Pieces	DIMENSIONS			Extension	Total Quantity	UNIT PRICE	LABOR	UNIT PRICE	MATERIALS	UNIT PRICE	TOTAL

ESTIMATE SHEET — Page No.

JOB . . . . DATE . . . .

ESTIMATED . . . . EXTENDED . . . . CHECKED . . . .

Figure 13.2   Stationery for Combined Takeoff and Pricing

- Complexity of the tasks involved—clearly, complicated work will be less productive than work that is more straightforward
- Location of the job—are the required materials available locally, or do they have to be trucked in? This is one of the number of questions that can arise regarding job location

Remodeling Labor and Management Factors

The labor and management factors that are particularly important to productivity on remodeling projects include:

- Quality of supervision—because much of the work is done by the builder's own workers, good supervision is very important in order to attain high productivity

- Motivation and morale of the workers—high productivity is possible only with good workers
- Good tools and equipment—productivity suffers if tools are too old or poorly maintained
- Experience of supervisors and workers—if supervisors and workers lack experience with remodeling jobs, high productivity will be difficult to achieve

The estimator needs to be aware of all these factors and make adjustments to productivity rates to take into account the particular circumstance of the project being considered.

Labor and Equipment Productivities

Records from past jobs are the best source of productivity information for accurate estimates. Some estimators keep track of the cost of operations on past jobs, they maintain a *price book* so that they can use the information to price estimates. There are at least two problems with this approach to pricing:

1. Without additional notes, the estimator may not know what the wage levels were when the price was calculated. For example, the *price book* may say it cost $15.00 per square foot for labor to construct a 4" slab-on-grade, but what were the wage rates at that time?
2. The composition of the crew that performed the work is also probably unknown. For example, was the $15.00 per square foot price achieved with a crew of one foreman and two laborers, or was it one carpenter and four laborers?

Clearly, if wage rates have changed, or a different combination of workers is used, the unit price will probably be different, but perhaps more importantly, there is no easy way to adjust the price to take into account these changes.

When records are kept in the form of crew productivities, they are much easier to adapt to different situations.

Where a contractor does not have its own productivity records available to use in estimating future projects, it can turn to publications such as Means Cost Data[1] that express crew productivity rates. However, there is no substitute for good historic records of productivities established on a contractor's own projects as discussed in Chapter 9.

A summary chart of historic productivity rates can be prepared from these data records. This chart lists work operations, shows the typical crew constituents for each operation, and indicates its historic productivity range. A sample chart is shown on Figure 13.3a, Figure 13.3b, Figure 13.3c, and Figure 13.3d for the kind of work encountered on remodeling projects. The productivity rates on this chart are used in pricing the examples that follow.

Pricing Materials

A wide range of materials may be required for a remodeling job. These can be summarized from the takeoff on a separate **bill of materials** or they may be priced with the labor and equipment on the recap all as described in Chapter 9.

---

1. Building Construction Cost Data, R. S. Means Company, Kingston, MA

PRODUCTIVITIES

CREWS	CREW MEMBERS	CREWS	CREW MEMBERS
CREW A	1.0 Foreman 2.0 Laborers	CREW E	1.0 Cement Finisher
		CREW F	1.0 Carpenter
CREW B	1.0 Equipment Operator		
		CREW G	1.0 Foreman 2.0 Bricklayers
CREW C	1.0 Oarp. Foreman 2.0 Carpenters		
		CREW H	1.0 Foreman 2.0 Painters
CREW D	1.0 Foreman 2.0 Laborers 1.0 Cement Finisher		

ITEM	OPERATION	CREW	EQUIP.	OUTPUT
1.0	**Demolition**			
1.1	Remove 1'-0" x 2'-0" Conc. Footing	A	Jackhammers	20 – 24 feet/hour
1.2	Remove Pad Footings	A	Jackahmmers	6 – 8 cu. ft./hour
1.3	Remove 8" Concrete Wall	A	Jackhammers	10 – 12 sq. ft./hour
1.4	Remove 8" Block Walls	A	Jackhammers	50 – 60 sq. ft./hour
1.5	Remove Asphalt Shingles	A	Hand Tools	150 – 250 sq. ft./hour
1.6	Remove 4" Conc. Slab-on-Grade (not reinf.)	A	Jackhammers	35 – 40 sq. ft./hour
1.7	Remove Stud Partition	A	Hand Tools	65 – 70 sq. ft./hour
1.8	Remove Kitchen Cabinets	A	Hand Tools	10 – 15 feet/hour
1.9	Remove Countertop	A	Hand Tools	25 – 30 feet/hour
1.10	Remove Vinyl Tiles	A	Hand Tools	80 – 125 sq. ft./hour
1.11	Cut Openings in Stud Walls	A	Hand Tools	10 – 20 sq. ft./hour
2.0	**Excavation and Backfill**			
2.1	Strip Topsoil	B	Skid-Steer Loader	50 – 100 cu. yd./hour
2.2	Excavate Trench	B	Small Backhoe	12 – 20 cu. yd./hour
2.3	Excavate Pit	B	Small Backhoe	8 – 12 cu. yd./hour
2.4	Backfill Trench	A & B	Backhoe & Compactors	10 – 14 cu. yd./hour
2.5	Backfill Pit	A & B	Backhoe & Compactors	6 – 8 cu. yd./hour
3.0	**Concrete Work**			
	Forming:			
3.1	Continuous Strip Footings	C	Hand Tools	30 – 50 sq. ft./hour
3.2	2x4 Keyways	C	Hand Tools	150 – 190 feet/hour
3.3	Isolated Footings and Pile Caps	C	Hand Tools	32 – 43 sq. ft./hour
3.4	Foundation and Retaining Walls	C	Hand Tools	27 – 35 sq. ft./hour
3.5	Columns — Rectangular	C	Hand Tools	55 – 80 sq. ft./hour
3.6	10" Dia. Columns	C	Hand Tools	12 – 15 feet/hour
3.7	Edges of Slab-on-Grade	C	Hand Tools	27 – 33 sq. ft./hour
3.8	Edges and Risers of Stairs	C	Hand Tools	20 – 27 sq. ft./hour
3.9	Edges of Sidewalks	C	Hand Tools	27 – 33 sq. ft./hour
3.10	Stripping Forms	A	Hand Tools	40 – 155 sq. ft./hour
	Placing concrete in:			
3.11	Continuous Strip Footings	D	Concrete Pump	5 – 10 cu. yd./hour
3.12	Isolated Footings and Pile Caps	D	Concrete Pump	3 – 6 cu. yd./hour
3.13	Foundation Walls	D	Concrete Pump	5 – 7 cu. yd./hour
3.14	Slab-on-Grade	D	Concrete Pump	6 – 11 cu. yd./hour
3.15	Slab Topping — Separate	D	Concrete Pump	2 – 4 cu. yd./hour
3.16	Sidewalks	D	Concrete Pump	8 – 12 cu. yd./hour
	Placing rebar in:			
3.17	Footings	A	Hand Tools	390 – 400 lbs./hour
3.18	Walls	A	Hand Tools	560 – 570 lbs./hour
3.19	Columns	A	Hand Tools	280 – 290 lbs./hour
3.20	Placing Mesh in Slabs	A	Hand Tools	400 – 500 sq. ft./hour
3.21	Finishing Slabs	E	Hand Tools	60 – 80 sq. ft./hour
3.22	Install Anchor Bolts — 1/2 to 3/4"	F	Hand Tools	40 – 80 no./hour
3.23	Rigid Insulation to Walls — 1"	C	Hand Tools	200 – 250 sq. ft./hour
3.24	— 2"	C	Hand Tools	180 – 225 sq. ft./hour
3.25	6mil Polyethylene Vapor Barrier	C	Hand Tools	900 – 1000 sq. ft./hour
4.0	**Masonry Work**			
4.1	8" Concrete Blocks	G	Hand Tools	22 – 50 Blks/hour
4.2	Brick Facings	G	Hand Tools	90 – 110 Brks/hour
4.3	Fireplace — Single Story	G	Hand Tools	40 – 60 hours

Figure 13.3a   Renovation Work Productivities

PRODUCTIVITIES

CREWS	CREW MEMBERS	CREWS	CREW MEMBERS
CREW A	1.0 Foreman 2.0 Laborers	CREW E	1.0 Cement Finisher
CREW B	1.0 Equipment Operator	CREW F	1.0 Carpenter
CREW C	1.0 Oarp. Foreman 2.0 Carpenters	CREW G	1.0 Foreman 2.0 Bricklayers
CREW D	1.0 Foreman 2.0 Laborers 1.0 Cement Finisher	CREW H	1.0 Foreman 2.0 Painters

ITEM	OPERATION	CREW	EQUIP.	OUTPUT
6.0	**Carpentry**			
	Rough Carpentry			
6.1	3" Diameter Telescopic Post	C	Hand Tools	1.50 – 2.50 No./hour
6.2	2 x 3 Plates	C	Hand Tools	50 – 54 BF/hour
6.3	2 x 4 Plates	C	Hand Tools	63 – 68 BF/hour
6.4	2 x 5 Plates	C	Hand Tools	90 – 95 BF/hour
6.5	2 x 3 Studs	C	Hand Tools	79 – 92 BF/hour
6.6	2 x 4 Studs	C	Hand Tools	100 – 115 BF/hour
6.7	2 x 6 Studs	C	Hand Tools	110 – 125 BF/hour
6.8	2x8 in Built-Up Beam I	C	Hand Tools	150 – 260 BF/hour
6.9	2x10 in Built-Up Beam	C	Hand Tools	200 – 310 BF/hour
6.10	2x8 Joists	C	Hand Tools	150 – 180 BF/hour
6.11	2x10 Joists	C	Hand Tools	160 – 185 BF/hour
6.12	2x4 Blocking	C	Hand Tools	20 – 35 BF/hour
6.13	Trusses 24'-40' Span	C	Hand Tools	2 – 3 No./hour
6.14	2x2 Cross-Bridging	C	Hand Tools	30 – 34 Sets/hour
6.15	1x3 Rafter Ties (Ribbons)	C	Hand Tools	25 – 35 BF/hour
6.16	2x4 Rafters	C	Hand Tools	60 – 110 BF/hour
6.17	2x6 Rafters	C	Hand Tools	65 – 120 BF/hour
6.18	2x4 Lookouts	C	Hand Tools	30 – 50 BF/hour
6.19	2x4 Rough Fascia	C	Hand Tools	30 – 50 BF/hour
6.20	2x6 Rough Fascia	C	Hand Tools	35 – 60 BF/hour
9.21	T & G Floor Sheathing	C	Hand Tools	150 – 160 sq. ft./hour
6.22	Wall Sheathing	C	Hand Tools	130 – 150 sq. ft./hour
6.23	Roof Sheathing	C	Hand Tools	160 – 180 sq. ft./hour
6.24	½" Ply Soffit	C	Hand Tools	40 – 60 sq. ft./hour
6.25	Aluminum Soffit	C	Hand Tools	40 – 60 sq. ft./hour
6.26	1x6 Fascia Board	C	Hand Tools	15 – 20 BF/hour
6.27	6" Aluminum Fascia	C	Hand Tools	30 – 40 ft./hour
	Finish Carpentry			
6.28	3'-0" Wide Stair with 13 Risers	C	Hand Tools	0.4 – 0.6 units/hour
6.29	2x2 Handrail	C	Hand Tools	12 – 20 feet/hour
6.30	Wood Railings	C	Hand Tools	8 – 12 feet/hour
	Exterior Doors with Frame and Trim:			
6.31	— 2'-8" x 6'8" x 1³/₄"	C	Hand Tools	0.75 – 1.25 units/hour
6.32	— 3'-0" x 6'8" x 1³/₄"	C	Hand Tools	0.65 – 1.00 units/hour
	Interior Doors with Frame and Trim:			
6.33	— 1'-6" x 6'8" x 1³/₈"	C	Hand Tools	0.90 – 1.35 units/hour
6.34	— 2'-0" x 6'8" x 1³/₈"	C	Hand Tools	0.85 – 1.30 units/hour
6.35	— 2'-6" x 6'8" x 1³/₈"	C	Hand Tools	0.80 – 1.25 units/hour
6.36	— 2'-8" x 6'8" x 1³/₈"	C	Hand Tools	0.75 – 1.20 units/hour
	Bifold Closet Doors with Jambs and Trim:			
6.37	— 2'-0" x 6'8" x 1³/₈"	C	Hand Tools	0.85 – 1.00 units/hour
6.38	— 3'-0" x 6'8" x 1³/₈"	C	Hand Tools	0.80 – 0.85 units/hour
6.39	— 4'-0" x 6'8" x 1³/₈"	C	Hand Tools	0.70 – 0.90 units/hour
6.40	— 5'-0" x 6'8" x 1³/₈"	C	Hand Tools	0.65 – 0.85 units/hour
6.41	Attic Access Hatch	C	Hand Tools	1.00 – 2.00 units/hour

Figure 13.3b   Renovation Work Productivities (Continued)

PRODUCTIVITIES

CREWS	CREW MEMBERS		CREWS	CREW MEMBERS
CREW A	1.0 Foreman 2.0 Laborers		CREW E	1.0 Cement Finisher
			CREW F	1.0 Carpenter
CREW B	1.0 Equipment Operator			
			CREW G	1.0 Foreman 2.0 Bricklayers
CREW C	1.0 Oarp. Foreman 2.0 Carpenters			
			CREW H	1.0 Foreman 2.0 Painters
CREW D	1.0 Foreman 2.0 Laborers 1.0 Cement Finisher			

ITEM	OPERATION	CREW	EQUIP.	OUTPUT
6.0	**Carpentry**			
	Finish Carpentry (Continued)			
	Finish Hardware to Doors:			
6.42	— Passage Sets	F	Hand Tools	1.0 – 1.5 No./hour
6.43	— Privacy Sets	F	Hand Tools	1.0 – 1.5 No./hour
6.44	— Dead Bolts	F	Hand Tools	0.8 – 1.3 No./hour
6.45	— Key-in-Knob Lock/Latch	F	Hand Tools	0.5 – 1.5 No./hour
	Window Trim			
	Windows: Installing Manufactured Units			
6.46	— 3'-0" x 1'-0"	C	Hand Tools	1.5 – 3.2 units/hour
6.47	— 2'-0" x 3'-6"	C	Hand Tools	1.0 – 3.0 units/hour
6.48	— 4'-0" x 3'-6"	C	Hand Tools	1.0 – 2.8 units/hour
6.49	— 6'-0" x 3'-6"	C	Hand Tools	1.0 – 2.5 units/hour
6.50	— 4'-0" x 4'-0"	C	Hand Tools	0.9 – 2.5 units/hour
6.51	— 4'-0" x 6'-0"	C	Hand Tools	0.9 – 2.3 units/hour
6.52	— 4'-0" x 8'-0"	C	Hand Tools	0.9 – 2.2 units/hour
6.53	Base Board	F	Hand Tools	25.0 – 30.0 feet/hour
6.54	Floor Cabinets Not Including Countertop	C	Hand Tools	6.0 – 8.0 feet/hour
6.55	1'-0" x 2'-6" Wall Cabinets	C	Hand Tools	7.0 – 9.0 feet/hour
6.56	1'-0" x 1'-10" Wall Cabinets	C	Hand Tools	7.5 – 9.5 feet/hour
6.57	2'-0" Wide Countertop	C	Hand Tools	6.0 – 8.0 feet/hour
6.58	2'-0" x 2'-6" Bathroom Vanities	C	Hand Tools	2.0 – 5.5 feet/hour
6.59	12" Wide Closet Shelves	F	Hand Tools	8.0 – 14.0 feet/hour
6.60	16" Wide Closet Shelves	F	Hand Tools	8.0 – 13.0 feet/hour
6.61	Adjustable Closet Rods	F	Hand Tools	2.0 – 3.0 No./hour
	Bathroom Accessories:			
6.62	— Toilet Roll Holders	F	Hand Tools	3.0 – 4.0 No./hour
6.63	— 2'-0" x 3'-0" Mirrors	F	Hand Tools	1.5 – 2.0 No./hour
6.64	— Medicine Cabinets	F	Hand Tools	1.0 – 1.5 No./hour
6.65	— Shower Curtain Rod	F	Hand Tools	2.0 – 3.0 No./hour
7.0	**Exterior Finishes**			
	Sprayed Damp Proofing:			
7.1	— 1 Coat	A	Paint Spray	200 – 220 sq. ft./hour
7.2	— 2 Coats	A	Paint Spray	120 – 140 sq. ft./hour
7.3	1/2" Parging	E	Hand Tools	9 – 11 sq. ft./hour
7.4	Aluminum Siding	C	Hand Tools	60 – 65 sq. ft./hour
7.5	Stucco	E	Hand Tools	10 – 12 sq. ft./hour
7.6	Building Paper	C	Hand Tools	850 – 900 sq. ft./hour
7.7	4" Wide Flashing	C	Hand Tools	90 – 130 sq. ft./hour
7.8	5" Eaves Gutter	C	Hand Tools	20 – 30 sq. ft./hour
7.9	3" Downspouts	C	Hand Tools	30 – 48 sq. ft./hour
7.10	210# Asphalt Shingles	C	Hand Tools	125 – 138 sq. ft./hour
7.11	Ridge Cap	C	Hand Tools	90 – 100 sq. ft./hour
7.12	Drip-Edge Flashing	C	Hand Tools	90 – 100 sq. ft./hour
7.13	6mil Poly Eaves Protection	C	Hand Tools	800 – 900 sq. ft./hour

Figure 13.3c Renovation Work Productivities (Continued)

PRODUCTIVITIES

CREWS	CREW MEMBERS	CREWS	CREW MEMBERS
CREW A	1.0 Foreman 2.0 Laborers	CREW E	1.0 Cement Finisher
CREW B	1.0 Equipment Operator	CREW F	1.0 Carpenter
CREW C	1.0 Oarp. Foreman 2.0 Carpenters	CREW G	1.0 Foreman 2.0 Bricklayers
CREW D	1.0 Foreman 2.0 Laborers 1.0 Cement Finisher	CREW H	1.0 Foreman 2.0 Painters

ITEM	OPERATION	CREW	EQUIP.	OUTPUT
7.0	**Exterior Finishes (Continued)**			
	Precast Concrete Steps:			
7.14	— 2 Risers	A	Hand Tools	0.4 – 0.6 No./hour
7.15	— 4 Risers	A	Hand Tools	0.3 – 0.5 No./hour
7.16	Wrought-Iron Railing	C	Hand Tools	2.0 – 4.0 feet/hour
7.17	Paint Iron Railing	H	Hand Tools	10.0 – 14.0 feet/hour
8.0	**Interior Finishes**			
8.1	32 oz. Carpet	C	Hand Tools	180 – 230 sq. ft./hour
8.2	Vinyl Flooring	C	Hand Tools	100 – 180 sq. ft./hour
8.3	1/2" Drywall Ceilings	C	Hand Tools	90 – 95 sq. ft./hour
8.4	Textured Ceiling Finish	H	Hand Tools	180 – 200 sq. ft./hour
8.5	R35 Loose Cellulose Insulation	A	Hand Tools	150 – 160 sq. ft./hour
8.6	6 mil. Poly Vapor Barrier	C	Hand Tools	850 – 900 sq. ft./hour
8.7	Insulation Stops	C	Hand Tools	20 – 30 sq. ft./hour
8.8	R20 Batt Insulation	C	Hand Tools	260 – 340 sq. ft./hour
8.9	R12 Batt Insulation	C	Hand Tools	300 – 400 sq. ft./hour
8.10	1/2" Drywall Walls	C	Hand Tools	110 – 120 sq. ft./hour
8.11	Spray Paint Walls and Ceilings	H	Hand Tools	375 – 425 sq. ft./hour
8.12	Roller Paint Walls and Ceilings	H	Hand Tools	150 – 300 sq. ft./hour
8.13	Paint Exterior Doors	H	Hand Tools	40 – 45 sq. ft./hour
8.14	Stain Doors	H	Hand Tools	50 – 55 sq. ft./hour
8.15	Paint Attic Access Hatch	H	Hand Tools	3 – 4 No./hour
8.16	Paint Handrail	H	Hand Tools	90 – 95 feet/hour
8.17	Paint Wood Railing	H	Hand Tools	12 – 14 feet/hour
8.18	Paint Shelves	H	Hand Tools	60 – 64 sq. ft./hour

Figure 13.3d Renovation Work Productivities (Continued)

Pricing General Expenses

The general expenses on a remodeling project will be similar to those on a housing job. The estimator should use the General Expense Sheet as a checklist and tick off the items required for the job as demonstrated in the example below, Figure 13.8.

The Summary and Bid

The estimate is summarized and the bid prepared in accordance with the principles discussed in Chapter 12. See Figure 13.9 for an example.

Example of a Remodeling Estimate

The example estimate involves the remodeling of a kitchen. Assume that the estimator has visited the site, and Figure 13.4 is a sketch of the kitchen layout as it exists before the work begins. Figure 13.5 shows the proposed layout of the new kitchen together with some notes on requirements obtained from the owner. The work comprises the following tasks:

- Remove the wall on two sides of the kitchen.
- Remove old cabinets and install new kitchen cabinets including a new island cabinet.
- Replace the vinyl tile floor in the kitchen and also in the entrance hallway.
- Install a new 4'-0" × 3'-0" window to the outside wall of the kitchen.
- The owner is going to keep the existing appliances but temporarily remove them to the adjacent garage for the duration of the work.
- Repaint all walls and ceilings on the whole of the main floor of the house.

The estimate begins with the takeoff of the work shown on Figure 13.6a and Figure 13.6b.

Comments on Remodeling Takeoff Shown in Figure 13.6a and Figure 13.6b

1. Moving appliances are enumerated.
2. Note that part of the electrical work is adding a duplex outlet in the new location of the refrigerator.
3. Cabinets to be removed and new cabinets are enumerated, but the counter-top is measured in linear feet.
4. The area of stud wall to be removed is measured over the two openings in the wall (no deductions).
5. The wall did not pass through the ceiling, but there will still be patching required where the wall was attached to the ceiling and to the walls.
6. A double 2 × 8 lintel has been allowed over the opening for the new window.
7. If the wall is neatly cut out, there will be little patching and repairs required.
8. An aluminum J-mold trim has been allowed around the window on the outside.
9. A quick method of calculating the interior wall areas is used. Wall areas are usually roughly three times the floor area. This approximation is close enough for assessing the painting area.
10. A subcontractor will do the electrical work on the project.

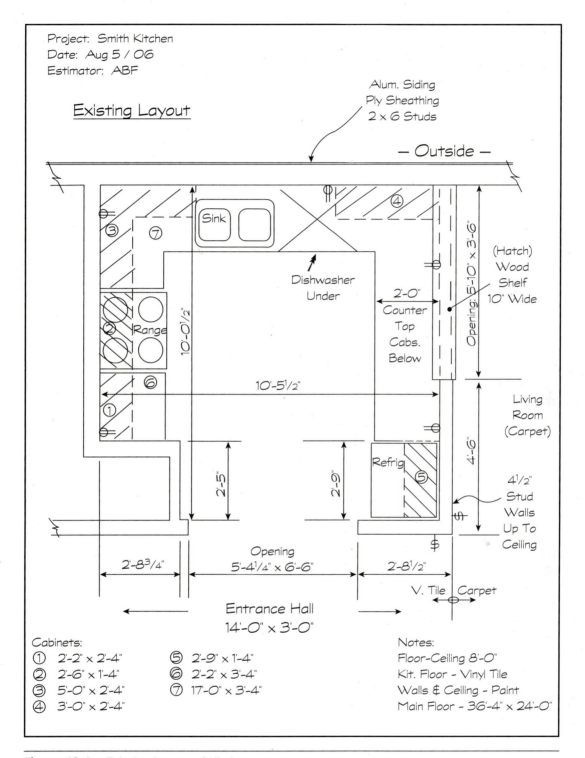

Project: Smith Kitchen
Date: Aug 5 / 06
Estimator: ABF

Existing Layout

Alum. Siding
Ply Sheathing
2 x 6 Studs

— Outside —

Sink

④

(Hatch)
Wood
Shelf
10" Wide

③   ⑦

Dishwasher
Under

Opening: 5'-10" x 3'-6"

2'-0"
Counter
Top
Cabs.
Below

10'-0½"

② Range

Living
Room
(Carpet)

⑥

10'-5½"

①

4'-6"

4½"
Stud
Walls
Up To
Ceiling

2'-5"

2'-9"

Refrig

⑤

V. Tile | Carpet

2'-8¾"

Opening
5'-4¼" x 6'-6"

2'-8½"

Entrance Hall
14'-0" x 3'-0"

Cabinets:
① 2'-2" x 2'-4"   ⑤ 2'-9" x 1'-4"
② 2'-6" x 1'-4"   ⑥ 2'-2" x 3'-4"
③ 5'-0" x 2'-4"   ⑦ 17'-0" x 3'-4"
④ 3'-0" x 2'-4"

Notes:
Floor-Ceiling 8'-0"
Kit. Floor - Vinyl Tile
Walls & Ceiling - Paint
Main Floor - 36'-4" x 24'-0"

Figure 13.4   Existing Layout of Kitchen

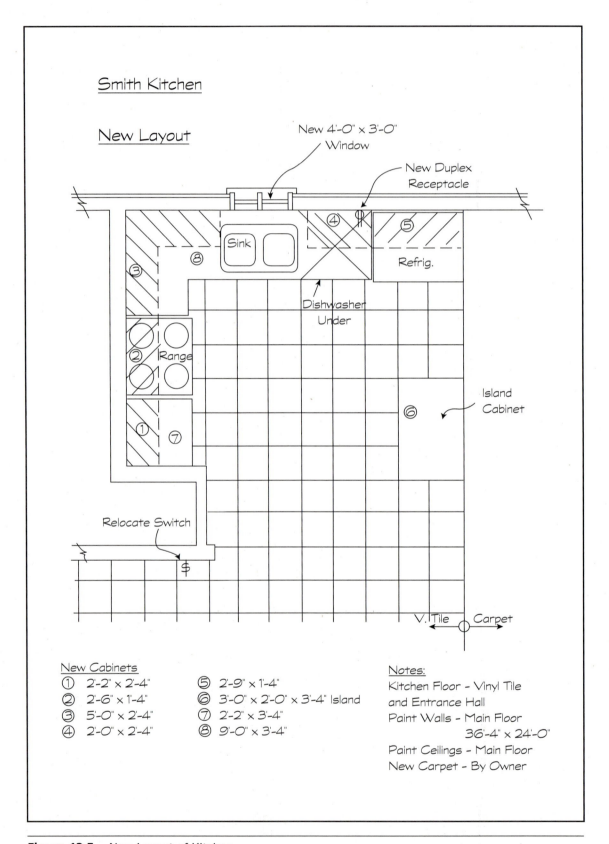

Figure 13.5 New Layout of Kitchen

QUANTITY SHEET   SHEET No.   1 of 2

JOB: Kitchen Remodeling Example   DATE: Aug 10 / 06

ESTIMATOR ABF   EXTENDED: EXT. CHKD:

DESCRIPTION	TIMES	Length	Width	Height			
			DIMENSIONS				
TEMP. RELOCATE AND							
RESTORE AT COMPLETION:							
RANGE		-	-	-	1	No.	
DISHWASHER		-	-	-	1	No.	
REFRIG.		-	-	-	1	No.	
REMOVE CABINETS:							
1-0 x 2-4 WALL		2.17	-	-	2		
1-0 x 1-4 WALL		2.50	-	-	3		
1-0 x 2-4 WALL		5.00	-	-	5		
1-0 x 2-4 WALL		3.00	-	-	3		
1-0 x 1-4 WALL		2.90	-	-	3		
2-0 x 3-4 BASE		2.00	-	-	2		
2-0 x 3-4 BASE		17.00	-	-	17		
					35	LF	
REMOVE AND REPLACE SINK		-	-	-	1	No.	
REMOVE COUNTERTOP		19.17	-	-	19	LF	
NEW CABINETS:							
1-0 x 2-4 WALL		2.17	-	-	2		
		5.00	-	-	5		
		2.00	-	-	2		
					9	LF	
1-0 x 1-4 WALL		2.50	-	-	3		
		2.75	-	-	3		
					5	LF	
2-0 x 3-4 BASE		3.00	-	-	3		
		2.17	-	-	2		
		9.00	-	-	9		
					14	LF	
2-0 COUNTERTOP		DITTO	-	-	14	LF	
REMOVE 4½" STUD WALL		18.41	-	8.00	147	SF	

5.83
4.50
2.75
5.33
18.41

Figure 13.6a   Remodeling Takeoff

QUANTITY SHEET   SHEET No. | 2 of 2

JOB: _____ Kitchen Remodeling Example _____   DATE: Aug 10 / 06

ESTIMATOR: **ABF** _____   EXTENDED: _____   EXT. CHKD: _____

DESCRIPTION	DIMENSIONS					
	TIMES	Length	Width	Height		
PATCH CEILING 4$^1/_2$" WIDE		18.41	—	—	18	LF
PATCH WALL 4$^1/_2$" WIDE		—	—	8.00		
		—	—	8.00		
					16	LF
REMOVE VINYL TILE						
FLOORING		10.42	10.42		109	
10.04 DDT		2.71	2.42		(7)	
0.38 (Entrance Hall)		14.00	30.00		420	
10.42					522	SF
NEW VINYL TILE FLOORING		DITTO			522	SF
CUT OPENING 4-0 x 3-0 IN		—	—	—	1	No.
EXTERIOR 2x6 STUD WALL						
with ALUM SIDING						
2 x 8 LINTEL	2	4.25			9	LF
PATCH DRYWALL AROUND		—	—	—	1	No.
4-0 x 3-0 OPENING						
NEW WINDOW 4-0 x 3-0		—	—	—	1	No.
ALUM. J - MOLD	4	4.00			16	LF
PAINT CEILINGS		36.33	24.00		872	SF
(Whole Main Floor)						
PAINT WALLS	3	36.33	24.00		2,616	SF
(Whole Main Floor)						
ELECTRICAL WORK		—	—	—	SUB	

Figure 13.6b   Remodeling Takeoff (Continued)

Example of Remodeling Recap

Figure 13.7 shows the priced recap for the example.

Comments on Remodeling Recap Shown in Figure 13.7

1. On this small project, the work is listed in just two categories: demolition work and new work, rather than listing each trade separately.
2. Relocating the range and refrigerator consists of just unplugging these appliances and moving them. Two laborers spending less than half an hour on each should be sufficient time for this.
3. The dishwasher and the sink require plumbing work, which will take extra time to deal with.
4. Again, because this is such a small project, most of the productivity rates used are from the low end of the scale.
5. The productivity removing the cabinets for crew A is 10 linear feet per hour, so the cost is calculated in this fashion:

   Crew A: 1 Labor foreman @ $26.00 = $26.00
             2 Laborers @ $23.00   = $46.00
                                         $72.00 per hour

   Per linear foot   = $\frac{\$72.00}{10}$

                                       = $7.20 per foot

6. The remaining demolition items use the same crew A with the following productivities:
   a. Remove countertop   -   25 linear feet/hour
   b. Remove 4½" stud wall   -   65 square feet/hour
   c. Remove vinyl tile   -   80 square feet/hour
   d. Cut opening 4'-0" × 3'-0" -   12 square feet/hour
7. Installing the 2 × 8 lintel is based on a productivity of 20 linear feet per hour for crew C:

   Crew C: 1 Carpenter foreman @ $35.00 = $35.00
             2 Carpenters @ $30.00   = $60.00
                                             $95.00 per hour

   Per linear foot   = $\frac{\$95.00}{20}$

                                           = $4.75 per foot

8. The following new items use the same crew C with the following productivities:
   a. 4'-0" × 3'-0" Window   -   1 unit/hour
   b. Aluminum "J" Mold   -   50 linear feet/hour
   c. Cabinets: 1'-0" × 2'-4"   -   8.5 linear feet/hour
                1'-0" × 1'-4"   -   8.5 linear feet/hour
                2'-0" × 3'-4"   -   7.6 linear feet/hour
   d. Countertop   -   7.6 linear feet/hour
9. Patching the wall is based on a productivity of 50 linear feet per hour for crew A:

   Crew A: 1 Labor foreman @ $26.00 = $26.00
             2 Laborers @ $23.00   = $46.00
                                         $72.00 per hour

   Per linear foot   = $\frac{\$72.00}{50}$

                                       = $1.44 per foot

PRICING SHEET

JOB.....*Kitchen Remodeling Example* .... DATE.................................…….

ESTIMATED.........*ABF.* ....

No.	DESCRIPTION	QUANTITY	UNIT	UNIT PRICE	LABOR $	UNIT PRICE	MATERIALS $	UNIT PRICE	EQUIP. $	SUBS. $	TOTAL $
	Demolition										
1	TEMP. RELOCATE AND										
	RESTORE AT COMPLETION:										
	RANGE	1	No.	$18.00	18	---	---	---	---	---	18
2	DISHWASHER	1	No.	$24.00	24	---	---	---	---	---	24
3	REFRIG.	1	No.	$18.00	18	---	---	---	---	---	18
4	REMOVE AND RESTORE SINK	1	No.	$80.00	80	---	---	---	---	---	80
5	REMOVE CABINETS	35	LF	$7.20	252	---	---	---	---	---	252
6	REMOVE COUNTERTOP	19	LF	$2.88	55	---	---	---	---	---	55
7	REMOVE 4½" STUD WALL	147	SF	$1.11	163	---	---	---	---	---	163
8	REMOVE VINYL TILE										
	FLOORING	522	SF	$0.90	470	---	---	---	---	---	470
9	CUT OPENING 4-0 x 3-0 IN										
	EXTERIOR 2x6 STUD WALL with										
	ALUM SIDING	1	No.	$72.00	72	---	---	---	---	---	72
	TOTAL DEMOLITION:				1,152	---	---	---	---	---	1,152
	New Work										
10	2 x 8 LINTEL	9	LF	$4.75	43	$1.10	10	---	---	---	53
11	4-0 x 3-0 WINDOW	1	No.	$95.00	95	$450.00	450	---	---	---	545
12	ALUM. J - MOLD	16	LF	$1.90	30	$2.75	44	---	---	---	74
13	CABINETS:										
14	1-0 x 2-4 WALL	9	LF	$11.18	101	$85.00	765	---	---	---	866
15	1-0 x 1-4 WALL	5	LF	$11.18	56	$70.00	350	---	---	---	406
16	2-0 x 3-4 BASE	14	LF	$12.50	175	$120.00	1,680	---	---	---	1,855
17	COUNTERTOP	14	LF	$12.50	175	$35.00	490	---	---	---	665
18	PATCH WALL 4½" WIDE	16	LF	$1.44	23	$0.55	9	---	---	---	32
19	PATCH CEILING 4½" WIDE	18	LF	$1.80	32	$0.55	10	---	---	---	42
20	PATCH DRYWALL AROUND										
	4-0 X 3-0 OPENING	1	No.	$25.00	25	---	---	---	---	---	25
21	VINYL TILE FLOORING	522	SF	$0.72	376	$3.60	1,879	---	---	---	2,255
22	PAINT CEILINGS	872	SF	$0.48	419	$0.10	87	---	---	---	506
23	PAINT WALLS	2616	SF	$0.48	1,256	$0.10	262	---	---	---	1,517
	TOTAL NEW WORK:				2,805		6,036		0	0	8,841

Figure 13.7   Remodeling Pricing Sheet

10. The following items of new work use the same crew A with the following productivities:
    a. Patch ceiling   40 linear feet/hour
    b. Vinyl tile   100 square feet/hour
    c. Paint ceilings   150 square feet/hour
    d. Paint walls   150 square feet/hour
11. Little patching around the window opening will be necessary if the opening is cut neatly. The $25.00 amount is included as a contingency amount; it is available should it be required.
12. Material prices are based on quotes from suppliers for the products specified by the owner. It is important to identify the particular type of new window, cabinets, paint, and vinyl tiles selected by the owner and used as the basis of the estimate.
13. There are no equipment costs included in the estimate; all work is to be completed using hand tools that are priced on the summary sheet under a *small tools* allowance.

Example of Remodeling General Expenses

Figure 13.8 shows the priced general expenses for the example.

Comments on Remodeling General Expenses Shown in Figure 13.8

1. This is a very small project, so there are few items of general expenses to consider here.
2. There is nothing here for a superintendent or foreman; supervision costs are allowed for in the direct labor prices for the work.
3. An allowance is made for first aid supplies consumed on this two-month project.
4. Telephone expenses are expected to be $65.00 per month.
5. The cost of a temporary dust partition has been included to isolate the work area from the rest of the house while the demolition work is in progress.
6. Final cleanup costs are estimated to be $0.15 per square foot of the main floor.

Example of Remodeling Summary

Figure 13.9 shows the summary of prices for the example.

Comments on Remodeling Summary Shown in Figure 13.9

1. The price of the builder's own work and subtrade work is listed separately here, together with the general expense prices.
2. "Small tools" is an allowance for the cost of hand tools used in the work. This is based on the labor amount at the rate of 6 percent.
3. "Payroll Additive" is an allowance for the payroll costs paid by the employer; it includes items such as *Social Security Tax* and *Unemployment Compensation Tax*. It is also calculated as a percentage of the labor amount.
4. A building permit is required for the job at the rate of $7.00 per $1,000 of the job cost.
5. Insurance for the work includes property insurance and liability insurance; the price of insurance coverage in this example is 5 percent of the job cost.
6. The fee is the builder's markup on the job.
7. A small adjustment is made to the final price to make it an even $15,900.
8. The final amount, $15,900, can now be entered onto the bid forms for the project.

PROJECT: *Kitchen Remodeling Example*

LOCATION: *Townsville*

DATE:

ESTIMATOR: *ABF*

GENERAL EXPENSES

	Quantity Unit	Unit Rate Labor	Labor $	Unit Rate Material	Material $	Unit Rate Equip.	Equip. $	Subtrade $	Total $
Schedule	---	---	---	---	---	---	---	---	---
Project Superintendent	---	---	---	---	---	---	---	---	---
Assistant Superintendent	---	---	---	---	---	---	---	---	---
Stakeout	---	---	---	---	---	---	---	---	---
Survey and Plot Plan	---	---	---	---	---	---	---	---	---
Safety and First Aid	2 Mo.	---	---	20.00	40.00	---	---	---	40
Rentals:– office trailer	---	---	---	---	---	---	---	---	---
– office equipment	---	---	---	---	---	---	---	---	---
– office supplies	---	---	---	---	---	---	---	---	---
– storage	---	---	---	---	---	---	---	---	---
– tool lockups	---	---	---	---	---	---	---	---	---
– toilets	---	---	---	---	---	---	---	---	---
Permanent Connections:	---	---	---	---	---	---	---	---	---
– Sewer Connection	---	---	---	---	---	---	---	---	---
– Underground Wiring	---	---	---	---	---	---	---	---	---
Municipal Charges	---	---	---	---	---	---	---	---	---
Temporary Site Services:	---	---	---	---	---	---	---	---	---
– Water services	---	---	---	---	---	---	---	---	---
– Electrical services	---	---	---	---	---	---	---	---	---
– Telephone	2 Mo.	---	---	65.00	130.00	---	---	---	130
Scaffolds	---	---	---	---	---	---	---	---	---
Hoardings	192 sf	0.30	57.60	0.25	48.00	---	---	---	106
Temporary Heating	---	---	---	---	---	---	---	---	---
Temporary Fire Protection	---	---	---	---	---	---	---	---	---
Site Access	---	---	---	---	---	---	---	---	---
Site Security	---	---	---	---	---	---	---	---	---
Equipment Rentals	---	---	---	---	---	---	---	---	---
Truck Rentals	---	---	---	---	---	---	---	---	---
Dewatering Excavations	---	---	---	---	---	---	---	---	---
Cleanup:	---	---	---	---	---	---	---	---	---
– General site cleanup	---	---	---	---	---	---	---	---	---
– Final site cleanup	872 sf	0.15	130.80	---	---	---	---	---	131
– Cleaning the furnace	---	---	---	---	---	---	---	---	---
– Cleaning windows	---	---	---	---	---	---	---	---	---
– Move-in cleanup	---	---	---	---	---	---	---	---	---
Snow Removal	---	---	---	---	---	---	---	---	---
Photographs	---	---	---	---	---	---	---	---	---
Project Signs	---	---	---	---	---	---	---	---	---
Soils Testing	---	---	---	---	---	---	---	---	---
New Home Warranty	---	---	---	---	---	---	---	---	---
TOTALS TO SUMMARY:			188.40		218.00		0.00	0.00	406

Figure 13.8 Remodeling Example of General Expenses

PROJECT: Kitchen Remodeling Example
LOCATION: Townsville
DATE:
ESTIMATOR: ABF

ESTIMATE SUMMARY

DESCRIPTION	LABOR $	MATL. $	EQIUP. $	SUBS. $	OTHER $	TOTAL $
OWN WORK:						
DEMOLITION WORK	1,152	---	---	---	---	1,152
NEW WORK	2,805	6,036	---	---	---	8,841
Subtotal 1:	3,957	6,036	---	---	---	9,993
SUBCONTRACTORS						
DEMOLITION	---	---	---	---	---	---
LANDSCAPING	---	---	---	---	---	---
PILING	---	---	---	---	---	---
REINFORCING STEEL	---	---	---	---	---	---
PRECAST	---	---	---	---	---	---
MASONRY	---	---	---	---	---	---
STRUCTURAL AND MISC. STEEL	---	---	---	---	---	---
CARPENTRY	---	---	---	---	---	---
MILLWORK	---	---	---	---	---	---
ROOFING	---	---	---	---	---	---
CAULKING AND DAMP PROOFING	---	---	---	---	---	---
DOORS AND FRAMES	---	---	---	---	---	---
WINDOWS	---	---	---	---	---	---
RESILIENT FLOORING	---	---	---	---	---	---
CARPET	---	---	---	---	---	---
DRYWALL	---	---	---	---	---	---
ACOUSTIC CEILING	---	---	---	---	---	---
PAINTING	---	---	---	---	---	---
SPECIALTIES	---	---	---	---	---	---
PLUMBING	---	---	---	---	---	---
HEATING AND VENTILATING	---	---	---	---	---	---
ELECTRICAL	---	---	---	250	---	250
CASH ALLOWANCES	---	---	---	---	---	---
Subtotal 2:	3,957	6,036	0	250	0	10,243
GENERAL EXPENSES	188	218	---	---	---	406
Subtotal 3:	4,145	6,254	0	250	0	10,649
ADD-ONS						
SMALL TOOLS: 4,145 x 6%	---	---	249	---	---	249
PAYROLL ADDITIVE: 4,145 x 25%	1,036	---	---	---	---	1,036
BUILDING PERMIT 15,900 x $7 / $1,000	---	---	---	---	111	111
PERFORMANCE BOND	---	---	---	---	---	---
INSURANCE 15,900x 5%	---	---	---	---	795	795
FEE	---	---	---	---	3,000	3,000
ADJUSTMENT	---	---	---	---	60	60
				BID TOTAL:		15,900

Figure 13.9 Remodeling Example of Bid Summary

SUMMARY

- The remodeling contractor has to deal more closely with the customer than the average homebuilder.
- There are usually no architects or designers on remodeling jobs, so drawings and specifications are rarely available for this work.
- Remodeling involves structures that are in place, so they can be physically measured in a takeoff. However, items that are concealed will be difficult to assess.
- There are three main types of contract used in remodeling:
  - Lump sum—these contracts are usually awarded on the basis of low bid price, so an accurate estimate is required.
  - Cost plus—the builder is paid all costs plus a fee. An estimate may be required to determine a maximum price or provide a target budget.
  - Unit price—used when the amount of work is difficult to predict on a project. The builder quotes a series of unit prices that are used to price work once it is completed.
- There are three key steps to preparing an accurate estimate for remodeling work:
  - Clearly identify all the work to be done
  - Measure the work to be done
  - Price the work
- It can be difficult with remodeling jobs to define the scope of work; good communication with the owner and the use of outline sketches will help.
- The estimator needs a thorough knowledge of construction methods to prepare an estimate for remodeling projects.
- Remodeling work can involve all the same trades as homebuilding plus demolition.
- Demolition work is generally measured in the same way and using the same units as new work is measured.
- Sketches and specification notes should be prepared to indicate what has been allowed for in the takeoff.
- Much of the takeoff can be compiled on site by measuring the existing structures.
- An orderly takeoff sequence should be followed to ensure that all work items are included in the takeoff.
- The estimate is usually recapped and priced following a trade breakdown.
- Job factors that should be taken into account when pricing the work include:
  - Weather conditions
  - Access to and around the work area
  - Size of the project
  - Complexity of the tasks involved
  - Location of the job
- Labor and material factors to account for include:
  - Quality of supervision
  - Motivation and morale of the workers
  - Good tools and equipment
  - Experience of supervisors and workers
- The best sources of labor and equipment productivities are records from past projects.
- Pricing materials, general expenses, and summaries on remodeling jobs follow the principles discussed in previous chapters.
- A complete estimate of a remodeling project is demonstrated.

RECOMMENDED RESOURCES

Information	Web Page Address
■ There is a great deal of information on the web for remodeling contractors. **Remodeling Online** is one example.	http://www.remodeling.hw.net/
■ Remodelers have their own association: **The National Association of the Remodeling Industry**	http://www.nari.org/
■ Information about electronic tape measures can be obtained by entering the key words **electronic tape measures** into your Internet search engine.	
■ Information about laser tape measures can be obtained by entering the key words **laser tape measures** into your Internet search engine.	

REVIEW QUESTIONS

1. How does the remodeling business differ from the housing business?
2. What is the most common type of contract used on remodeling projects, and what kind of an estimate is required for this type of contract?
3. Under what circumstances would an owner choose a unit price contract to use on their remodeling project?
4. What are the three key steps in the process of estimating a remodeling project?
5. What should the estimator have at the end of the first step in the estimating process?
6. If a builder, as part of a renovation project, had to remove a basement consisting of a concrete foundation wall on a concrete footing and a concrete slab-on-grade, how should this work be measured?
7. When can an estimator make use of sketches on a remodeling project?
8. Describe some recent innovations that help when measuring work on site.
9. Identify four major job factors that influence the price of remodeling work.
10. What is the best source of information about work productivities?
11. Why is it not recommended that the estimator keep a price book for pricing estimates?
12. What recap format is most often used for remodeling estimates?

PRACTICE PROBLEMS

1. Calculate a labor price and an equipment price to remove an 8-inch thick concrete wall 40 feet long and 8 feet high based on the following data:
   a. Crew wages: Foreman $26.00 per hour
                       Laborers $23.00 per hour
   b. Compressor and jackhammers: $200.00 per hour
   c. Productivity is at the high end of the scale

PRICING SHEET Page No. [1 of 1]

JOB.....*Replacing Detached Garage* .... DATE...................................…….

ESTIMATED..

No.	DESCRIPTION	QUANTITY	UNIT	UNIT PRICE	LABOR $	UNIT PRICE	MATERIALS $	UNIT PRICE	EQUIP. $	SUBS. $	TOTAL $
	Demolition										
	REMOVE EXISTING GARAGE	1	Lsum	---	---	---	---	---	---	1,500.00	
	REMOVE 4" SLAB-ON-GRADE	240	SF			---	---	$0.25		---	
	TOTAL DEMOLITION:										
	New Work										
	FORM SLAB-ON-GRADE EDGES	48	SF			$10.50			---	---	---
	STRIPPING FORMS	48	SF			$0.50			---	---	---
	CONC. SLAB-ON-GRADE	9	CY			$150.00		$12.50		---	
	WELDED WIRE MESH	528	SF			$0.15		---	---	---	
	VAPOR BARRIER	528	SF			$0.06		---	---	---	
	FINISH SLAB	480	SF			---	---	---	---	---	
	ANCHOR BOLTS	32	No.			$1.25		---	---	---	
	2 x 4 PLATES	192	BF			$0.50		---	---	---	
	2 x 4 STUDS	512	BF			$0.50		---	---	---	
	2 x 8 LINTELS	24	BF			$0.55		---	---	---	
	2 x 10 BEAMS	100	BF			$0.70		---	---	---	
	WALL SHEATHING	608	SF			$0.54		---	---	---	
	2" RIGID INSULATION	608	SF			$0.65		---	---	---	
	20' SPAN TRUSSES	13	No.			$75.00		---	---	---	
	2 x 4 ROUGH FASCIA	32	BF			$0.50		---	---	---	
	ALUM. FASCIA	48	LF			$2.50		---	---	---	
	ROOF SHEATHIG	505	SF			$0.55		---	---	---	
	210# SHINGLES	505	SF			$0.50		---	---	---	
	ALUM. SOFFIT	96	SF			$2.50		---	---	---	
	STUCCO	608	SF			$0.90		---	---	---	
	4-0 x 3-0 WINDOW	2	No.			$350.00		---	---	---	
	18' OVERHEAD DOOR	1	No.	---	---	---	---	---	---	$2,400.00	
	2'-8" EXTR. DOOR	1	No.			$290.00		---	---	---	
	LOCK SET	1	No.			$85.00		---	---	---	
	EAVES GUTTER	48	LF			$2.50		---	---	---	
	DOWN SPOUTS	20	LF			$1.10		---	---	---	
	TOTAL NEW WORK:										

Figure 13.10   Remodeling Pricing Sheet

2. Calculate a labor price to place concrete in a continuous footing for a new wall based on the following data:
   a. Footing size is 110 feet long, 2 feet wide, and 1 foot deep
   b. Crew wages as in question 1 plus cement finisher $27.00 per hour
   c. Productivity is expected to be on the low end of the scale
3. Figure 13.10 shows the pricing sheet for a renovation project that involves replacing an old detached garage. Complete the pricing sheet using the wage rates shown below and assuming the crews involved will have low outputs because the project is so small.

   Crew wages: Labor foreman: $26.00 per hour
                  Laborer: $23.00 per hour
                  Cement finisher: $27.00 per hour
                  Carpentry foreman: $35.00 per hour
                  Carpenter: $30.00 per hour

14

COMPUTER ESTIMATING

OBJECTIVES

After reading this chapter and completing the review questions, you should be able to:

- Describe the process of estimating using ICE software.
- Identify the five steps in an ICE estimate.
- Explain the function of the *Project Information* window.
- Describe the four takeoff methods available with this software.
- Use **Construction Systems** and *Miscellaneous Items* to prepare a takeoff.
- Complete an estimate of a project using MC^2 ICE software.

KEY TERMS

assemblies	ICE software	primary quantity
construction systems	job information	project information
drill down navigation bar	mark field	takeoff notes
estimate maintenance	markups	unit price catalogue
estimate summary	miscellaneous items	wizards

Introduction

Many homebuilders now use specialized software to produce their estimates, so, to demonstrate the power of computer estimating, this chapter presents a complete estimate of a small 4-unit building using one of these specialized estimating programs. The particular software we will be using is Interactive Cost Estimating (ICE) from MC^2 and their *Residential Knowledge Base*, which has been prepared specifically for residential construction. Drawings for this example project are shown in Figure 14.1.

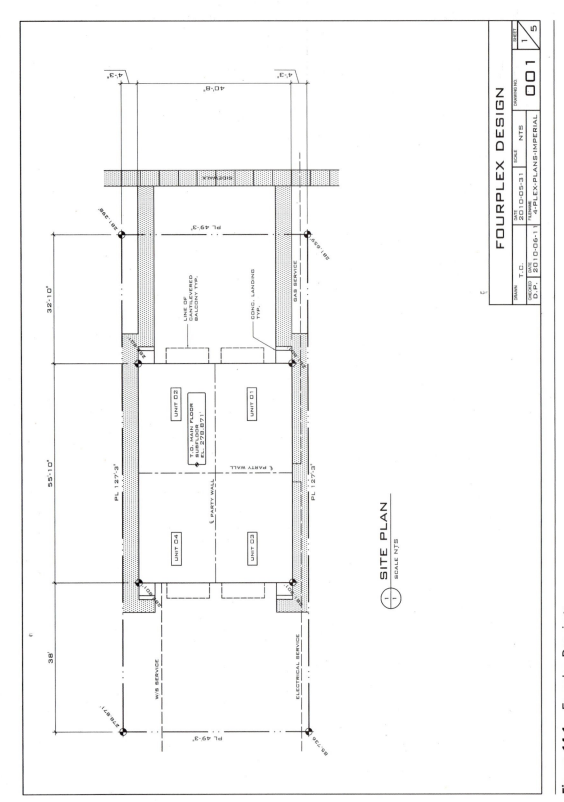

Figure 14.1 Fourplex Drawings

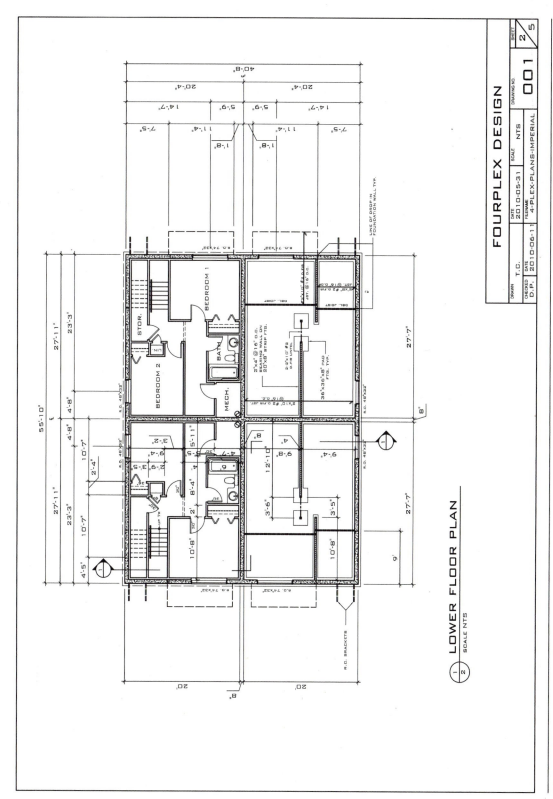

Figure 14.1 Fourplex Drawings (continued)

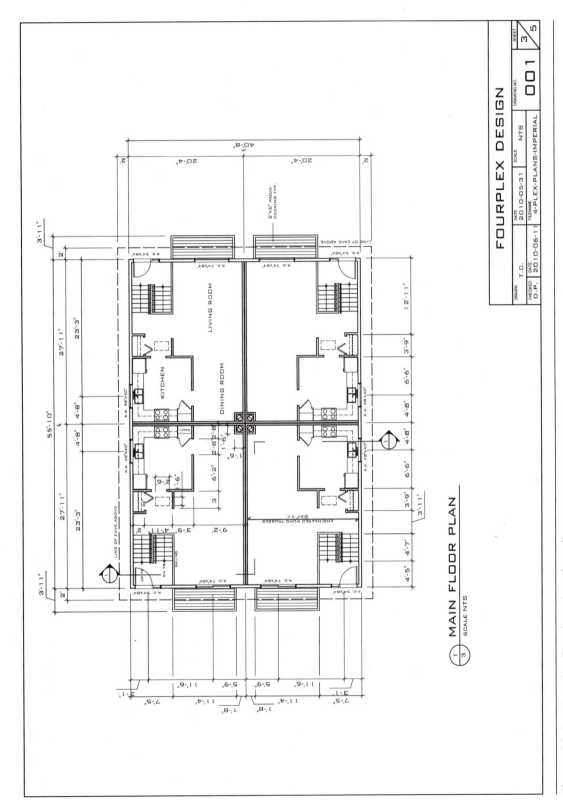

Figure 14.1   Fourplex Drawings (continued)

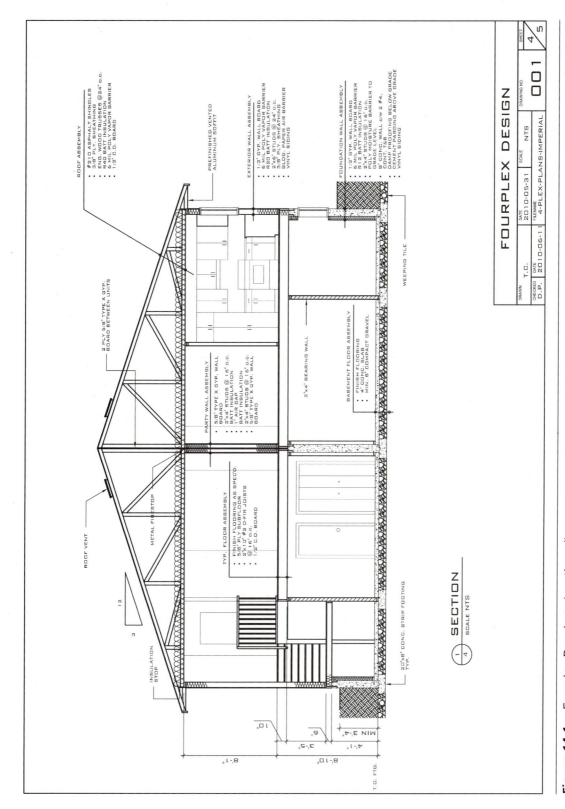

ROOF ASSEMBLY
• #210 ASPHALT SHINGLES
• 3/8" PLY. SHEATHING
• ENG. WOOD TRUSSES @24" O.C.
• R40 BATT INSULATION
• 6 MIL POLY VAPOR BARRIER
• 1/2" C.D. BOARD

PREFINISHED VENTED
ALUMINUM SOFFIT

EXTERIOR WALL ASSEMBLY
• 1/2" GYP. WALL BOARD
• 6 MIL POLY VAPOR BARRIER
• R20 BATT INSULATION
• 2"x6" STUDS @ 24" O.C.
• 3/8" PLY. SHEATHING
• BLDG. PAPER AIR BARRIER
• VINYL SIDING

FOUNDATION WALL ASSEMBLY
• 1/2" GYP. WALL BOARD
• 6 MIL POLY VAPOR BARRIER
• R12 BATT INSULATION
• 2"x4" STUDS @ 16" O.C.
• 8" CONC. WALL C/W 2 #4.
  CONT. T&B
• DAMP PROOFING BELOW GRADE
• CEMENT PARGING ABOVE GRADE
• VINYL SIDING

2 PLY 5/8" TYPE X GYP.
BOARD BETWEEN UNITS

PARTY WALL ASSEMBLY
• 5/8" TYPE X GYP. WALL
  BOARD
• 2"x4" STUDS @ 16" O.C.
• BATT INSULATION
• 1" AIR GAP
• BATT INSULATION
• 2"x4" STUDS @ 16" O.C.
• 5/8" TYPE X GYP. WALL
  BOARD

2"x4" BEARING WALL

BASEMENT FLOOR ASSEMBLY
• FINISH FLOORING
• 4" CONC. SLAB
• MIN. 8" COMPACT GRAVEL

TYP. FLOOR ASSEMBLY
• FINISH FLOORING AS SPEC'D.
• 5/8" PLY. SUBFLOOR
• 2"x10" #2 D-FIR JOISTS
  @ 16" O.C.
• 1/2" C.D. BOARD

ROOF VENT

METAL FIRESTOP

WEEPING TILE

INSULATION
STOP

12
3

20"x8" CONC. STRIP FOOTING
TYP.

10"

6"

MIN 3'-4"

3'-5"

4'-1"

8'-1"

8'-10"

T.O. FTG.

1 SECTION
4 SCALE NTS

FOURPLEX DESIGN

| DRAWN | T.C. | DATE | 2010-05-31 | SCALE | NTS | DRAWING NO. | 001 | SHEET | 4 |
| CHECKED | D.P. | DATE | 2010-06-11 | FILENAME | 4-PLEX-PLANS-IMPERIAL | | | | 5 |

Figure 14.1   Fourplex Drawings (continued)

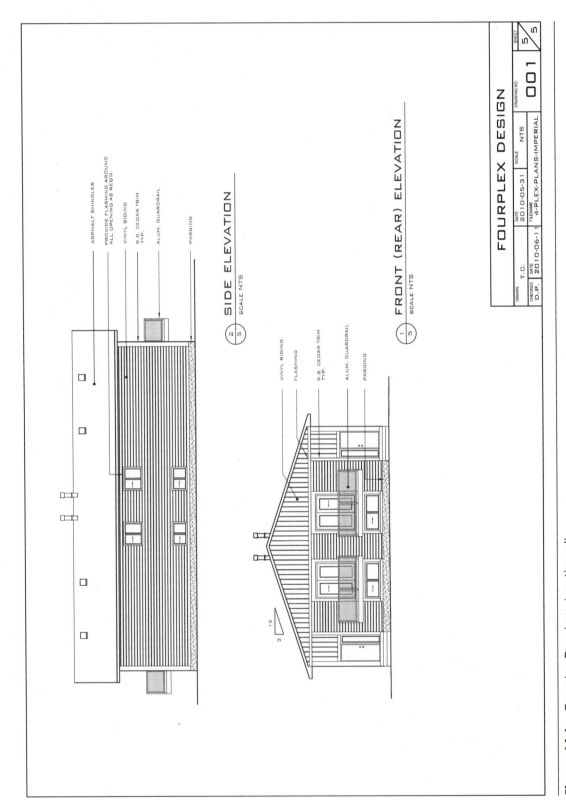

Figure 14.1   Fourplex Drawings (continued)

Estimating with ICE

There are five steps in the ICE estimating process:

1. Setting up the estimate—Create the estimate from a template by imputing **project information.**
2. Performing the takeoff—Measure the work of the project by selecting building components and entering dimensions.
3. Reviewing the estimate—Make adjustments to quantities and/or prices in the estimate.
4. Printing reports—After adjusting the estimate, print one or more of the many different types of ICE reports.
5. Analyzing totals—Insert add-ons and **markups** to complete the estimate and prepare a bid price.

Setting Up the Estimate

With the MC² ICE program installed on your computer, you begin an estimate by starting the program as described below in a step-by-step description of the process. Once the program begins, the software will load the *Project Information* window; to the left of this screen is a vertical menu bar offering *Project Information*, **Miscellaneous Items**, *Add Item*, and other options (see Figure 14.2). Because this menu bar contains options that can be easily accessed on the horizontal tool bar and menu bar at the top of the page, I usually switch off this feature to give more space to work within the window. To hide the navigation bar, click on the blank gray button to the top right of this feature.

The *Project Information* window contains many items beginning with General, followed by Rates, then Crews, and so on. At the beginning of an estimate, we can enter some general information, and later in the estimating process, we can return to this window to enter more data as the need arises.

The Job Information Window

Under the *General* heading in the list of items you will find *Job Information.* In this category, the estimator can enter information about the project to be estimated including location, project web page, date of the estimate, the name of the

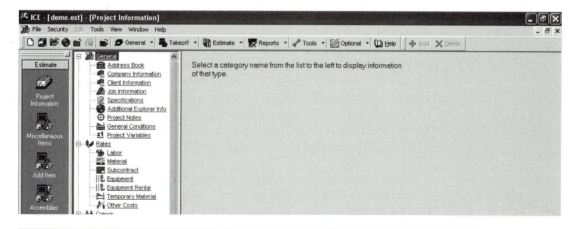

Figure 14.2 *Project Information* Window

estimator, and so on. For this example, we will enter the names of the project and the estimator and identify the kind of estimate. Following this, we will go to the *Specifications* window to enter the project floor area. This particular information is in the **estimate summary** to calculate the price per square foot of the final amount.

The Takeoff Process

There are basically four methods of preparing a takeoff using **ICE software**:

1. **Add Items (or Miscellaneous Items)**—Take off one item at a time by either selecting an item from the database or compiling a completely new item. This is the simplest way to takeoff, but it is time-consuming and, thus, fails to take advantage of the full potential of computer estimating.
2. **Construction Systems**—Measure a complete component of the work by entering dimensions and selecting specifications from a pre-set menu. This is the method of takeoff used most often in the estimating process; it allows the estimator to deal with a number of items all at once by sharing a set of dimensions with a group of work activities that relate to a particular component of the project such as a footing, a wall, or a slab.
3. **Assemblies**—Use a preprogrammed function to measure a group of work items by entering the minimum amount of data in response to a set of prompts. This takeoff method has been especially useful for measuring the work of specialized trades such as plumbing and electrical work.
4. **Wizards**—Prepare a complete conceptual estimate by means of a preprogrammed function that uses a relatively small set of questions to define the work of the job. This technique can be employed before detailed drawings are available to obtain an approximate estimate for a proposed project.

Estimate Example

Here we will work through the process of estimating a bid price for a 2,271 square feet 4-unit bi-level project: the *Fourplex*.

Starting the Estimate

In ICE an estimate begins by creating an estimate file:

1. Start the ICE program and click on the *New Estimate* button. This is found on the left end of the tool bar.
   a. The *New Estimate* window will now open to allow you to enter the file name: *Fourplex*. Be careful not to use characters other than letters, numbers, spaces, and underscores in your filename. See Figure 14.3.
   b. When you click **Open,** the *Create New Estimate* window will open. See Figure 14.4.
   c. We will be using the *Company Standard* that comes with this software; this is the default selection on this window. The *Company Standard* contains information such as wage rates, materials prices, crew make-ups, and many more items that are needed to produce an estimate.
   d. You will also notice that the *Default Unit Price Catalogue* has been automatically selected as the cost database for this estimate.

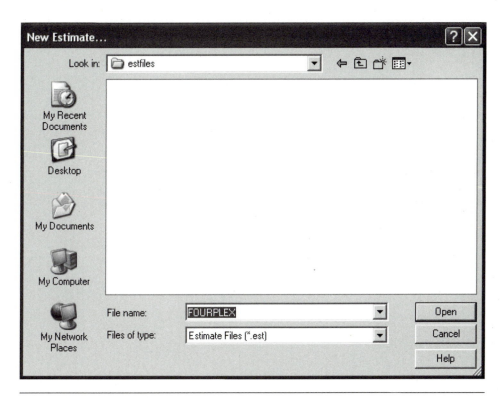

Figure 14.3   New Estimate Window

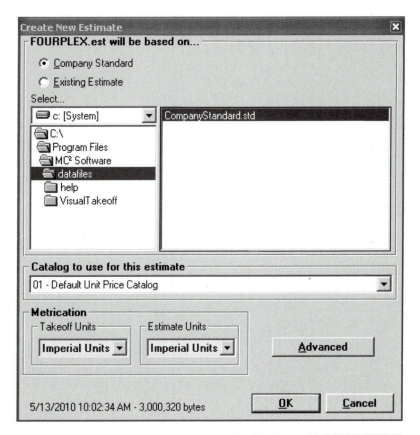

Figure 14.4   Create New Estimate Window

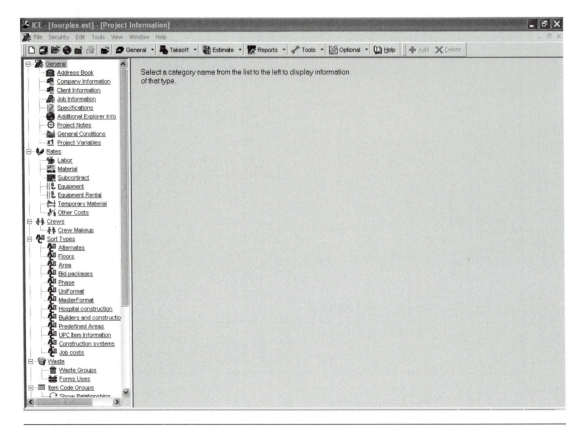

Figure 14.5 *Project Information* Window

2. Click OK to continue with the estimate.
   a. The *Project Information* screen will now open ready for you to enter general information, rates, crew information, and so on for this estimate. See Figure 14.5.
   b. In this example, we will restrict ourselves to the basics, so at this stage we will access only the *Job Information* and *Specifications* screens.
3. Click on *Job Information* listed under the *General* heading on the left.
   a. Enter the name of the project in the field at the top of this screen.
   b. Identify yourself in the *Estimator* under *Estimate* toward the middle of this page.
   c. Against *Estimate Type* select *Detailed document*. See Figure 14.6.
4. Click on *Specifications*, which is also listed under the *General* heading on the left.
   a. Here we enter the area of the building, which is 55'-10" × 40'-8" × 2 levels = 4,541 square feet, as the **Primary Quantity**. This amount will be displayed on the *Estimate Summary* and will facilitate the calculation of amounts per square foot of gross floor area.

   HINT: Often when windows are opened, they do not fill the whole screen; if you click on the maximize button (to the left of the X at the top right), you will find it easier to work with the full window.

   We can now close the *Project Information* window before starting the takeoff:
   b. Click on the X (red) on the right above the *Imperial Units* label on the *Project Information* window.

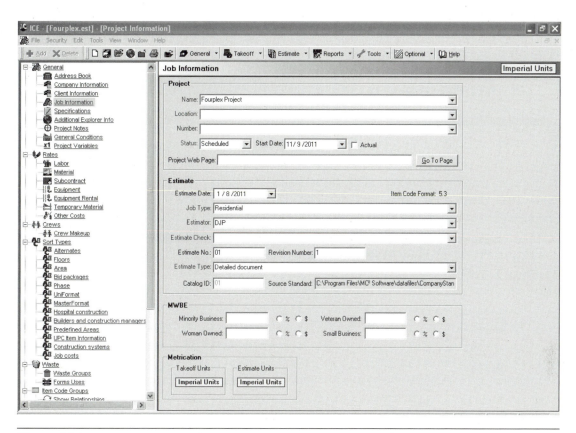

Figure 14.6 *Job Information* Window

Excavation Takeoff

1. The first items in the takeoff are strip topsoil and basement excavation. Because these are stand-alone items, we will use *Miscellaneous Item* takeoff for this part of the work:

   a. On the toolbar, click on the down arrow just to the right of the *Takeoff* button and select *Miscellaneous Item*s.

   b. When the *Miscellaneous Items* window opens, the curser will be on the *Item Code* cell. Click on the button with three dots to the right of this. This opens up the *Unit Price Catalogue* where we can pick the item we want to measure. After you have used the program for a while, you will get to know your way around this catalogue, but as a beginner, you can use a shortcut to find an item.

   c. Click on the Find button; it is the bottom of three buttons up on the right of this window.

   d. The first item to measure is *Strip Topsoil*, so if we type the word *topsoil* here, the find facility will list all the items that contain that word.

   e. On your list should be the item *Strip and haul away topsoil*; this is what we need, so double click on this item.

   f. The fields on the *Miscellaneous Items* window will prefill and the curser will be on the *Quantity field*.

   g. At this stage you need to calculate the quantity for this item (127.25 × 49.25 × 0.5 divided by 27 = 116 CY).

   h. Enter the 116 into the *Quantity field*.

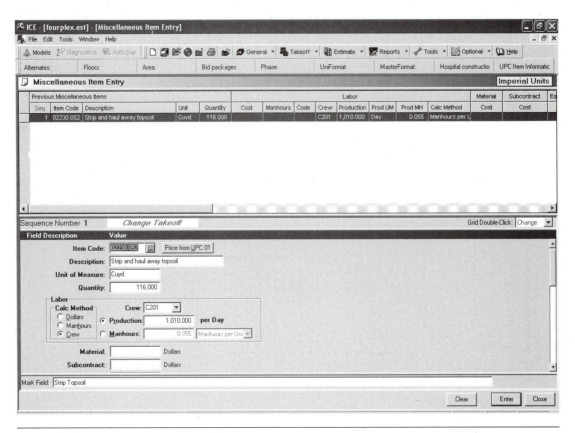

Figure 14.7   Miscellaneous Item Entry

    i. Click on the **Mark Field** at the bottom of this window and type *Strip Top-soil*. This is a note that helps identify what was measured in this item.

    j. Now press enter to place the item into the estimate and your *Miscellaneous Items* window will reflect this. See Figure 14.7.

2. We propose to use a subcontractor to excavate and backfill the basement. To measure a one-off subtrade, we can again use the *Miscellaneous Item* takeoff. This time we will be adding an item to the database for this work, so we need to give it a new item code.

    a. Click on the button with three dots to the right of the *Item Code* field. This opens the *Unit Price Catalogue* so that we can find a suitable number for this new item. I like to list subtrades at the beginning of each section, so, as this item is to go into the *Earthwork* section of Division 2—Sitework, an appropriate number would be 2300.001.

    b. Close the *Unit Price Catalogue* window, then back in the *Miscellaneous Item* window, type 2300.001 in the *Item Code* field and complete the following fields:

    c. Description: Excavation Subtrade

    d. Unit of measurement: Lsum

    e. Quantity: 1

    f. Mark Field: Excavation & Backfill Subtrade

    g. At the time of the takeoff, prices may not be available so leave the subtrade price field blank at this time.

    h. Press enter to complete this item. See Figure 14.8.

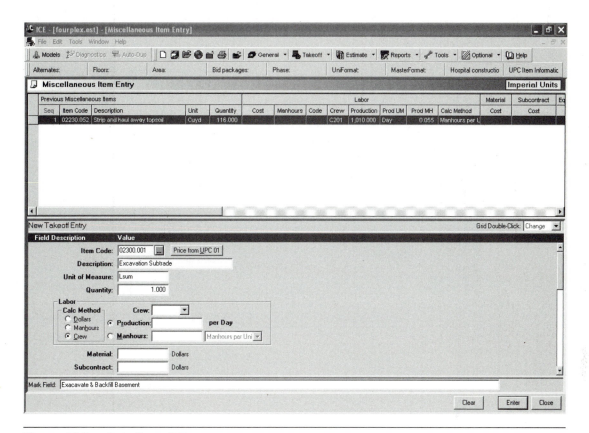

Figure 14.8   Strip Topsoil Entry

The sitework takeoff is now complete; to see the status of the estimate so far, click the *Estimate* button on the toolbar. This takes you to the ***Estimate Maintenance*** window (see Figure 14.9). As you can see on this page, the items we measured are now listed in the estimate, and labor, material, and equipment prices have been applied automatically. At any stage in the process, we could view the ongoing estimate summary or print a report of what we have measured, but we will examine these options in more detail later.

Footings and Walls Takeoff

Figure 14.10 shows the hand notes that accompany this takeoff. The software we are using will record all the amounts that are entered such as lengths, widths, and so on. Many of these amounts have first to be calculated, and because the figures entered into a calculator are lost, it is useful to have a record that shows where these amounts came from when they are not obtained directly from the drawings. Therefore, we have made use of **Takeoff Notes** to facilitate this.

For the footings and walls takeoff, we will use a **construction system** from the *Residential Contractor* database:

1. On the toolbar, click on *Takeoff* and select construction system.
   a. Double click on *Residential Contractor*.
   b. From the list of systems that pops up, double click on 03.200 *Footings and Walls*.

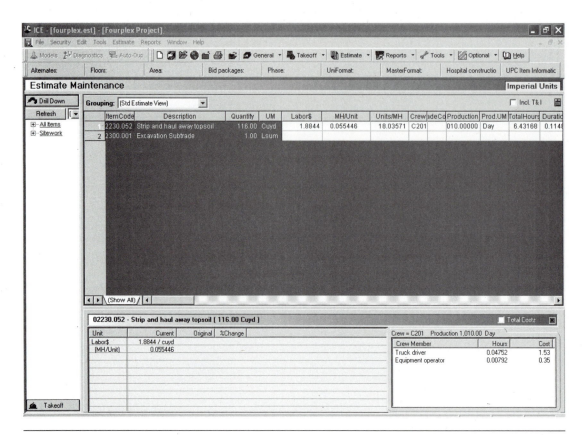

Figure 14.9 *Estimate Maintenance* Window

The construction system window is set up so that dimensions and/or quantities are entered on the left and specifications of how and what to measure with the dimensions are listed on the right of this window.

2. Starting on the left, type 365.47 in the B dimension field and, using your down key or your tab key, move down to dimension C.
   a. Enter 1.67 and tap the down key again.
   b. Enter 0.67 and tap the down key again. (Note that tapping the enter key at any stage will close off this window, so do not hit enter yet.)
   c. Lastly, go to field A and enter 1 as the quantity.
3. In the right-hand column, we enter the specifications that indicate how the dimensions we have entered are to be handled.
   a. In field 9, enter 1 to indicate that these are concrete footings.
   b. In field 8, enter 1 to indicate that 2,500 psi concrete is required.
   c. The two vertical surfaces of this footing require formwork, so enter 3 into field 7.
   d. There are no excavations, wall finishes, reinforcement, or insulation, so all the remaining values are 0.
   e. In the *Mark Field* type: Continuous Footing.
   f. See Figure 14.11 for all the selections for this item.
4. Hit the enter key and the Takeoff Answers will be processed (see Figure 14.12).

TAKEOFF NOTES   SHEET No. **1 of 9**

PROJECT: *Fourplex*   DATE:

ESTIMATOR: *DJP*

NOTES					DIMENSIONS		
				TIMES			
Excavation							
		Strip Topsoil			127.25	49.25	0.50
		Excavate and Backfill - subtrade					
Foundations							
	Perimeter	Length:	Perimeter	190.32			
2 x 55.83	111.66		Party walls	52.49			
2 x 40.67	81.34		2 x 18.33	36.66			
Less 4 x 0.67	-2.68		Partitions				
	190.32		4 x 10.67	42.68			
Less	55.83		4 x 10.83	43.32			
2 x 0.67	-1.34			365.47			
2 x 0.50	-1.00	**Continuous Footings**			365.47	1.67	0.67
	52.49						
Less	20.33						
1 x 0.67	-0.67	**2 x 4 Keyway**			365.47		
1 x 0.33	-0.33						
2 x 0.50	-1.00						
	18.33	**Pad Footings**		2 x 4	3.00	3.00	0.67
9.00		**Walls**			190.32		
7.42					53.49		
16.42				2 x 19.33	38.66		
					282.47	0.67	8.00
		(deduct drops @ landings)		-4	16.42	0.67	3.92
		190.32'/2.00 **2 x 8 (Ladder) Sill Plate**		2	190.32 =	381	
				95	0.67 =	64	
						445	LF
2 x 3.92 =	7.84						
	16.42						
	24.26	**Form Wall Bulkheads**		4	24.26	0.67	65
2 x 4.00 =	8.00	(windows)		4	13.34	0.67	36
2 x 2.67 =	5.34						101 SF
	13.34						

Figure 14.10   Fourplex Takeoff Notes

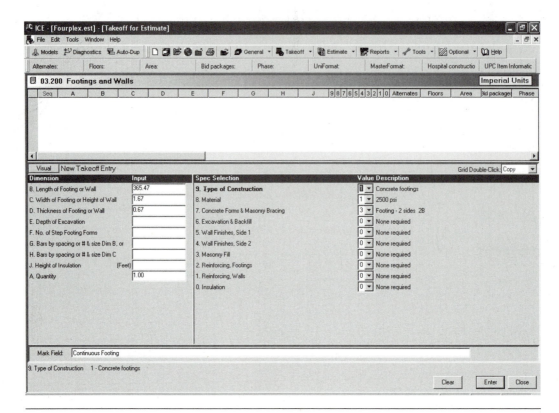

Figure 14.11 Continuous Footing Entry

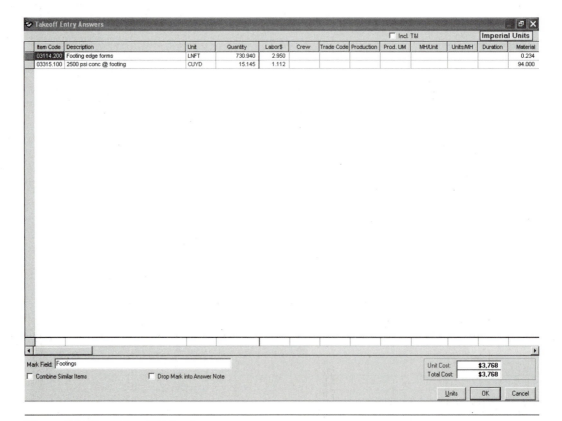

Figure 14.12 Takeoff Entry Answers

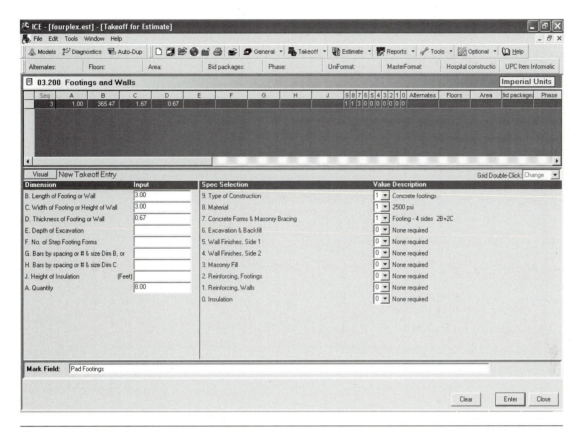

Figure 14.13   Pad Footing Entry

5. If the answers look okay, press OK and you will be returned to the construction system window with the takeoff item you have just measured listed as sequence 3.

6. Because the footing keyway was not included in this takeoff, it needs to be added as a miscellaneous item. Using the *find* facility as we did above, the item number is 02830.501, the quantity is 365.47 feet and in the *Mark Field* type: 2 × 4 Keyway.

Next we consider the pad footings.

7. Return to the Footings and Walls construction system to enter dimensions and specifications for the pad footings required to support columns in the units. Figure 14.13 shows the entries for this takeoff.

8. To measure the basement walls, we again use the Footings and Walls construction system, so, in the construction system window, enter dimensions and specifications for the walls as shown in Figure 14.14, which shows the entries for this takeoff.

   a. Note that the entry for rebar is 4.04, which is interpreted as 4 pieces of #4 bars with length of B dimension. Over on the right, we select option 4 in cell 2 to indicate that the bars are measured by pieces.

   b. Press enter when you have completed this item.

   c. Press OK to accept the Takeoff Entry Answers.

   d. The wall drops to the level of the landing at each corner, so adjustments have to be made to account for these cutouts. This can be done by measuring the cutouts as a wall takeoff but using a negative quantity. See Figure 14.15 for the entries for these cutouts.

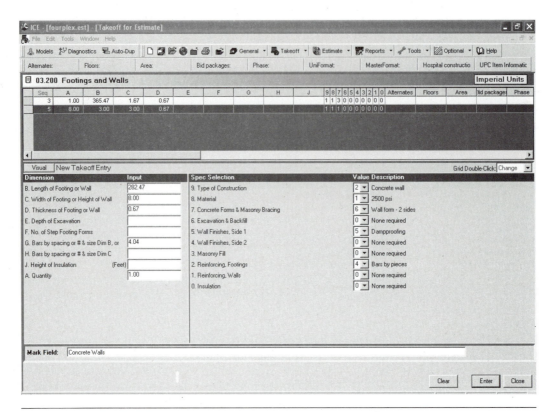

Figure 14.14 Concrete Walls Takeoff

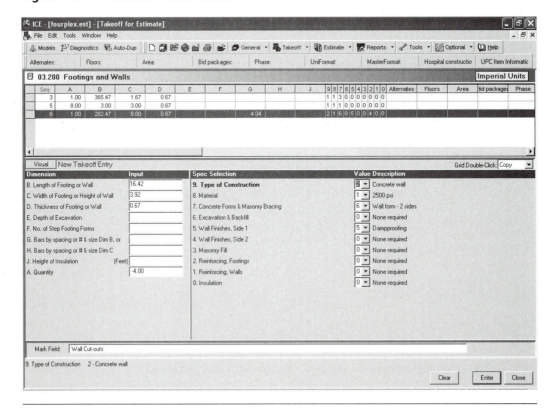

Figure 14.15 Adjustment for Cutouts

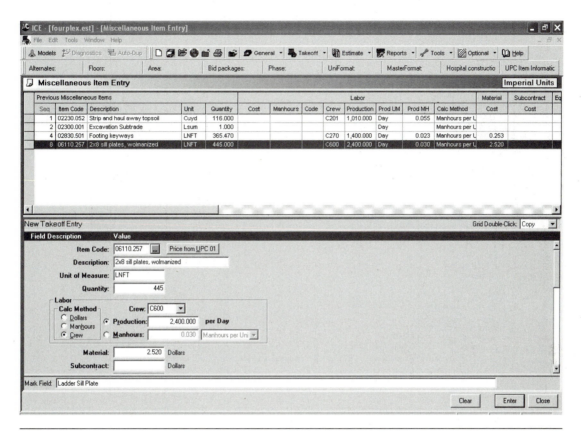

Figure 14.16   Sill Plates Entry

e. The drawings indicate a double sill plate cast into the top of the wall (ladder sill plate); this can be measured using miscellaneous item code 06110.257. See Figure 14.16 for the data entries for this item.

f. Bulkheads are needed on the walls to form the corner cutouts. These bulkheads can also be measured using a miscellaneous item code 3111.412. See Figure 14.17 for the data entries for this item.

9. There is a weeping tile noted on the drawings to be placed in the backfill outside the footings. This is to be a part of the excavation subcontractor's work; to ensure that they include this in their price, we can place a note on their takeoff item. To do this, on the estimate window click on the 2 next to the item. A window will open allowing you to enter the note. See Figure 14.18.

This concludes the foundations takeoff; next is the slab-on-grade.

Slab-on-Grade Takeoff

1. Select a new construction system: 03.210 Slabs and Thickening Slabs.
2. See the takeoff notes in Figure 14.19 for the dimensions.
3. Enter the dimensions and specifications as shown in Figure 14.20.
   a. Press enter when you have completed this window.
   b. Press OK to accept the Takeoff Entry Answers.

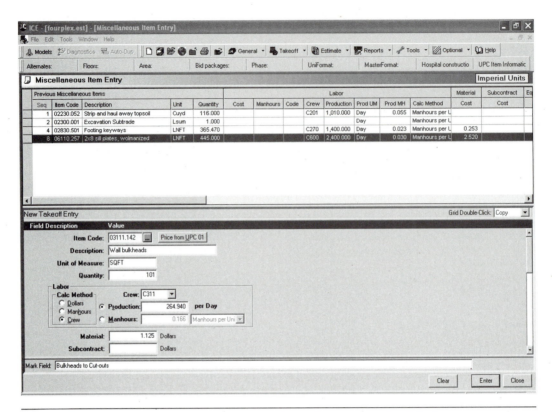

Figure 14.17 Wall Bulkheads Entry

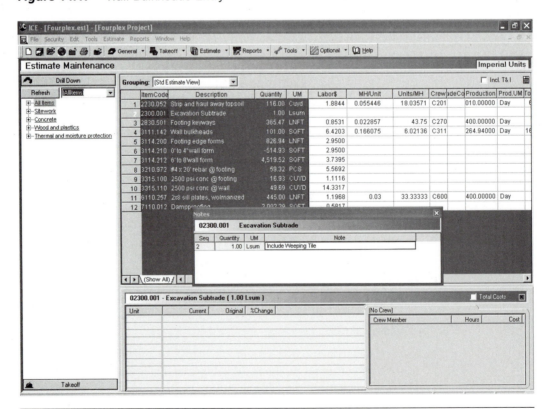

Figure 14.18 Adding a Note

TAKEOFF NOTES

PROJECT: *Fourplex*

DATE:

ESTIMATOR: *DJP*

NOTES		TIMES	DIMENSIONS					
	Conc Slab-on-Grade	4	26.91		19.33		0.33	
27.58 x 20.00								
Less -0.67 -0.67								
26.91 19.33								
	Gravel under	4	26.91		19.33		0.67	
Floor System								
	4 x 4 Posts	4	8.00	=	32			
					43	BF		
	2 x 10 in Beams	4 x 2	3.50	=	28			
					37	BF		
	2 x 10 Joists	4 x 13	20.00	=	1040			
12.83		4 x 5	14.00	=	280			
Less -3.50		4 x 4	8.00	=	128			
16.33		4 x 11	12.00	=	528			
27.58		4 x 2	27.58	=	221			
Less -7.33		4 x 2	20.00	=	160			
20.25		4	11.33	=	45			
Less -16.33					2402			
3.33 /1.33 = 3 + 2 = 5					3203	BF		
11.33	*2 x 8 Joists*	4 x 5	8.00	=	160			
1.33	*(landing)*	4 x 2	4.00	=	32			
12.66 /1.33 = 10 + 1 = 11					192			
					256	BF		
	5/8" Ply Floor Sheathing		55.83		40.67	=	2271	
	(Balconies)	4	11.33		3.92	=	178	
18.58 /1.33 = 14							2448	SF
	2 x 4 Bridging	4 x 2	40	=	320	PCS		
20.25 /1.33 = 15								
14.00/1.33 = 11								

Figure 14.19   Fourplex Takeoff Notes (continued)

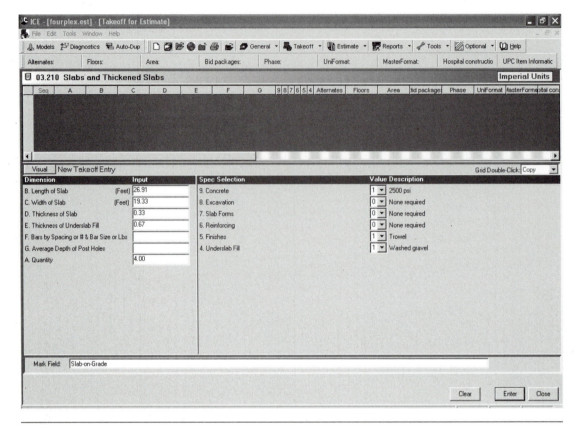

Figure 14.20 Slab-on-Grade Entry

Carpentry Floor System Takeoff

The carpentry takeoff begins with the floor system:

1. There is no construction system available for the kind of floor required here, so the components are taken off as miscellaneous items using quantities calculated on the takeoff notes (Figure 14.19).
2. For the 4 × 4 posts, use a new code of 06111.099. Because this item is new to the database, you have to complete the fields on the *Miscellaneous Item* takeoff window for this item. See Figure 14.21 for entries.
3. The remaining miscellaneous items are all in the database, so use the following data to complete the floor system:

Item	Item Code	Quantity
a. 2 × 10 in Beams	6111.033	47 BDFT
b. 2 × 10 Joists	6111.132	3,203 BDFT
c. 2 × 8 Joists	6111.124	256 BDFT
d. 5/8" Ply Floor Sheathing	6160.127	2,448 SQFT
e. 2 × 2 Bridging	6110.103	320 PCS

Carpentry Wall System Takeoff

The takeoff of the carpentry wall systems, including exterior walls and the partitions, begins with the basement walls. See Figure 14.22 for dimensions and other notes.

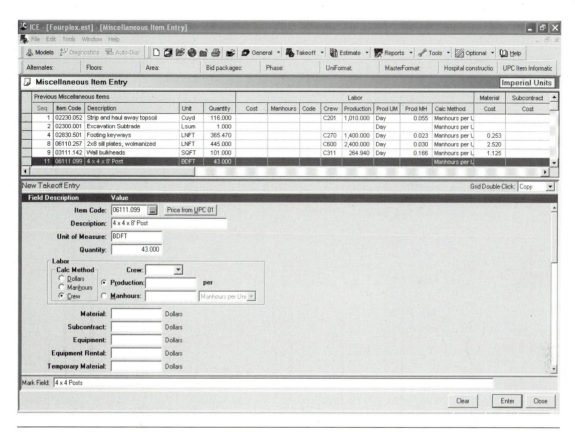

Figure 14.21   Wood Posts Entry

We use the construction system for partitions for the stud wall takeoff:

- Figure 14.23 shows the entries for the pony walls to the corners of the basement.
- In the record of the takeoff items, note that we moved the **mark field** to just after the sequence number. This is done by clicking on the column heading (mark field) and dragging it to the new location on the left. In this location, it is easier to keep track of multiple construction system pages.
- Figure 14.24 shows the entries for the exterior walls in the basement.
- Note that the stud spacing is entered as 12" rather than the true spacing of 16". This provides a quick way of allowing for the extra studs to openings and corners.
- Figure 14.25 shows the entries for the interior partitions in the basement.
- Figure 14.26 shows the entries for the main floor exterior walls.
- Figure 14.27 shows the entries for the party walls on the main floor.
- Figure 14.28 shows the entries for the main floor partitions.

Carpentry Roof System Takeoff

Dimensions and notes for the carpentry roof system takeoff are shown in Figure 14.29. For this roof system takeoff, we can use the *Prefabricated Trusses Construction System.*

- Figure 14.30 shows the entries for the roof system.
- The remaining items listed in Figure 14.29 are entered using miscellaneous items. See Figure 14.31 for a summary of these items.

TAKEOFF NOTES

PROJECT: *Fourplex*

DATE:

ESTIMATOR: *DJP*

NOTES		TIMES	DIMENSIONS		
Wall System					
(Basement Pony Walls at Landings)					
2 x 6 Plates		4 x 3	9.00		
2 x 6 Studs (4')					
3/8" Ply Wall Sheathing		4	9.00	3.42	
6" Insulation					
Anchor Bolts					

Partitions:

Exterior	Interior			
27.92	14.33 (L - R)			
-1.00	2 x 2.33 4.67			
26.92	10.67			
14.75	12.58			
19.33	16.42			
21.00	12.58 (T - B)			
82.00	2 x 6.00 12.00			
	9.33			
	6.17			
	98.75			

NOTES		TIMES	DIMENSIONS		
2 x 4 Plates (extr)		4 x 3	82.00		
(intr)		4 x 3	98.75		
6.42	2 x 4 Studs (8')				
4.25					
10.67	2 x 10 Headers	4 x 2	5.33		
/ 2 = 5.33 ave. width of windows					
5 x 2.75 = 13.75	2 x 6 Headers	4 x 8	2.81		
3.25					
3.75					
1.75					
22.50					
/ 8 = 2.81 ave. width of doors					

Figure 14.22a Fourplex Takeoff Notes (continued)

TAKEOFF NOTES

PROJECT: *Fourplex*

DATE:

ESTIMATOR: *DJP*

NOTES			TIMES	DIMENSIONS		
(Main Floor - Exterior Walls)						
2 x 55.83 *111.66*						
2 x 40.67 *81.34*						
203.00						
	2 x 6 Plates		*3*	*203.00*		
203.00 / 2.00 = 102						
	2 x 6 Studs		*102*			
		extras	*48*			
6.17	*3/8" Ply Wall Sheathing*			*203.00*	*8.00*	
4.00						
10.17	*2 x 10 Lintels*	*(doors)*	*4*	*4.50*		
/2 = 5.08 ave width of windows		*(windows)*	*4 x 2*	*5.08*		
	½" Gyproc & 6" Insulation					
(Party Walls)						
27.92						
20.33	*2 x 4 Plates*		*4 x 3*	*48.25*		
48.25						
	2 x 4 Studs		*4 x 37*			
48.25 / 1.33 = 37		*extras*	*4 x 12*			
	5/8" Gyproc & 4" Insulation					
(Interior Partitions) *- Length:*	*3.50 (L - R)*					
	3.00					
	11.50					
	2.00 (T - B)					
	10.67					
(corner)	*3.00*					
	33.67					
	2 x 4 Plates		*4 x 3*	*33.67*		
3.00						
3.00	*2 x 4 Studs*		*4 x 34*			
2.67						
8.67	*2 x 6 Headers*	*(doors)*	*4 x 3*	*2.89*		
/2 = 2.89 ave width of openings						
	½" Gyproc					

Figure 14.22b   Fourplex Takeoff Notes (continued)

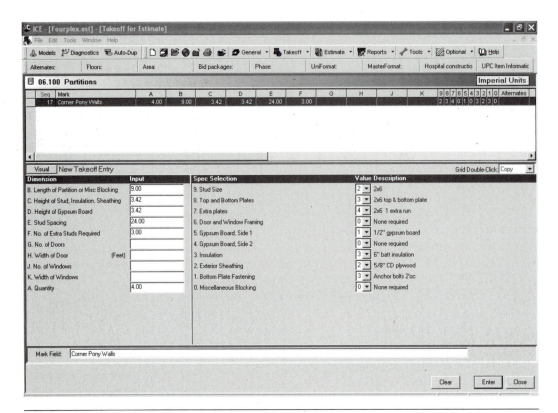

Figure 14.23 Basement Pony Walls Entry

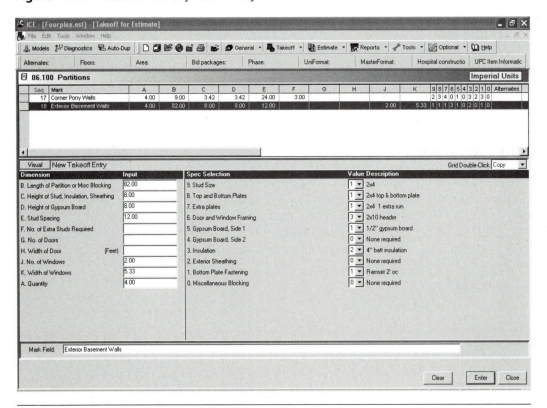

Figure 14.24 Exterior Basement Walls Entry

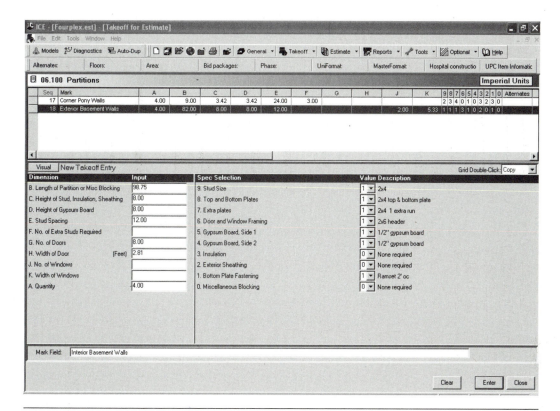

Figure 14.25 Interior Basement Partitions Entry

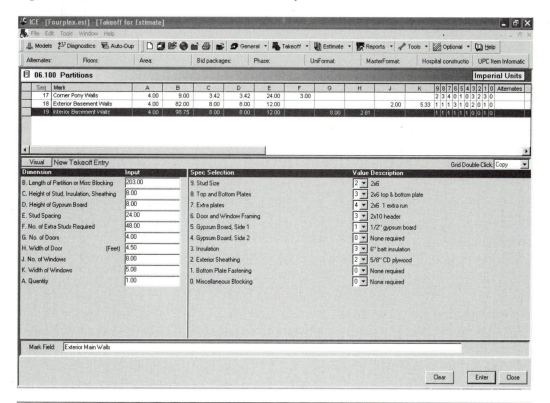

Figure 14.26 Main Floor Exterior Walls Entry

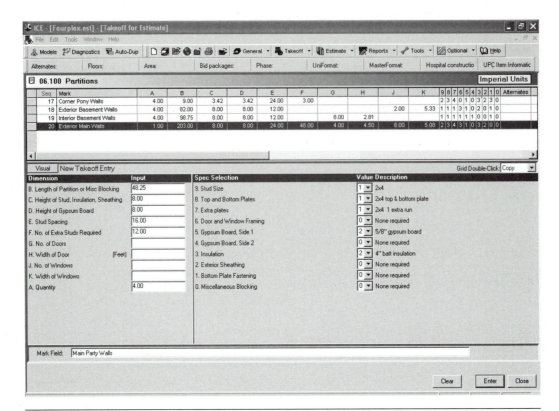

Figure 14.27   Main Floor Party Walls Entry

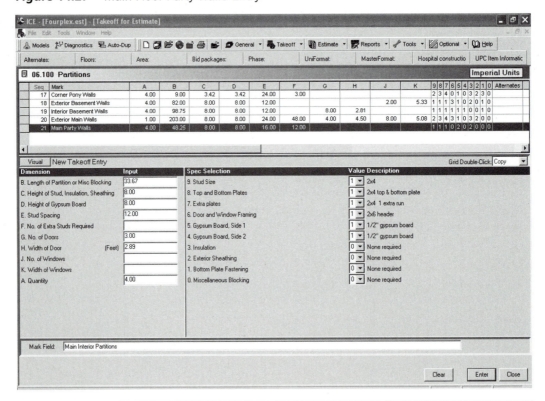

Figure 14.28   Main Floor Partitions Entry

TAKEOFF NOTES

PROJECT: *Fourplex*   DATE:

ESTIMATOR: *DJP*

NOTES			TIMES	DIMENSIONS			
Roof System	Half Trusses - 20' Span		2	56.00			
	½" Roof Sheathing						
	Asphalt Shingles						
	Sheet Metal Flashings		2 x 4	4.50	=	36	
			4	6.67	=	27	
			4	5.00	=	20	
			2	40.67	=	81	
						164	LF
	Eaves Gutter		2	63.83			
22.33² = 498.63	Downspouts		2	14.00			
5.58² = 31.14							
sq. root 529.77	Gable Ends - 40'-8" Span		4				
= 23.02							
	2 x 4 Barge Rafters		4	23.02	=	92	LF
	2 x 4 Blocking						
23.02 / 2.00 = 12		Lookouts	4 x 12	3.33	=	160	
		Rough fascia	2	59.83	=	120	
(around vents etc.)	Misc. blocking		2 x 4 x 3 x 2	2.00	=	96	
	Eaves blocking		4 x 12	2.00	=	96	
			2 x 4	2.00	=	16	
					488	=	325 BFM
	1 x 4 Ribbons		6	55.83	= 335	=	112 BFM
	(Truss ties)						
	Alum. Soffit		2	59.83	2.00	239	
			4	23.02	2.00	184	
						423	SF
	6" Alum. Fascia		2	59.83	=	120	
			4	23.02	=	92	
						212	LF
	1/2" Wall Sheathing		2 x ½	44.67	5.58	249	SF
	(gable)					8	SHTS
	5/8" Drywall (4 x 12 shts)			Ditto		249	
	(in attic between units)		2	59.83	23.02	2755	
						3004	SF
						63	SHTS
	3/8" Ply Soffit		4	4.00	11.33	181	SF
	(under balconies)					6	SHTS

Figure 14.29   Fourplex Takeoff Notes (continued)

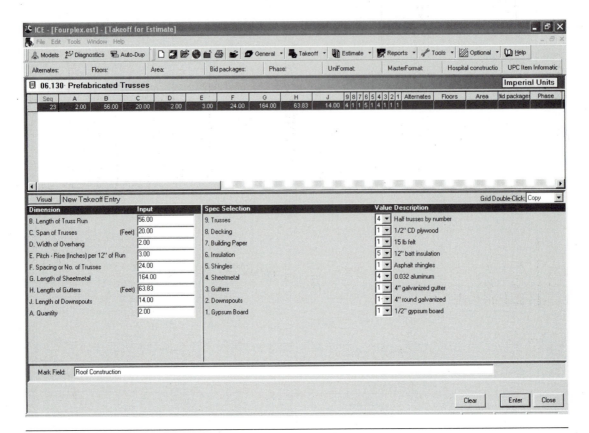

Figure 14.30   Roof System Entry

	Item Code	Description	Unit of Measure	Quantity	Mark Field	
New*	6110.412	Gable End - 40'-8" Span	EACH	4	Gable Ends	
	6110.308	2 x 4 x 24 rafter	PCS	4	Barge Rafters	
	6110.635	2 x 4 x RL blocking	BDFT	325	Misc. Blocking	
	6110.566	1 x 4 x RL blocking	BDFT	112	Truss Ties	
	7460.056	Aluminum vented soffit	SQFT	423	Eaves Soffit	
	7460.083	6" Aluminum fascia	LNFT	212	Fascia	
	6160.100	½" 4 x 8 CD plywood	SHTS	8	Gable Sheathing	
	9250.411	⅝" 4 x 12 gyp bd @ wall	SHTS	63	Drywall to Attic	
	7460.052	⅜" exterior grade plywood	SQFT	181	Soffit to Balconies	

Figure 14.31   Roof System—Miscellaneous Items

Finish Carpentry and Exterior Takeoff

1. Dimensions and notes for the finish carpentry takeoff are shown in Figure 14.32. We start with the doors; these can be measured using the 08.100 *Residential Doors* construction system.
   a. Figure 14.33 shows the entries for the entrance doors.
   b. The remaining doors listed in Figure 14.32 are entered in a similar fashion.
   c. The attic access door is not listed in the construction system, so this item is measured using *Miscellaneous Item* 08310.010 Access Panel.
2. Measure windows using the construction system 08.200 *Residential Windows*.
   a. Figure 14.34 shows the entries for the 48" × 40" windows. Note that the size of the windows is entered as feet and inches. So this window entry is 40.34 indicating 4'-0" × 3'-4".
   b. The remaining windows listed in Figure 14.32 are entered in a similar fashion.
3. Measure cabinets listed in Figure 14.32 using the construction system 11.100 *Residential Cabinets*. See Figure 14.35.
4. Measure the stair using the construction system 06.160 *Stairs*. See Figure 14.36.
5. The iron railing, closet shelves, exterior siding, and the remaining items listed in Figure 14.32 are taken off using miscellaneous items. See Figure 14.37 for a summary of these items.

Finishes and Subcontractors Takeoff

1. Dimensions and notes for the finishes are shown in Figure 14.38.
2. The *Floor and Ceiling Joists* construction system number 06.110 is used to measure the ceiling drywall. See Figure 14.39 for entries.
3. The ceiling, floor, and wall finishes are measured using miscellaneous items. We can also use this method to take off the remaining subcontractors for the site paving, plumbing, HVAC, and electrical work. See Figure 14.40 for a summary of these items.
4. You can close the miscellaneous items window now; the software should return you to the *Estimate Maintenance* window.

Pricing the Estimate

Most of the items measured in the takeoff processes using MC² ICE software were automatically priced from the unit price catalogue as the takeoff proceeded. Any of the rates, prices, crews, and productivities that are priced using the default settings in the database can be modified; here we will consider a number of ways these changes can be made. However, we added a number of new items using the *Miscellaneous Item* takeoff without entering their prices. We will enter prices for these below.

Labor Rates

The most efficient way to make changes that apply to the entire estimate such as wage rates and material prices is to use the pricing codes found in the *Project Information* window.

1. In the *Estimate Maintenance* window, tap the *General* button on the toolbar to open the *Project Information* window. Here we are going to make the required changes starting with wage rates.

TAKEOFF NOTES   SHEET No. | 6 of 9 |

PROJECT: *Fourplex*   DATE:

ESTIMATOR: *DJP*

NOTES		TIMES	DIMENSIONS		
Finish Carpentry					
(Doors)	36" x 80" Extr. Door & Frm.	4			
	with 12" side light				
	30" x 80" Intr. Door & Frm.	4 x 5			
	18" x 80" Intr. Door & Frm.	4			
	36" x 72" Bifold	4 x 2			
	42" x 72" Bifold	4			
	74" x 84" Slinding Patio Door & Frm.	4			
	30" x 18" Attic Access	4			
(Windows)	48" x 40" Window	4			
	74" x 32" Window	4			
	48" x 32" Window	4			
(Cabinets)	24" Base Cabinet	4	12.00		
8.67					
3.33	12" Wall Cabinet	4	9.00		
12.00		4	12.00		
2.00					
2.67	24" x 24" Pantry	4	pcs		
3.00					
1.33	24" Vanity	4	4.00		
9.00					
(Misc Items)					
	3" Wide Stair w. 7-risers	4 x 2	7.00	(diagonal)	
4.25 [triangle 7.00 / 5.58]					
7.00					
3.75	*Iron Railing*	4	19.75	=	79 LF
9.00					
19.75					

Figure 14.32a   Fourplex Takeoff Notes (continued)

TAKEOFF NOTES

PROJECT: *Fourplex*

DATE:

ESTIMATOR: *DJP*

NOTES		TIMES	DIMENSIONS		
Shelf Unit		4			
(linen closet)					
Closet Rod & Shelves		2 x 4	3.00 =	24	
		4	6.00 =	24	
				48	LF
1 x 3 Base Board	(basement)	4 x 2	15.00 =	120	
		4 x 2	3.42 =	27	
		4 x 2	26.92 =	215	
		4 x 2	14.75 =	118	
		4 x 2	2.33 =	19	
		4 x 2	3.42 =	27	
		4 x 2	2.33 =	19	
		4 x 2	12.58 =	101	
		4 x 2	9.33 =	75	
		4 x 2	10.67 =	85	
		4 x 2	11.00 =	88	
		4 x 2	6.00 =	48	
		4 x 2	2.00 =	16	
		4 x 2	8.33 =	67	
		4 x 2	6.00 =	48	
		4	4.58 =	18	
		4	5.92 =	24	
	(main floor)	4 x 2	27.25 =	218	
		4 x 2	20.00 =	160	
		4 x 2	3.50 =	28	
		4 x 2	2.00 =	16	
		4 x 2	3.00 =	24	
		4 x 2	10.67 =	85	
		4 x 2	10.33 =	83	
				1729	LF

Figure 14.32b   Fourplex Takeoff Notes (continued)

TAKEOFF NOTES

PROJECT: *Fourplex*

DATE:

ESTIMATOR: *DJP*

NOTES			TIMES	DIMENSIONS		
Exterior Finishes						
	Alum Siding	*- Subtrade*				
	PC Conc steps	*- Subtrade*				
2 x 4.00 =	*8.00*					
	11.33					
	19.33	*Balcony Rails*	*4*	*19.33*	*=*	*77* LF
	Neoprene to Balcanies		*4*	*4.00*	*11.33 =*	*181* SF
	4" Cedar Trim		*2*	*40.67 =*	*81*	
			2 x 6	*13.33 =*	*160*	
			2 x 2	*4.50 =*	*18*	
			2 x 4	*6.17 =*	*49*	
					309 LF	
	10" Cedar Trim		*2 x 4*	*13.33 =*	*107*	
			2 x 2	*6.17 =*	*25*	
			4 x 2	*4.00 =*	*32*	
			4	*11.33 =*	*45*	
					209 LF	

Figure 14.32c Fourplex Takeoff Notes (continued)

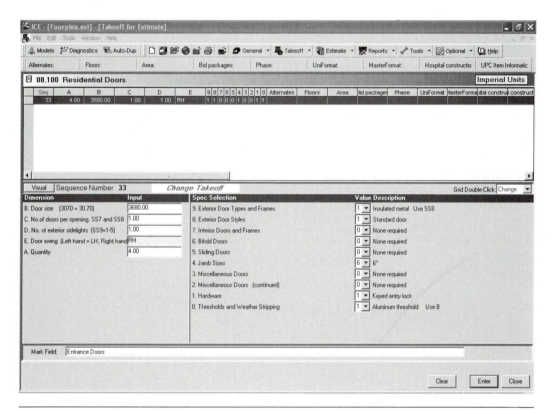

Figure 14.33   Entry for Doors

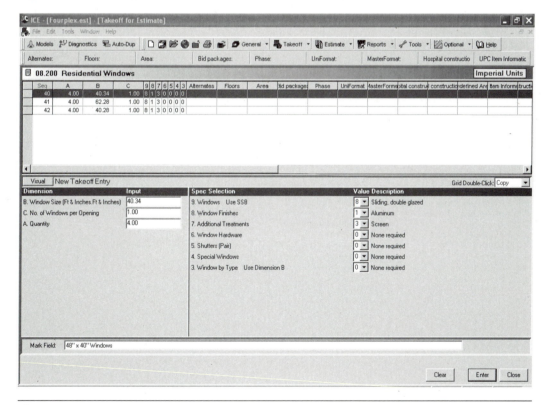

Figure 14.34   Entry for Windows

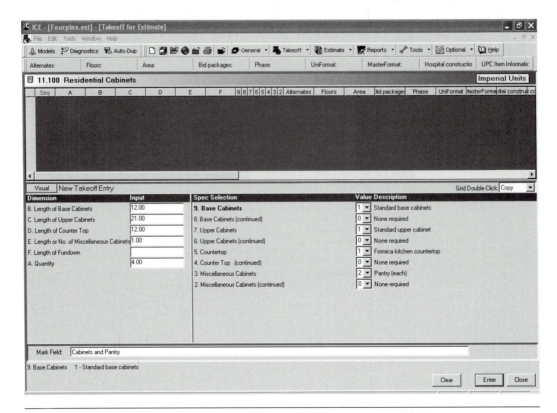

Figure 14.35 Entry for Cabinets

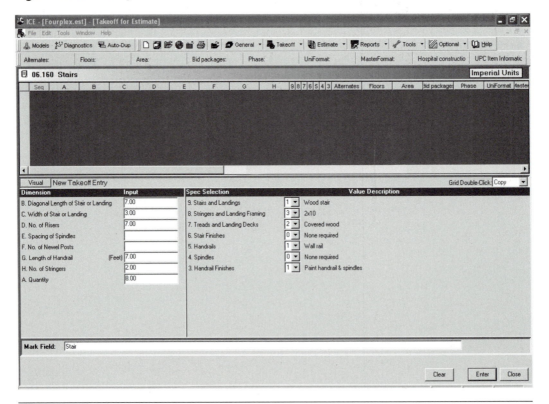

Figure 14.36 Entry for Stairs

	Item Code	Description	Unit of Measure	Quantity	Mark Field	
	5520.019	Metal Railing	LNFT	79	Iron Railing	
	10670.050	Shelf Unit	EACH	4	Linen Shelves	
	6270.130	Closet rod and shelf	LNFT	48	Closet Shelves	
	6450.150	Wood Base Board	LNFT	1729	1 x 3 Base Board	
new*	7405.001	Aluminum Siding Subtrade	Lsum	1	Siding Subtrade	
new*	3420.001	Precast Concrete Steps	EACH	4	Entrance Steps	
	7560.027	Neoprene Deck Finish	SQFT	181	Balcony Deck Finish	
	6220.134	1 x 4 Cedar trim	LNFT	309	4" Trim	
	6220.131	1 x 10 Cedar trim	LNFT	209	10" Trim	
	5520.019	Metal railing	LNFT	77	Balcony Rails	

Figure 14.37   Finish Carpentry and Siding—Miscellaneous Items

2. From the menu on the left, under *Rates* click on *Labor*.
3. To see a list showing just the trades associated with the work you have taken off, check the box *Rates Used*.
4. We are going to add payroll burden as a percentage of labor on the estimate summary instead of allowing fringes here, so the fringe value in each case will be zero. Double click on the fringe amount, type 0, and press enter.
5. In the same manner, enter the required basic rates for each trade as shown in Figure 14.41.

Material Prices

Material prices can also be updated on the *Project Information* window:

1. Under *Rates* click on *Material*.
2. To change the price of 2,500 psi concrete to $107.00 per cubic yard, double click on the price displayed for this material and type in 107.00, then press enter (see Figure 14.42). Note that the date and time when the prices were updated are recorded against each item you have changed.
3. To return to the *Estimate Maintenance* window, on the toolbar click on Estimate. Here a message will appear asking you if you want to recost the estimate; in order to apply the new prices to the estimate, you have to click yes, then click OK to confirm that you want the changes made.

Using the Drill Down Navigation Bar

You can change many of the values shown on the *Estimate Maintenance* window by simply typing over the values shown in each cell. For example, to change the quantity for washed gravel fill to 52.00 cubic yards, double click on the number shown, type in 52, and press enter. In the same way, you could change the labor unit price for an item, the crew selected for an item, the productivity rate, and so forth.

This window is currently set up to show all the items measured in the takeoff. To review and make changes more effectively, it can be useful to focus on specific parts

TAKEOFF NOTES

PROJECT: *Fourplex*

DATE:

ESTIMATOR: *DJP*

NOTES			DIMENSIONS		
		TIMES			
Interior Finishes					
½" Drywall Ceiling	(basement)	4	27.92	20.33	
Paint Ceilings	(bsmt & main)	2 x 4	27.92	20.33	= 4541 SF
Carpet		2 x 4	27.92	20.33	= 4541
	Bathroom DDT	-4	8.33	6.00	= -200
	Furnace DDT	-4	5.92	4.58	= -434
	Storage DDT	-4	15.00	3.42	= -205
	Kitchen DDT	-4	10.67	10.33	= -441
					3261
					= 362 SY
Vinyl Flooring		4	8.33	6.00	= 200
		4	5.92	4.58	= 434
		4	15.00	3.42	= 205
		4	10.67	10.33	= 441
					1280
					= 142 SY
Paint Walls			1729.00	8.00	= 13 832 SF
Subtrades:					
	- Surface Paving				
	- Plumbing				
	- HVAC				
	- Electrical				

Figure 14.38 Fourplex Takeoff Notes (continued)

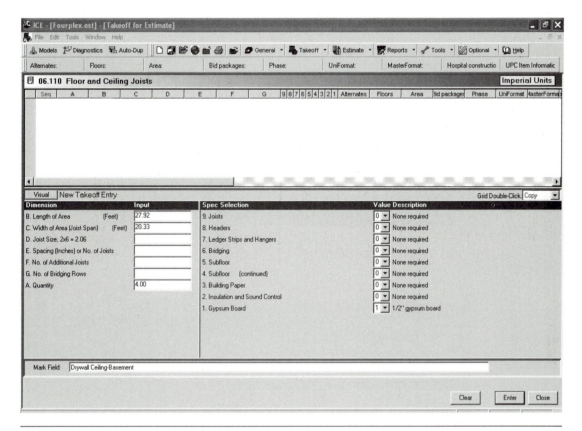

Figure 14.39 Entry for Drywall Ceiling-Basement

	Item Code	Description	Unit of Measure	Quantity	Mark Field	
	9910.110	Paint gypsum board ceiling	SQFT	4,541	Paint Ceilings	
	9680.015	Carpeting	SQYD	362	Carpet	
	9650.065	Sheet vinyl floor	SQYD	362	Vinyl Flooring	
	9910.101	Paint gypsum board wall	SQFT	13,832	Paint Walls	
new*	2700.001	Paving Subtrade	Lsum	1	Surface Paving	
new*	15400.001	Plumbing Subtrade	Lsum	1	Plumbing	
new*	15700.001	HVAC Subtrade	Lsum	1	HVAC	
new*	16000.001	Electrical Subtrade	Lsum	1	Elecctrical	

Figure 14.40 Finishes and Subtrades—Miscellaneous Items

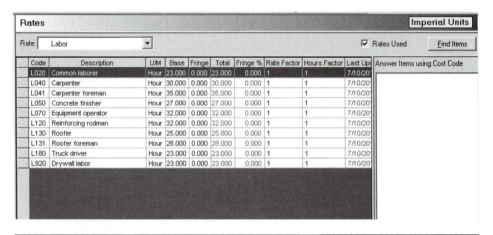

Figure 14.41 Wage Rates

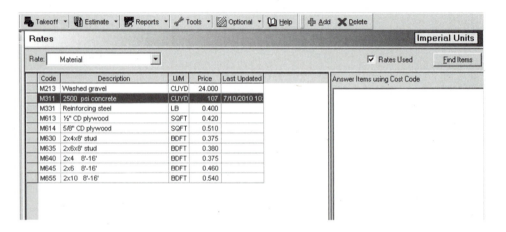

Figure 14.42 Material Prices

of the estimate; the pane on the left of this window with the heading *Drill Down* can be used to achieve this. Presently listed in the left window is CSI breakdown of the estimate. Here you can see *Sitework* from division 1, followed by *Concrete*, then *Metals*, and so on. If you want to specifically check on your concrete takeoff, you can obtain a summary of the concrete items by clicking on *Concrete* on the list in this window.

To obtain a different breakdown of the estimate, click on the *Drill Down* bar at the top of the left pane. This pane now shows a list of the miscellaneous items and construction systems used in the takeoff. For instance, if you want to review the items measured with the continuous footing construction system, you would click on *Continuous Footings* in this pane as shown in Figure 14.43. To return to the original CSI breakdown, click again on the *Drill Down* bar.

Adding Prices

A large number of pricing columns are displayed on the right-hand side of the *Estimate Maintenance* window. We can set up another view of this window that shows

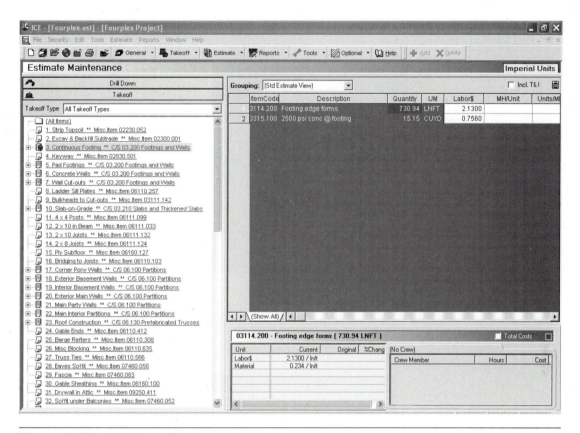

Figure 14.43   Drill Down Data

only the columns in which we need to attach prices to the new items we added in the takeoff process.

1. At the bottom of the Description column, there is a tab labeled "(Show All)." There is a slider next to this tab so that you can open a space. See Figure 14.44.
2. Right click in this space and select *Add View* from the menu.
3. On the window that pops up, type *Pricing* as the name of this view.
4. Insert the following columns by double clicking on the name of the column in the list on the left:
   a. ItemCode
   b. Description
   c. Quantity
   d. UM
   e. Labor$
   f. Material
   g. Subcontract
   h. TotalCost
5. These choices will now be listed on the *Columns To Show* window. See Figure 14.45.
6. Click OK on this popup window. Now, back on the *Estimate Maintenance* window, there will be a new tab at the bottom of the Description column labeled Pricing.

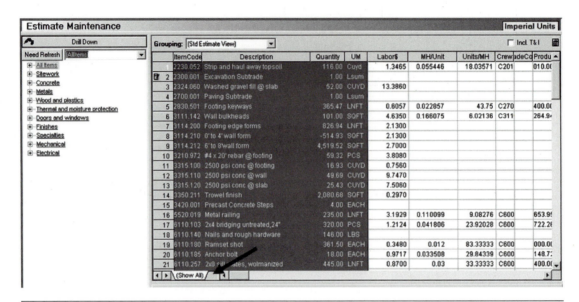

Figure 14.44 Slider Next to (Show All)

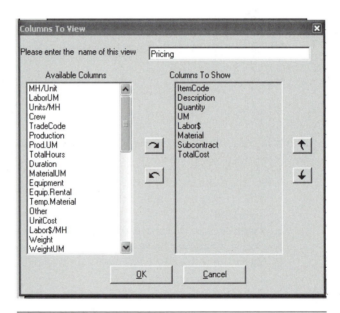

Figure 14.45 Adding a View

7. Click on the Pricing tab and only the selected columns will be displayed in this window.
8. It is now easy to see from the blank cells in these columns which items are still in need of pricing.
9. Enter the prices shown in Figure 14.46 into the appropriate blank cells.
10. This completes the pricing of the work, but we still need to consider the general expenses for this project.

Item Code	Description	Unit of Measure	Unit Prices		
			Labor$	Material	Subcontract
2230.052	Strip & haul away topsoil	CuYd	---	---	$1.00
2300.001	Excavation Subtrade	Lsum	---	---	$1,980.00
2700.001	Paving Subtrade	Lsum	---	---	$5,250.00
3420.001	Precast Concrete Steps	EACH	---	---	$690.00
6110.412	Gable End - 40'-8" Span	EACH	$28.00	$190.000	---
6111.099	4 x 4 x 8" Post	BDFT	$0.25	$0.375	---
7405.001	Aluminum Siding Subtrade	Lsum	---	---	$7,450.00
15400.001	Plumbing Subtrade	Lsum	---	---	$24,975.00
15700.001	HVAC Subtrade	Lsum	---	---	$22,705.00
16000.001	Electrical Subtrade	Lsum	---	---	$20,500.00

Figure 14.46   Pricing New Miscellaneous Items

Pricing General Expenses

ICE software allows the estimator to assess and account for general expenses in many different ways. The simplest way is to select items required from the list that can be accessed in the *Project Information* window.

1. On the toolbar click on the *General* button to open the *Project Information* window.
2. Click on *General Conditions* under the *General* heading of items. A list of general expense items will appear in the window. See Figure 14.47.
3. Enter a quantity against an item on this list by double clicking in the quantity field to bring that general expense item into your estimate. See Figure 14.48 for the items required for this estimate. (To generate a list that shows only the items you have selected, click on the *Omit Zero Quantity* box.)
4. To return to the full estimate window, click on *Estimate* on the toolbar and all the items of general expense for this project are now listed in the *Estimate Maintenance* window. Prices can be adjusted here. For instance, the superintendent and the timekeeper may be working only part-time on this project, so the weekly amounts can be reduced to $750.00 and $400.00. See Figure 14.48.

Estimate Summary

All that remains to be done to complete the bid price for this project is to set up the markups in the summary. The MC² ICE summary window is basically a spreadsheet that has been set up to show a recap of the dollar amounts in each of the estimate cost categories and to allow the estimator to insert markups (add-ons).

1. On the toolbar click the arrow to the right of *General* and choose *Summary* from the options listed. This will open the summary window. You can adjust

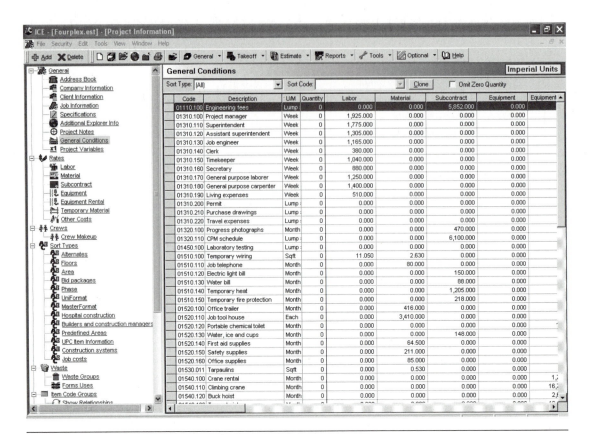

Figure 14.47 General Conditions

Item Code	Description	Quantity	Unit of Measure
1310.110	Superintendent	20	Week
1310.150	Timekeeper	20	Week
1510.110	Job telephone	5	Month
1520.120	Portable chemical toilet	5	Month
1520.140	First aid supplies	5	Month
1540.190	Air compressor	3	Month
1540.230	Generators	3	Month
1540.240	Pickup truck rental	5	Month
1740.100	Job clean up	4,541	Sqft

Figure 14.48 General Expenses

the size of the Summary screen by accessing the *Zoom* button on the toolbar. See Figure 14.49.

(Note: Markups are *not* set up by adding percentage values to cells in the Summary window.)

2. Click on the *Factors* button on the toolbar. Here the estimator can set up markups for the estimate. Enter the markups listed below and tap OK when all entries are complete.

Builder's Risk (Insurance)	0.60%
Labor Burden (Payroll Additive)	25%
Materials Sales Tax	12%
Miscellaneous 1 (Building Permit)	0.70%
Overhead Labor (Small Tools)	3%
Profit (Overall)	10%

3. Before we leave the Summary window, let us add labels to clarify a couple of markups we have set up:
   a. In cell C22, type: *Small Tools* to explain what the Labor Overhead is for.
   b. In cell A29, type: *Misc. 1 – Building Permit*.

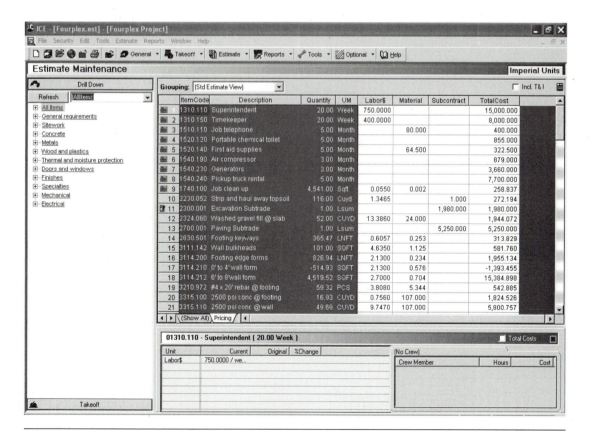

Figure 14.49   *Adjusting Prices*

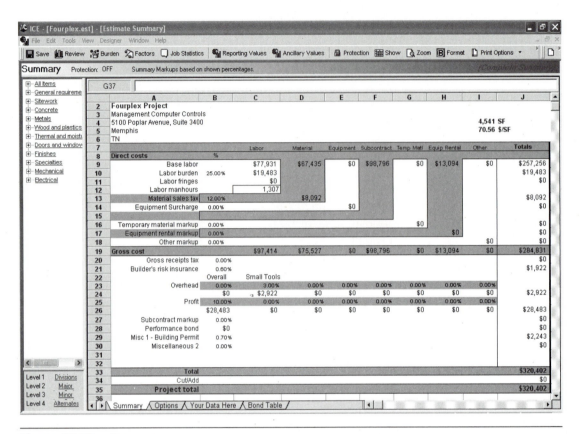

Figure 14.50   Summary

4. We can also use the data on this window to calculate the cost per SF of this bid:
   a. In cell I5, type the formula: *=J35/I4*
   b. In cell I6, type: *$/SF*
5. See Figure 15.50 for the finished Summary window.

Printing Reports

ICE software allows you to print many different types of reports in an almost unlimited variety of formats. Here our choice of report is the Unit Cost Estimate report:

1. Close the Summary window and click on the *Reports* button that will now be listed on the toolbar.
2. The Estimate Report window will open; in the box on the top left of this window select: *11 - Est Detail - 198 - Unit Cost*
3. Down the left of this window are a number of buttons that allow you to switch on or off features on this printout. Switch on the following:
   a. Detail Report
   b. Print Summary
   c. Show Code and Desc
   d. Combine by Sort Codes
   (Leave all remaining buttons down the left blank.)

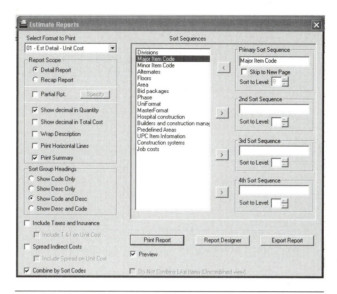

Figure 14.51   Estimate Reports

4. By default the *Preview* button will be checked (this allows you to review the printout before it is sent to the printer). See Figure 14.51.
5. Click on *Print Report*, and then at the top of the next window, click on the *Print Setup* tab. Here select *Landscape* and tap *OK*.
6. Review the printout; if it looks acceptable, print a copy. See Figure 14.52 for a copy of the printout.

Materials Reports

Because this estimate is in an electronic format, it allows the estimator to quickly conduct many types of price analysis. For example, we may receive a quote offering a price to supply rough carpentry materials for this project. We can generate a special report showing the price of rough carpentry materials in the estimate so that we can make a comparison with the supplier's quote:

1. With the *Estimate Maintenance* window open, click on the down arrow to the right of the *Reports* button on the toolbar and a menu of different reports will open.
2. Select *MC² Misc. Reports* from this menu.
3. In the Miscellaneous Report window select: *Material Detail*.
4. Down the left of this window are a number of buttons that allow you to switch on or off features on this printout. Switch on the following:
   a. Detail Report
   b. Show decimal in Total Cost
   c. Show Costs
   d. Show Description only
   e. Preview
   Then, on the right, set the Primary Sort Sequence to *Major Item Code*.
5. If you look at the report now it would list all the materials required for the project, but here we are interested only in the rough carpentry, so we will restrict the scope of the printout to the range 6,100–6,200.

Estimate Detail - Fourplex Project

Detail - Without Taxes and Insurance

Group 1: Major ItemCode Groups

Estimator: DJP
Project Size: 4541 SF

ItemCode	Description	Quantity	UM	Lab.Unit	Mat.Unit	Eqp.Unit	Sub.Unit	Other Unit	Tot.UnitCost	TotalCost
01300.000 Administrative requirements										
01310.110	Superintendent	20.00	Week	750.0000					750.000	15,000
01310.150	Timekeeper	20.00	Week	400.0000					400.000	8,000
	*** Total 01300.000 Administrative requirement**									**23,000**
01500.000 Temporary facilities and controls										
01510.110	Job telephone	5.00	Month		80.000				80.000	400
01520.120	Portable chemical toilet	5.00	Month						171.000	855
01520.140	First aid supplies	5.00	Month		64.500				64.500	323
01540.190	Air compressor	3.00	Month						293.000	879
01540.230	Generators	3.00	Month						1,220.000	3,660
01540.240	Pickup truck rental	5.00	Month						1,540.000	7,700
	*** Total 01500.000 Temporary facilities and controls**									**13,817**
01700.000 Execution requirements										
01740.100	Job clean up	4,541.00	Sqft	0.0550	0.002				0.057	259
	*** Total 01700.000 Execution requirements**									**259**
02200.000 Site preparation										
02230.052	Strip and haul away topsoil	116.00	Cuyd	1.3465			1.000		2.347	272
	*** Total 02200.000 Site preparation**									**272**
02300.000 Earthwork										
02300.001	Excavation Subtrade	1.00	Lsum				1,980.000		1,980.000	1,980
02324.060	Washed gravel fill @ slab	52.00	CUYD	13.3860	24.000				37.386	1,944
	*** Total 02300.000 Earthwork**									**3,924**
02700.000 Bases, ballasts, and pavement										
02700.001	Paving Subtrade	1.00	Lsum				5,250.000		5,250.000	5,250
	*** Total 02700.000 Bases, ballasts, and pavement**									**5,250**
02800.000 Site improvements and amenities										
02830.501	Footing keyways	365.47	LNFT	0.6057	0.253				0.859	314
	*** Total 02800.000 Site improvements and amenities**									**314**
03100.000 Formwork										
03111.142	Wall bulkheads	101.00	SQFT	4.6350	1.125				5.760	582
03114.200	Footing edge forms	826.94	LNFT	2.1300	0.234				2.364	1,955
03114.210	0' to 4' wall form	-514.93	SQFT	2.1300	0.576				2.706	-1,393
03114.212	6' to 8' wall form	4,519.52	SQFT	2.7000	0.704				3.404	15,385
	*** Total 03100.000 Formwork**	4,831.53	SQFT	2.7601	0.661				3.421	16,528
03200.000 Concrete reinforcing										
03210.972	#4 x 20' rebar @ footing	59.32	PCS	3.8080	5.344				9.152	543
	*** Total 03200.000 Concrete reinforcing**									**543**
03300.000 Cast in place concrete										
03315.100	2500 psi conc @ footing	16.93	CUYD	0.7560	107.000				107.756	1,825
03315.110	2500 psi conc @ wall	49.69	CUYD	9.7470	107.000				116.747	5,801
03315.120	2500 psi conc @ slab	25.43	CUYD	7.5060	107.000				114.506	2,912
03350.211	Trowel finish	2,080.68	SQFT	0.2970					0.297	618
	*** Total 03300.000 Cast in place concrete**									**11,155**

C:\Program Files\MC² Software\estfiles\Fourplex.est   Page 1   7/12/2010 01:53 PM

Figure 14.52a Estimate Printout

Estimate Detail - Fourplex Project

Detail - Without Taxes and Insurance

Group 1: Major ItemCode Groups

Estimator: DJP
Project Size: 4541 SF

ItemCode	Description	Quantity	UM	Lab.Unit	Mat.Unit	Eqp.Unit	Sub.Unit	Other Unit	Tot.UnitCost	TotalCost
03400.000 Precast concrete										
03420.001	Precast concrete steps	4.00	EACH				690.000		690.000	2,760
	* Total 03400.000 Precast concrete									**2,760**
05500.000 Metal fabrications										
05520.019	Metal railing	235.00	LNFT	3.1929	31.250				34.443	8,094
	* Total 05500.000 Metal fabrications									**8,094**
06100.000 Rough carpentry										
06110.103	2x4 bridging untreated, 24"	320.00	PCS	1.2124	0.501				1.713	548
06110.140	Nails and rough hardware	146.00	LBS		1.088				1.088	159
06110.180	Ramset shot	361.50	EACH	0.3480	0.375				0.723	261
06110.185	Anchor bolt	18.00	EACH	0.9717	0.448				1.420	26
06110.257	2x8 sill plates, wolmanized	445.00	LNFT	0.8700	2.520				3.390	1,509
06110.308	2x4x24 rafter	4.00	PCS	7.5518	6.126				13.678	55
06110.410	Half truss 3/12 pitch 20' "	48.00	EACH	69.6000	4.052				73.652	3,535
06110.412	Gable End - 40'-8" Span	4.00	EACH	28.0000	190.000				218.000	872
06110.450	2x4x92-5/8" stud	1,170.43	PCS	1.7861	2.010				3.796	4,443
06110.460	2x6x92-5/8" stud	174.50	PCS	2.4617	3.040				5.502	960
06110.483	2x6x16 stud	8.50	PCS	5.1086	7.360				12.469	106
06110.510	2x4x16 plate	199.67	PCS	6.1173	4.020				10.137	2,024
06110.511	2x6x16 plate	47.35	PCS	6.8114	7.360				14.171	671
06110.530	2x6x16 header	15.58	PCS	14.2892	7.360				21.649	337
06110.532	2x10x16 header	12.66	PCS	15.2979	14.429				29.727	376
06110.566	1x4xRL blocking	112.00	BDFT	1.0735	0.375				1.449	162
06110.635	2x4xRL blocking	325.00	BDFT	0.9116	0.375				1.287	418
06111.033	2x10x14 beam	47.00	BDFT	0.3036	0.375				0.679	32
06111.099	4 x 4 x 8' Post	43.00	BDFT	0.2500	0.375				0.625	27
06111.124	2x8x16 joist	256.00	BDFT	0.2277	0.375				0.603	154
06111.132	2x10x12 joist	3,203.00	BDFT	0.2240	0.375				0.599	1,919
06160.100	½" 4x8 CD plywood	8.00	SHTS	6.5153	13.440				19.955	160
06160.100	½" 4x8 CD plywood	79.39	SHTS	6.5153	13.440				19.955	1,584
06160.101	5/8" 4x8 CD plywood	54.60	SHTS	7.4037	16.320				23.724	1,295
06160.127	5/8" T&G plywood @ subfloor	2,448.00	SQFT	0.2314	0.640				0.871	2,133
	* Total 06100.000 Rough carpentry									**23,767**
06200.000 Finish carpentry										
06220.131	1x10 cedar trim	209.00	LNFT	1.4130	3.162				4.575	956
06220.134	1x4 cedar trim	309.00	LNFT	1.1220	1.280				2.402	742
06270.130	Closet rod and shelf	96.00	LNFT	6.5250	8.961				15.486	1,487
	* Total 06200.000 Finish carpentry									**3,185**
06400.000 Architectural woodwork										
06410.030	Standard base cabinet	48.00	LNFT	7.5000	33.283				40.783	1,958
06410.060	Vanity base cabinet	16.00	LNFT	7.5000	33.283				40.783	653
06410.230	Standard upper cabinet	84.00	LNFT	7.5000	32.003				39.503	3,318
06410.410	Pantry cabinet	4.00	EACH	30.0000	192.019				222.019	888
06415.140	Formica kitchen countertop	48.00	LNFT	3.7500	8.001				11.751	564
06415.160	Formica vanity countertop	16.00	LNFT	3.7500	8.001				11.751	188
06430.123	2x10x14 stair framing	8.00	PCS	7.6629	12.625				20.288	162
06430.124	2x10x16 stair framing	9.00	PCS	8.7642	14.429				23.193	209

Figure 14.52b   Estimate Printout (continued)

Estimate Detail - Fourplex Project

Detail - Without Taxes and Insurance

Group 1: Major ItemCode Groups

Estimator: DJP
Project Size: 4541 SF

ItemCode	Description	Quantity	UM	Lab.Unit	Mat.Unit	Eqp.Unit	Sub.Unit	Other.Unit	Tot.UnitCost	TotalCost
06430.140	1x8x12 stair framing	14.00	PCS	2.4565					2.457	34
06431.147	Wall rail	56.00	LNFT	6.4200	4.480				10.900	610
06450.150	Wood base board	3,458.00	LNFT	0.7959	2.230				3.026	10,464
	* Total 06400.000 Architectural woodwork									**19,048**
	07100.000 Waterproofing and dampproofing									
07110.012	Dampproofing	2,002.29	SQFT	0.4200	0.128				0.548	1,097
	* Total 07100.000 Waterproofing and dampproofing									**1,097**
	07200.000 Thermal protection									
07210.091	3½" batt insulation	4,168.00	SQFT				0.250		0.250	1,042
07210.092	6" batt insulation	1,747.12	SQFT				0.400		0.400	699
07210.095	12" batt insulation	2,240.00	SQFT				0.800		0.800	1,792
	* Total 07200.000 Thermal protection									**3,533**
	07300.000 Shingles, tiles, and roof coverings									
07310.100	Asphalt shingles	25.40	SQS	33.9048	30.710				64.615	1,641
07310.140	15 lb felt 432 sqft roll	5.88	ROLL	16.0615	12.481				28.543	168
07310.800	Roofing nails	25.40	LBS		1.267				1.267	32
	* Total 07300.000 Shingles, tiles, and roof coverings									**1,842**
	07400.000 Roof and siding panels									
07405.001	Aluminum Siding Subtrade	1.00	Lsum				7,450.000		7,450.000	7,450
07460.052	3/8" exterior grade plywood	181.00	SQFT	0.5090	0.360				0.869	157
07460.056	Aluminum vented soffit	423.00	SQFT	0.9532	3.260				4.213	1,782
07460.083	6" aluminum fascia	212.00	LNFT	0.7681	3.120				3.888	824
	* Total 07400.000 Roof and siding panels									**10,214**
	07500.000 Membrane roofing									
07560.027	Neoprene deck	181.00	Sqft	2.1576	1.330				3.488	631
	* Total 07500.000 Membrane roofing									**631**
	07600.000 Flashing and sheetmetal									
07620.020	.032 aluminum sheetmetal	328.00	LNFT	1.5250	1.344				2.869	941
	* Total 07600.000 Flashing and sheetmetal									**941**
	07700.000 Roof specialties and accessories									
07710.020	4' galvanized gutter	127.66	LNFT	1.8374	1.651				3.489	445
07710.120	4" round galvanized downspout	28.00	LNFT	1.3627	1.216				2.579	72
	* Total 07700.000 Roof specialties and accessories									**518**
	08100.000 Metal doors and frames									
08111.501	Insulated metal door		****							
08111.502	3680RH,6", standard dr	4.00	EACH		185.619				185.619	742
08111.508	Side lite	4.00	EACH		113.931				113.931	456
08160.014	7484 sliding glass door	4.00	EACH	75.0000	232.983				307.983	1,232
	* Total 08100.000 Metal doors and frames	8.00	EACH	37.5000	266.267				303.767	2,430
	08200.000 Wood and plastic doors									
08220.010	Flush masonite hollow core door		****							
08220.011	1880,4", door and frame	4.00	EACH	45.8571	79.368				125.225	501
08220.011	3080,4", door and frame	20.00	EACH	45.8571	79.368				125.225	2,505

Figure 14.52c   Estimate Printout (continued)

Estimate Detail - Fourplex Project

Detail - Without Taxes and Insurance

Estimator: DJP
Project Size: 4541 SF

Group 1: Major ItemCode Groups

ItemCode	Description	Quantity	UM	Lab.Unit	Mat.Unit	Eqp.Unit	Sub.Unit	Other Unit	Tot.UnitCost	TotalCost
	* Total 08200.000 Wood and plastic doors	24.00	EACH	45.8571	79.368				125.225	3,005
	08300.000 Specialty doors									
08310.010	Access panel	4.00	EACH	47.5556	57.606				105.162	421
08350.060	Raised panel masonite bi-fold	****	****							
08350.061	3672,4" door and frame	8.00	EACH	61.1429	131.853				192.996	1,544
08350.061	4272,4" door and frame	4.00	EACH	61.1429	131.853				192.996	772
	* Total 08300.000 Specialty doors	16.00	EACH	57.7461	113.291				171.037	2,737
	08500.000 Windows									
08520.017	4028 Sliding DG aluminum window	4.00	EACH		82.474				82.474	330
08520.017	4034 Sliding DG aluminum window	4.00	EACH		103.093				103.093	412
08520.017	6228 Sliding DG aluminum window	4.00	EACH		127.148				127.148	509
08580.020	4028 Screen	4.00	EACH		25.261				25.261	101
08580.020	4034 Screen	4.00	EACH		31.577				31.577	126
08580.020	6228 Screen	4.00	EACH		38.944				38.944	156
	* Total 08500.000 Windows									1,634
	08700.000 Hardware									
08710.010	Keyed entry lock	4.00	EACH		18.946				18.946	76
08710.030	Privacy lock	20.00	EACH		11.841				11.841	237
08710.040	Passage lock	4.00	EACH		8.321				8.321	33
08720.010	3680 aluminium threshold	4.00	EACH	15.1059	12.801				27.907	112
	* Total 08700.000 Hardware									458
	09200.000 Plaster and gypsum board									
09250.360	Drywall tape and joint cement	14,390.00	SQFT	0.3930	0.013				0.406	5,839
09250.401	½" 4x12 gyp bd @ ceiling	47.30	SHTS		9.217				9.217	436
09250.410	½ 4x12 gyp bd @ wall	267.63	SHTS		8.641				8.641	2,313
09250.411	5/8" 4x12 gyp bd @ wall	63.00	SHTS		10.753				10.753	677
09250.411	5/8" 4x12 gyp bd @ wall	32.17	SHTS		10.753				10.753	346
09250.500	Nails for gypsum board	166.60	LBS		0.768				0.768	128
09250.520	Hang gypsum board @ ceiling	2,270.45	SQFT	0.2453					0.245	557
09250.525	Hang gypsum board @ wall	14,390.00	SQFT	0.1897					0.190	2,730
09250.550	Finish gypsum board @ ceiling	2,270.45	SQFT	0.3505					0.351	796
09250.551	Finish gypsum board @ wall	14,390.00	SQFT	0.3067					0.307	4,413
	* Total 09200.000 Plaster and gypsum board									18,235
	09600.000 Floors									
09650.065	Sheet vinyl floor	142.00	SQYD				9.950		9.950	1,413
09680.015	Carpeting	362.00	SQYD				13.000		13.000	4,706
	* Total 09600.000 Floors									6,119
	09900.000 Paints and coatings									
09910.101	Paint gypsum board wall	13,832.00	SQFT				0.180		0.180	2,490
09910.110	Paint gypsum board ceiling	4,541.00	SQFT				0.190		0.190	863
09910.125	Paint posts and spindles	8.00	EACH				3.500		3.500	28
09910.130	Paint handrail	56.00	LNFT				0.500		0.500	28
	* Total 09900.000 Paints and coatings									3,409
	10670.000 Storage shelving									

C:\Program Files\MC² Software\estfiles\Fourplex.est   Page 4   7/12/2010 01:53 PM

Figure 14.52d Estimate Printout (continued)

Estimate Detail - Fourplex Project

Detail - Without Taxes and Insurance

Group 1: Major ItemCode Groups

Estimator: DJP
Project Size: 4541 SF

ItemCode Description	Quantity	UM	Lab.Unit	Mat.Unit	Eqp.Unit	Sub.Unit	Other Unit	Tot.UnitCost	TotalCost
10670.050 Closet shelf unit	8.00	EACH	16.0615	28.803				44.865	359
* Total 10670.000 Storage shelving									**359**
15400.000 Plumbing fixtures and equipment									
15400.001 Plumbing Subtrade	1.00	Lsum				24,975.000		24,975.000	24,975
* Total 15400.000 Plumbing fixtures and equipment									**24,975**
15700.000 HVAC equipment									
15700.001 HVAC Subtrade	1.00	Lsum				22,705.000		22,705.000	22,705
* Total 15700.000 HVAC equipment									**22,705**
16000.000 Power distribution equipment									
16000.001 Elecrical Subtrade	1.00	Lsum				20,500.000		20,500.000	20,500
* Total 16000.000 Power distribution equipment									**20,500**
Total Estimate									**257,256**

Figure 14.52e Estimate Printout (continued)

Estimate Summary

Fourplex Project

Management Computer Controls
5100 Poplar Avenue, Suite 3400
Memphis
TN

Bid date

4,541 SF
70.56 $/SF

Direct Costs	%	Labor	Material	Equipment	Subcontract	Temp Matl	Equip Rental	Other	Totals
Base labor		$77,931	$67,435	$0	$98,796	$0	$13,094	$0	$257,256
Labor burden	25.00%	$19,483							$19,483
Labor fringes		$0							$0
Labor manhours		1,307							
Material sales tax	12.00%		$8,092						$8,092
Equipment Surcharge	0.00%			$0					$0
Temporary material markup	0.00%					$0			$0
Equipment rental markup	0.00%						$0		$0
Other markup	0.00%							$0	$0
Gross cost		$97,414	$75,527	$0	$98,796	$0	$13,094	$0	$284,831
Gross receipts tax	0.00%								
Builder's risk insurance	0.60%								$1,922
	Overall	Small Tools							
Overhead	0.00%	3.00%	0.00%	0.00%	0.00%	0.00%	0.00%	0.00%	
	$0	$2,922	$0	$0	$0	$0	$0	$0	$2,922
Profit	10.00%	0.00%	0.00%	0.00%	0.00%	0.00%	0.00%	0.00%	
	$28,483	$0	$0	$0	$0	$0	$0	$0	$28,483
Subcontract markup	0.00%								$0
Performance bond	0.00%								$0
Misc 1 - Building Permit	0.70%								$2,243
Miscellaneous 2	0.00%								$0
Total									**$320,402**
Cut/Add									$0
Project total									**$320,402**

Figure 14.52f   Estimate Printout (continued)

6. Switch on the *Partial Rpt.* button on the left and click on *Specify*.
7. Double click in the *Begin Item* cell and enter 6100.
8. Double click in the *End Item* cell, enter 6200 and then press OK.
9. Now, when you press print, only the prices of rough carpentry materials will be listed. See Figure 14.53.

RECOMMENDED RESOURCES

Information	Web Page Address
■ Management Computer Controls, Inc.—The web page for MC² provides information on ICE software and also lots of general estimating data.	http://www.mc2-ice.com/
■ More information can be obtained on software for estimating by using the key works: **computer estimating** in **Web Search Engines**.	

REVIEW QUESTIONS

1. Identify the five steps in the ICE estimating process.
2. Where can you set up wage rates and crews in the estimate?
3. What do you have to set up in order to have the computer calculate the cost per square foot on the summary?
4. Explain the difference between a *Miscellaneous Item* takeoff and a *Construction System* takeoff.
5. What is contained in the *Company Standard*?
6. Where do you enter the name of the project and name of the estimator for an estimate?
7. What is the function of the *Mark Field* in the takeoff?
8. Explain the use of hand-written *Takeoff Notes* to accompany a computer estimate.
9. How can you change the price of concrete in one place so that it updates that price of that concrete throughout the estimate?
10. How can you set up the *Estimate Maintenance* window so that only a select number of columns are displayed?
11. What is a simple way to take off general expense items for an estimate?
12. Describe how markups are added to the estimate summary.

Material Detail - Fourplex Project

Estimator: DJP
Project Size: 4541 SF

Partial Report

Group 1: Major ItemCode Groups

Item Code	Description	Quantity	Cost Quantity	Unit Price	Total Material
Rough carpentry					
06110.103	2x4 bridging untreated, 24"	320 PCS			
	Lump Sum Price			0.501	160.32
06110.140	Nails and rough hardware	146 LBS			
	Lump Sum Price			1.088	158.86
06110.180	Ramset shot	362 EACH			
	Lump Sum Price			0.375	135.56
06110.185	Anchor bolt	18 EACH			
	Lump Sum Price			0.448	8.06
06110.257	2x8 sill plates, wolmanized	445 LNFT			
	Lump Sum Price			2.520	1,121.40
06110.308	2x4x24 rafter	4 PCS			
	Lump Sum Price			6.126	24.50
06110.410	Half truss 3 /12 pitch 20' "	48 EACH			
	Lump Sum Price			4.052	194.48
06110.412	Gablre End - 40'-8" Span	4 EACH			
	Lump Sum Price			190.000	760.00
06110.450	2x4x92-5/8" stud	1,170 PCS			
	M630 - 2x4x8' stud		6273.50 BDFT	0.375	2,352.56
06110.460	2x6x92-5/8" stud	175 PCS			
	M635 - 2x6x8' stud		1396.00 BDFT	0.380	530.48
06110.483	2x6x16 stud	9 PCS			
	M645 - 2x6 8'-16'		136.00 BDFT	0.460	62.56
06110.510	2x4x16 plate	200 PCS			
	M640 - 2x4 8'-16'		2140.44 BDFT	0.375	802.66
06110.511	2x6x16 plate	47 PCS			
	M645 - 2x6 8'-16'		757.64 BDFT	0.460	348.51
06110.530	2x6x16 header	16 PCS			
	M645 - 2x6 8'-16'		249.20 BDFT	0.460	114.63
06110.532	2x10x16 header	13 PCS			
	M655 - 2x10 8'-16'		338.28 BDFT	0.540	182.67
06110.566	1x4xRL blocking	112 BDFT			
	Lump Sum Price			0.375	42.00
06110.635	2x4xRL blocking	325 BDFT			
	Lump Sum Price			0.375	121.88
06111.033	2x10x14 beam	47 BDFT			
	Lump Sum Price			0.375	17.63
06111.099	4 x 4 x 8' Post	43 BDFT			
	Lump Sum Price			0.375	16.13
06111.124	2x8x16 joist	256 BDFT			
	Lump Sum Price			0.375	96.00
06111.132	2x10x12 joist	3,203 BDFT			
	Lump Sum Price			0.375	1,201.13
06160.100	½" 4x8 CD plywood	87 SHTS			
	Lump Sum Price			13.440	1,174.48
06160.101	5/8" 4x8 CD plywood	55 SHTS			
	M614 - 5/8" CD plywood		1747.12 SQFT	0.510	891.03
06160.127	5/8" T&G plywood @ subfloor	2,448 SQFT			
	Lump Sum Price			0.640	1,566.72
	* Total Rough carpentry				**$12,084**
	Total Estimate				**$12,084**

Figure 14.53 Rough Carpentry Material Printout

A

GLOSSARY OF TERMS

A

add-on—An estimate item (usually related to general expenses) calculated as a percentage of some part or the total of an estimate.

apartment—A building that contains a number of separate suites. (See also **low rise** and **high rise**)

architect—A professional person or company who, in residential construction, is responsible for preparing a design and constructing a building that meets the needs of the owner. (See also **prime consultant**)

assembly—A component of a project that consists of a collection of takeoff items.

B

BTUs—British thermal units. These are a measure of heat.

backflow preventor—A plumbing fitting installed on a sewer line to prevent material flowing back into a house from a sewer.

bank measure—A volume of excavation or backfill calculated using the actual dimensions of the hole to be excavated or filled with no allowance for swell or compaction of material.

bar-chart schedule—A graphic way of displaying a project schedule where tasks are represented by horizontal lines plotted against a time line.

bid bond—A form of guarantee issued by a surety company assuring the owner that the bidder who is awarded a contract will enter into the contract and provide the surety bonds required by the contract terms.

bid closing time—The time and date when all bids are required to be submitted on a project. Bids received after the specified time are usually rejected.

bid runner—The person, usually a junior estimator, who delivers a bid. This task often involves completing the bid forms at the place of the bid closing in accordance with last-minute information received by cell phone.

bill of materials—A list of the materials required for a project obtained from a takeoff.

board measure—A cubic measure of lumber equivalent to 144 cubic inches calculated using the nominal dimensions of the lumber.

bond beam—A reinforced concrete beam contained within "U"-shaped concrete blocks.

brick ties—Sometimes referred to as "strap anchors" or "wall ties"; these are wire or sheet metal devices inserted at regular intervals into horizontal masonry (brick or block) joints to attach one masonry wall to another or to attach a masonry wall to another part of the structure.

brick-on-edge course—A course of bricks laid down so that the brick ends are visible on the face of a wall and the long dimension of the end is vertical rather than horizontal, which is normally the case.

C

cash allowance—A sum of money included in a bid as an allowance for a certain project expenditure (sometimes called a PC Sum).

centerline—The length of a trench/footing/wall, etc., measured along its center rather than on the inside or on the outside.

compaction factor—An amount added to the volume of backfill material to account for the additional material required to increase the density of the backfill material when it is in place.

conceptual estimate—An estimate prepared when the project is no more than an idea under consideration; a time when very little specific detail is known about the design of the work involved.

condominium—An apartment building where residents own their apartments. (See also **apartment**)

consent of surety—A document issued by a bonding company asserting that the builder identified on the document is able to meet contract bonding requirements.

construction manager—The person or company charged with responsibility for the on-site activities of a construction project.

contingency allowance—A sum of money included in a bid as an allowance for non-specific project expenditures such as the cost of extra work in general.

cost-based pricing—Setting the price of a new home by first estimating the cost to construct the home.

cost plan—An estimate of project expenditures prepared in the early (feasibility) stage of the project and expressed as a collection of budgets, one for each element or major component of the project.

cost plus contract—A form of construction contract where the main terms allow for the builder to be paid all the costs incurred to complete the project plus an additional amount for profit.

CSI Master Format—A master list of numbers and descriptions for organizing information about construction requirements, products, and activities in specifications.

custom builders—Builders of custom homes. (Compare to **production builders**)

custom home—A house built to meet the owner's design.

D

depreciation—The allocation of the cost of an asset, such as an item of equipment, over the life of that asset.

designer—A person or company who is responsible for preparing the design of a building.

design-build—A project delivery method whereby an organization undertakes to both design and construct a project for an owner.

digitizer—Electronic device for measuring lengths, areas, and volumes directly from drawings.

duplex—A building comprising two residences, usually one located above the other.

E

engineered joists—Manufactured wood joists that meet a consistent structural standard.

extra over—The cost of providing an additional feature over and above the cost of the standard feature, for example, the extra cost of providing tinted glass rather than plain glass.

F

falsework—The temporary supports, scaffolding, and formwork required to carry concrete until it reaches sufficient strength to support itself.

fast tracking—Organizing a construction project so that the work begins before the design is complete in order to reduce the overall duration of the project.

feasibility estimate—A preliminary estimate prepared to assess the feasibility of a potential project.

firm price contract—A contract whereby a builder agrees to complete a project as defined in contract plans and specifications for a fixed sum of money.

form ties—Metal wires or rods used to produce the desired amount of separation between the sheets of form material on each side of a wall formwork system, and also to resist the pressures that result from filling the form system with liquid concrete.

G

general contractor—A builder hired, usually after a bid competition, to complete the work of a project. This builder assumes responsibility for construction of the entire project although subcontractors may perform some or all of the work involved.

gross floor area—The area of all floors in a building measured to the outside of exterior walls. No deductions are made for openings inside the building such as stairways and elevator shafts.

H

high rise—A building that has more than twelve floors above grade.

hoardings—A temporary fence or wall to a project under construction built to provide security or to enable the workspace to be heated.

I

instructions to bidders—Usually found at the beginning of the bid documents, they describe the project in general terms and provide information such as the required time and place of the bid closing. They will usually also include a clause that calls for the bidder to leave the bid open for acceptance for a specified period of time.

J

joist hanger—A metal device used to attach the end of a wood joist to a beam or a wall.

L

ladder reinforcing to masonry—This is a masonry wall reinforcing system that is laid in horizontal joints. It comprises two wires that run the length of the wall and short wires that pass between the two long wires at intervals so that the complete assembly resembles a ladder.

low rise—Building that has up to three floors above grade.

lump sum contract—A contract between an owner and a builder whereby the builder agrees to perform a defined scope of work for a stipulated sum. As a consequence, lump sum contracts are applicable only when the work can be well defined in the contract documents.

M

masonry ties—See brick ties.

medium rise—A building that is between four and twelve floors above grade.

mensuration—This is the science and principles of measurement. In this text, it applies to measuring lengths, areas, and volumes encountered in the estimating process.

method of measurement—A uniform basis for measuring construction work that provides rules relating to how work is described and measured in the estimating process.

N

net in place—Quantities of work measured using the dimensions obtained from drawings with no adjustment for factors such as waste.

O

order units and order quantities—The units of measurement used to order materials and the quantities of materials measured in these units. Order units often differ from the units of measurement used in the takeoff. (See **takeoff units**)

P

PC sum—Prime Cost Sum. (See **cash allowance**)

parallam beam—A manufactured wood component made by gluing together a number of thin boards to form a structural member bigger than the trees from which the boards were sawn.

pilaster—A column attached to a wall.

plot plan—The plan of the site showing the building location together with site topography, landscaping, and existing structures.

pre-hung—This term applies to doors that have been installed in their frames in a factory so that they arrive on site as a door and frame unit ready for installation.

preliminary estimate—An estimate prepared in the early stages of a project when little detail is known about the specific design of the project.

prime consultant—Usually an architect who takes on the responsibility of designing and supervising the construction of a building.

production builder—A builder who builds production homes. (Compare to **custom builder**)

production home—One of a series of houses built to the builder's standard design.

Q

qualified bids—A bid that has a *qualification* or some condition attached to it.

query list—A list of questions compiled by an estimator in the takeoff process. These are generally sent to the design consultants to obtain clarification in the form of *addenda* to bid documents.

R

R-value—This is a measurement of the resistance of a material to heat flow.

recap—A summary listing of takeoff items and quantities that gathers like items together in a prescribed order to facilitate ease of pricing.

row housing—Houses attached side by side.

royalty—A payment for the use of resources paid to the owner of those resources.

running bond—A method of laying bricks where the bricks are laid in courses with their long edge exposed and vertical joints staggered from one course to the next.

S

semi-custom home—Starting with a standard home plan, buyers make changes and choices so that the house that is built meets the homebuyer's needs.

Système International d'Unités—This is the official name adopted for the metric system of units of measurement (abbreviation SI).

slump—A measure of how liquid is a mix of concrete. This is determined in a standard test carried out with a conical mould 12 inches (300 mm) high. In the test, the mould is first filled with concrete. The mould is then lifted free, and the height of the cone of concrete formed is measured to determine the amount it is lower than 12 inches (the slump). Thus, a large slump signifies a more fluid mix of concrete.

soldier course—A course of bricks laid side by side on their ends so that the long dimension of the brick edge exposed is vertical rather than horizontal.

spec builder—(Speculative builder)—A builder who builds spec homes. (Compare to **custom builder**)

spec home—(Speculative home)—A house built before the builder has found a buyer for the house.

specifications—A written description of the standard of quality required for project workmanship and components.

standard bricks—The *Common Brick Manufacturers Association* adopted a standard brick size with the nominal dimensions of $2^1/4"$ by $3^3/4"$ by $8"$.

superplasticizer—A concrete additive that reduces the amount of water needed for a concrete mix while at the same time producing a more fluid concrete that flows more easily.

surety bond—A form of guarantee offered by a bonding company that states that the bonding company will accept liability in the event a contractor or subcontractor (the party bonded) defaults on their contractual obligations.

swell factor—The amount by which a volume of excavated material expands after it has been extracted from the ground in the excavation process.

T

taking off—The process of measuring construction work.

takeoff—A document that shows the measurements and quantities of construction work.

takeoff units and takeoff quantities—The units of measurement used in the takeoff process and the quantities obtained using these units.

telepost—Adjustable steel post used to support beams.

topsoil—The top layer of vegetable soil on a site.

truss reinforcing to masonry—This is a masonry wall reinforcing system that is laid in horizontal joints. It comprises two wires that run the length of the wall and a third wire that zigzags between the two straight wires so that the complete assembly resembles a truss.

U

unit price contract—A contract between an owner and a builder whereby the builder agrees to perform a defined scope of work on the basis of so-much-per-unit of work completed. For example, the builder may offer a price per cubic yard for trench excavation, a price per cubic yard for 3,000 psi concrete in footings, and so on. Subsequently, the builder is paid for the agreed quantity of each item completed multiplied by the unit price for that item.

V

value analysis—A systematic and creative method of obtaining the desired function from a component of a construction project for the least expenditure. In the process, the value of a component to the owner or the end user of a project is assessed, and an effort is then made to obtain the desired component at the best price.

value-based pricing—Setting the price of a new home by first determining the market price for similar homes.

W

wall ties—See **brick ties.**

waterstops—A feature introduced into a concrete joint to prevent the flow of water through that joint.

weep holes—Gaps left in a course of a masonry wall (usually by omitting mortar from some of the joints in that course) to allow water to drain from the inside of the wall.

B

THE METRIC SYSTEM AND CONVERSIONS

SI Units

These units belong to the International System of Units, which is abbreviated SI from the French *Le Système International d'Unités*. The SI is constructed from seven base units:

Base Units

Quantity	Unit Name	Symbol
length	meter	m
mass	kilogram	kg
time	second	s
electric current	ampere	A
amount of substance	mole	mol
luminous intensity	candela	cd

Units outside the SI that are acceptable for use with the SI

Unit Name	Symbol	Value in SI Units
minute (time)	min	1 min = 60 s
hour	h	1 h = 60 min
day	d	1 d = 24 h
liter	L	1 L = 10^{-3} m^3 (or 1 m^3 = 10^3 L)
metric ton	t	1 t = 10^3 kg
hectare	ha	1 ha = 10^4 m^2

Prefixes

Multiplication Factor	Prefix Name	Prefix Symbol
$1\ 000\ 000\ 000 = 10^9$	giga	G
$1\ 000\ 000 = 10^6$	mega	M
$1\ 000 = 10^3$	kilo	k
$100 = 10^2$	hecto	h
$10 = 10^1$	deka	da
$0.1 = 10^{-1}$	deci	d
$0.01 = 10^{-2}$	centi	c
$0.001 = 10^{-3}$	milli	m
$0.000\ 001 = 10^{-6}$	micro	μ
$0.000\ 000\ 001 = 10^{-9}$	nano	n

Conversion Factors

Length

English Units	→	Metric Units	Metric Units	→	English Units
1 inch	=	25.4 mm	1 mm	=	0.039 inches
1 inch	=	0.0254 m	1 m	=	39.37 inches
1 foot	=	0.305 m	1 m	=	3.28 feet
1 yard	=	0.914 m	1 m	=	1.094 yards
1 mile	=	1.609 km	1 km	=	0.621 miles

Area

English Units	→	Metric Units	Metric Units	→	English Units
1 square foot	=	$0.093\ m^2$	$1\ m^2$	=	10.764 square feet
1 square yard	=	$0.836\ m^2$	$1\ m^2$	=	1.196 square yards
1 acre	=	0.405 hectares	1 ha	=	2.471 acres
1 square mile	=	$2.590\ km^2$	$1\ km^2$	=	0.386 square miles

Volume

English Units	→	Metric Units	Metric Units	→	English Units
1 cubic foot	=	$0.028\ m^3$	$1\ m^3$	=	35.315 cubic feet
1 cubic yard	=	$0.765\ m^3$	$1\ m^3$	=	1.308 cubic yards
1 quart (liquid)	=	0.946 Liters	1 Liter	=	1.057 quarts
1 gallon	=	3.785 Liters	1 Liter	=	0.264 gallons

Mass

English Units	→	Metric Units	Metric Units	→	English Units
1 ounce	=	28.350 grams	1 gram	=	0.035 ounces
1 pound	=	0.454 kg	1 kg	=	2.205 pounds
1 ton (2000 lb)	=	907.185 kg	1 kg	=	0.001 tons
1 ton (2000 lb)	=	0.907 metric tons	1 t	=	1.102 tons

C

FORMULAE USED IN ESTIMATING

	Shape	Formulae
1. Rectangle	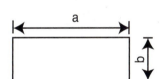	Perimeter = 2a + 2b Area = a x b
2. Parallelogram	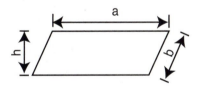	Perimeter = 2a + 2b Area = a x h
3. Trapezoid	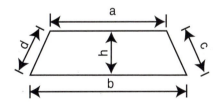	Perimeter = a + b + c + d Area = $\frac{(a+b) \times h}{2}$
4. Triangle	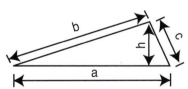	Perimeter = a + b + c Area = $\frac{a \times h}{2}$

5. Quadrilateral

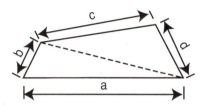

Perimeter = a + b + c + d
Area = (divide into triangles)

6. Circle

Circumference = $2\pi r$
Area = πr^2

7. Pythagoras

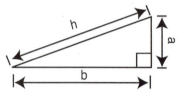

$h^2 = a^2 + b^2$
Therefore, length h = $\sqrt{a^2 + b^2}$

8. Pad Footing   Volume = length × width × depth

9. Column   Volume = length × width × height

10. Perimeter Footing Volume = centerline length × width × depth

11. Trenches   Volume = centerline length × average width × average depth

12. Volume of Basement Volume = average length × average width × average depth

13. Number of Trucks   Number of Trucks Required = $\left(\dfrac{\text{Unloading Time}}{\text{Loading Time}}\right) + 1$

14. Unloading Time   Unloading Time = Round-trip Travel Time + Time to Off-load the Truck

15. Loading Time   Loading Time = $\dfrac{\text{Truck Capacity}}{\text{Loader Output}}$

16. Average Annual
    Investment   Average Annual Investment = $\dfrac{\text{Total Initial Cost} + \text{Salvage Value}}{2}$

D

SAMPLE HOUSE DRAWINGS

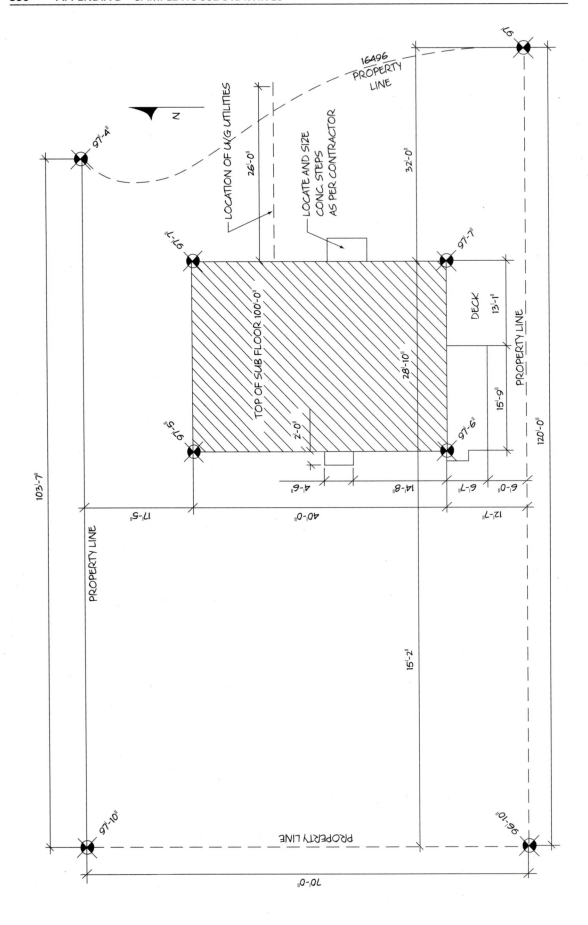

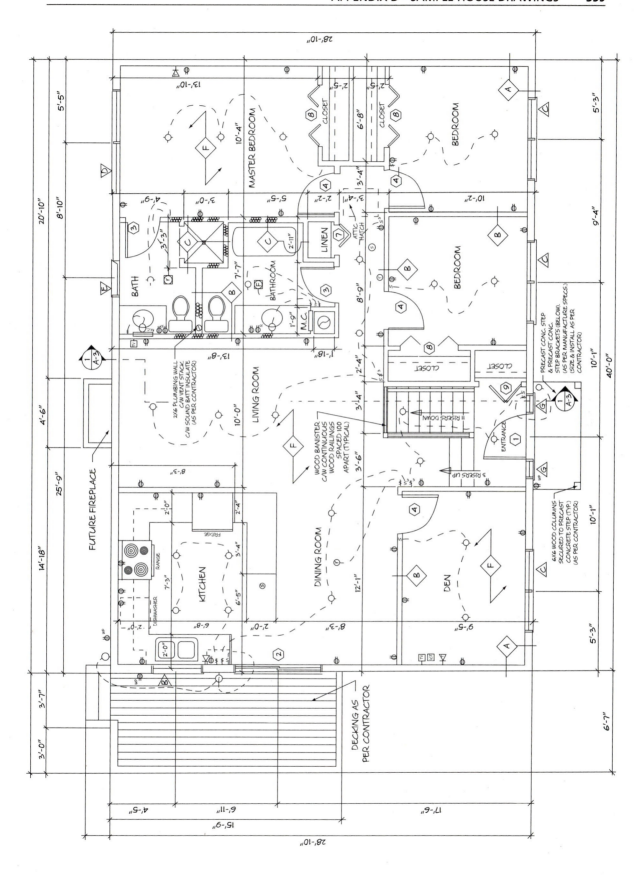

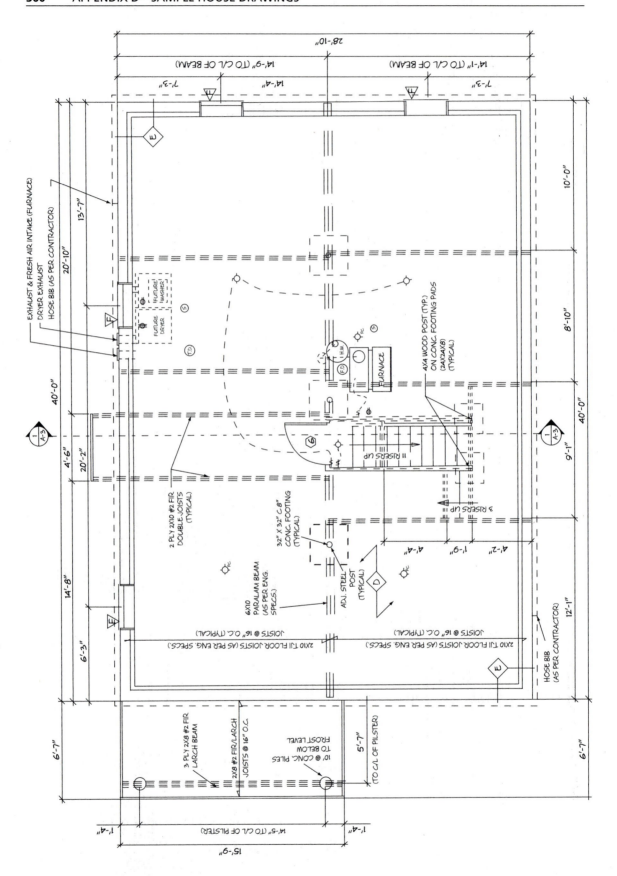

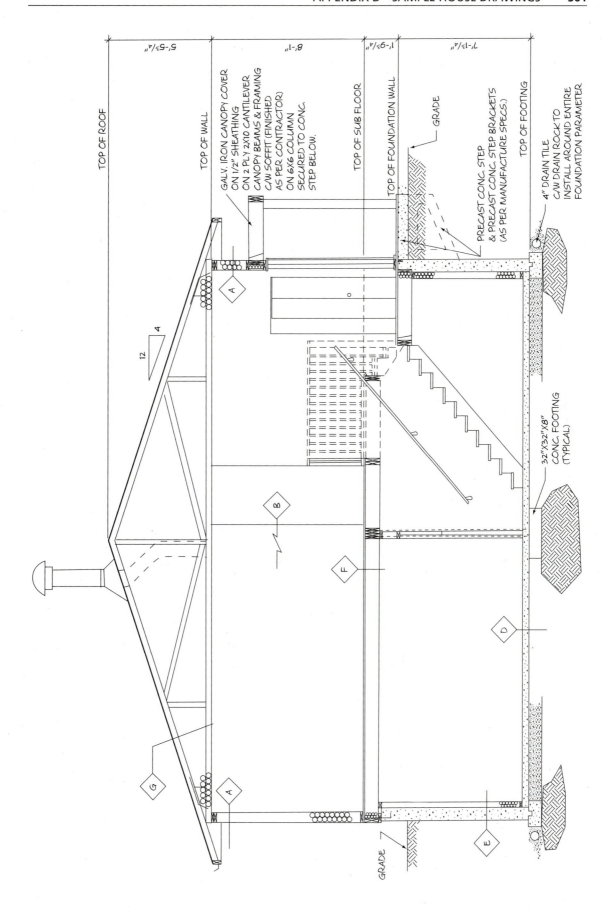

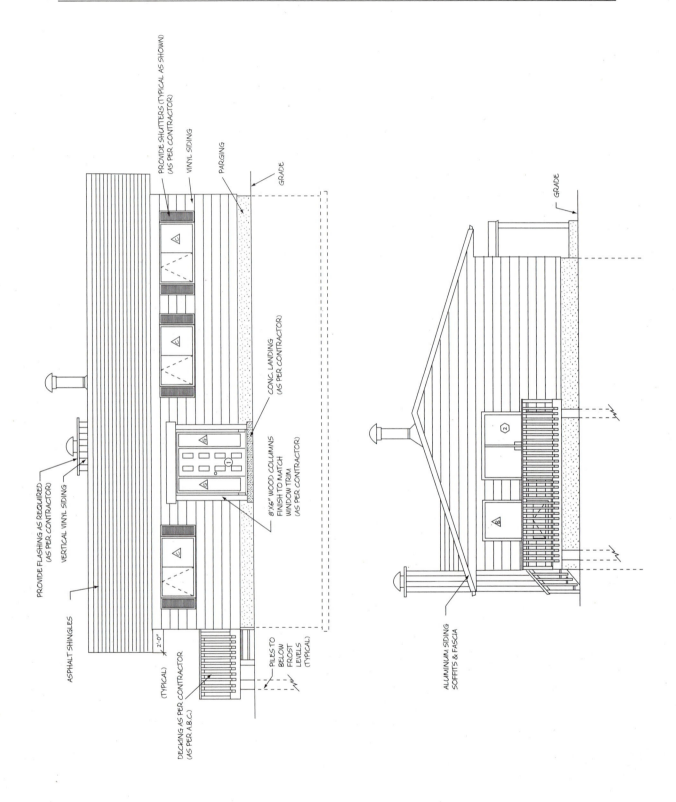

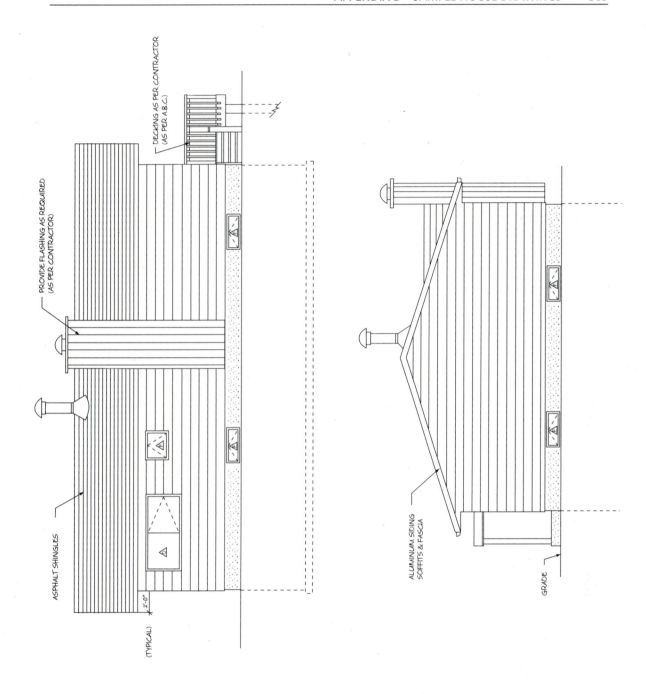

WINDOW SCHEDULE

UNIT:	UNIT SIZING:	R.O. (LXH):
A	NOT IN CONTRACT	
B	36"X50" (F)	42"X72"
	36"X15" (A)	
C X3	27"X30" (A-F)	64"X37"
D	36"X30" (A-F)	83"X37"
E	27"X22" (A)	33"X29"
F X4	36"X24" (A)	37"X13"
G X2	12"X6'4" (SEALED) LIGHTS	67"X84" WHOLE DOOR FRAME

UNIT:	DOOR SCHEDULE	LOCATION:
1	3'X6'8"X1" (C/W SIDE LIGHTS)	F-ENTRANCE
2	5'X6'8" PATIO DOOR	NOOK
3	2'4"X6'8"X1"	ENSUITE & BATH
4	2'6"X6'8"X1"	BEDROOMS
5	NOT IN CONTRACT	
6	2'10"X6'8"X1"	BASEMENT
7	2'0"X6'8"	BI-FOLD
8	4'0"X6'8"	BI-FOLD
9	3'0"X6'8"	BI-FOLD

ELEC. LEGEND

S	SINGLE POLE SWITCH
S^3	3-WAY POLE SWITCH
S_F	FURNACE SWITCH
$\ominus$	DUPLEX CONVENIENCE OUTLET
$\ominus_E$	ELECTRIC RANGE
$\ominus_D$	ELECTRIC CLOTHES DRYER
$\ominus_R$	RAZOR OUTLET
$\ominus_{WP}$	WEATHERPROOF OUTLET
$\Diamond$	CEILING LIGHT
$\Diamond_{PC}$	PULL CHAIN LIGHT
$\ominus$	WALL LIGHT
F	CEILING FAN
F	BATHROOM EXHAUST FAN
T	THERMOSTAT
S	SMOKE ALARM
ICT	INTERNET CABLE
TV	TELEVISION OUTLET
$\triangleright$	TELEPHONE JACK

G ROOF CONSTRUCTION
ASPHALT SHINGLES ON
1/2" O.S.B. SHEATHING c/w "H" CLIPS
ENG. APPROVED FINK TRUSSES @ 16" O.C.
R12 BATT INSUL. OR BETTER
6MIL. POLY. VAP. BARR.
1/2" DRYWALL CEILING
PROVIDE INSULATION STOPS AT WALL (TYP.)

F FLOOR CONSTRUCTION
FLOOR FINISH (AS PER CONTRACTOR)
UNDERLAY (WHERE REQUIRED AS PER CONTRACTOR)
3/4" T&G SUB FLOOR (GLUED & SECURED TO JOISTS)
2X10 TJI FLOOR JOISTS @ 16" O.C.
(OR AS PER ENG. SPECS.)
NOTE: INSTALL R12 INSULATION BATT.
     @ TOP OF FOUNDATION WALLS
        ALONG INSIDE FACE OF RIM JOISTS
        C/W 6MIL. VAPOUR BARRIER
ALL STRAPPING & CROSS BRIDGING
(AS PER ENG. SPECS.)
W/ 1/2" GYPSUM BOARD (BASEMENT CEILING)

A EXTERIOR WALL CONST. (NEW & EXISTING)
(SIDING BY CONTRACTOR)
(BUILDING PAPER/SHEATHING MEMBRANE)
1/2" O.S.B. SHEATHING
2X6 WALL STUDS @ 16" O.C.
R20 BATT. INSUL. OR BETTER
6 MIL. POLY. VAP. BARR.
1/2" DRYWALL

B INTERIOR WALL CONST.
2X4 STUDS @ 16" O.C.
1/2" DRYWALL BOTH SIDES

C INTERIOR BATHROOM WALL CONST.
2X4 STUDS @ 16" O.C.
1/2" WATERPROOF DRYWALL
(ALL AROUND TUB AREA
INCLUDING ANY MARKED
EXTERIOR WALLS)

D BASEMENT FLOOR SLAB
(AS PER CONTRACTOR, APPROVED BY ENG.)
4" MIN. CONC. SLAB
C/W 6MIL. POLY VAP. BAR.
ON 8" MIN. COMPACTED GRAVEL FILL
(REBAR OR WIRE MESH AS REQUIRED)

E FOUNDATION WALL CONST.
2X6 SILL PLATE (SECURED BY
LAG BOLTS INTO CONC. WALL,
C/W GASKETS TO MATCH) ON
8" WIDE CONCRETE WALL STRUCTURE:
C/W DAMPROOFING FROM ABOVE GRADE
ON OUTSIDE OF WALL
& PARGING ABOVE GRADE TO UNDER WALL SIDING
2-10MM REBAR TOP & BOTTOM
ON 16"X8" FOOTING (C/W KEYWAY)
W/ #4 REBAR TOP & BOTTOM
W/ WEEPING (DRAIN) TILE & ROCK
TO SURROUND EXTERIOR FACE OF FOOTINGS
(ENTIRE PERIMETER OF STRUCTURE)
.10MIL POLY MOISTURE BARRIER
     (OUTSIDE FACE) TO ABOVE GRADE
C/W INTERIOR FURRING WALL CONSTRUCTION:
     2X4 WALL STUDS @ 16" O.C.
     (INCLUDING TOP & BOTTOM PLATES)
     W/ R12 INSULATION

APPENDIX

E

SAMPLE DRAWINGS FOR A 4-UNIT APARTMENT BUILDING

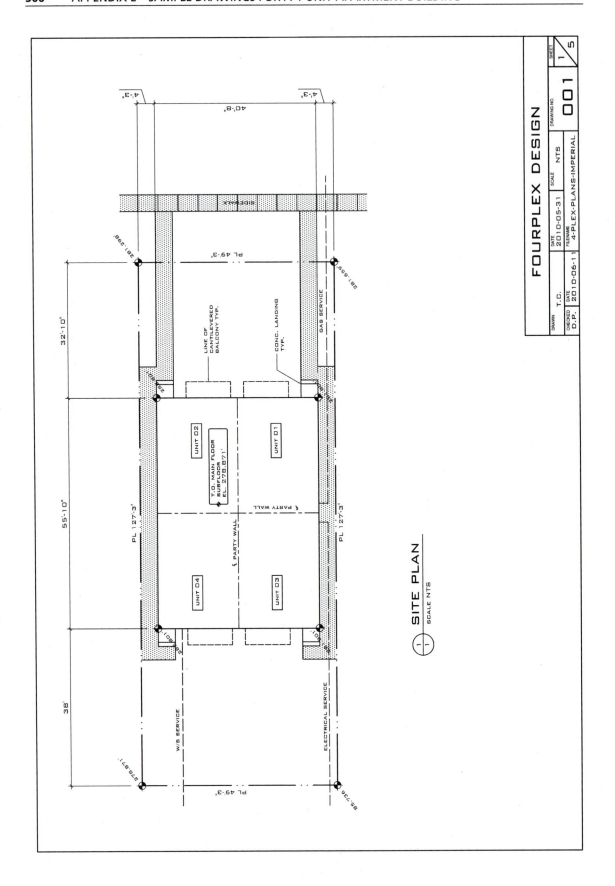

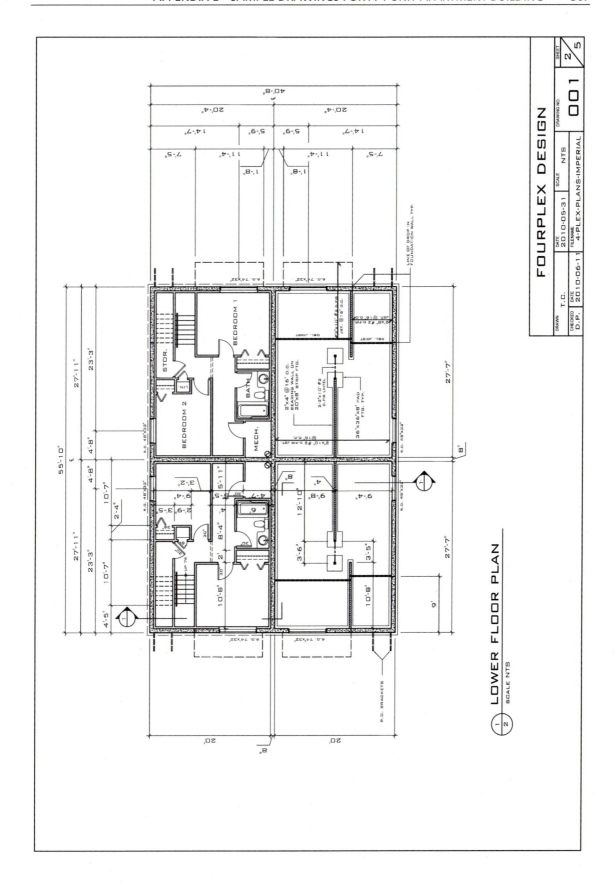

LOWER FLOOR PLAN
SCALE NTS

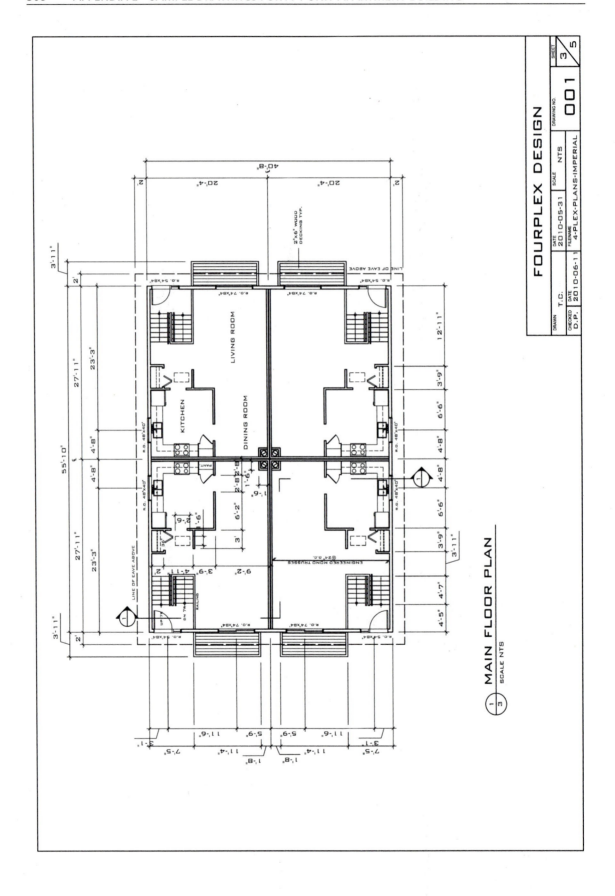

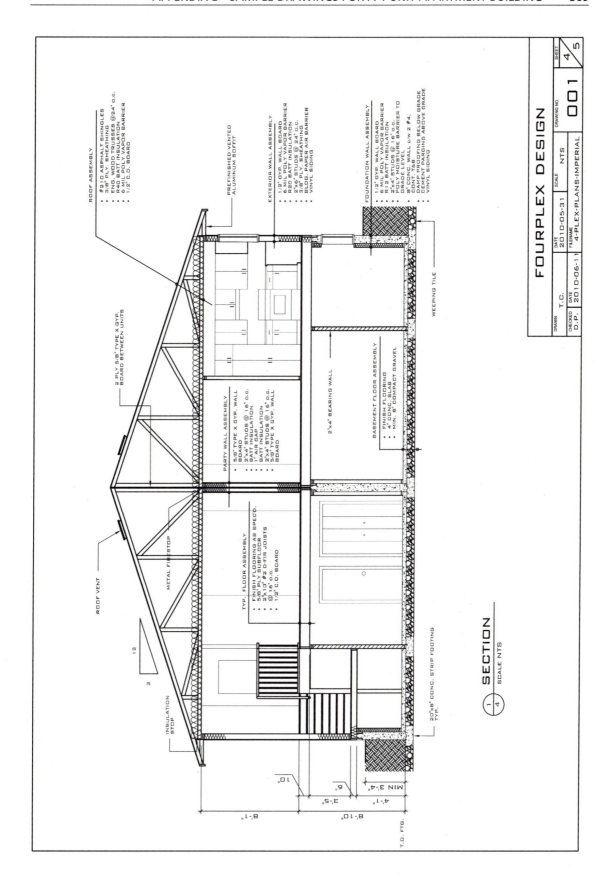

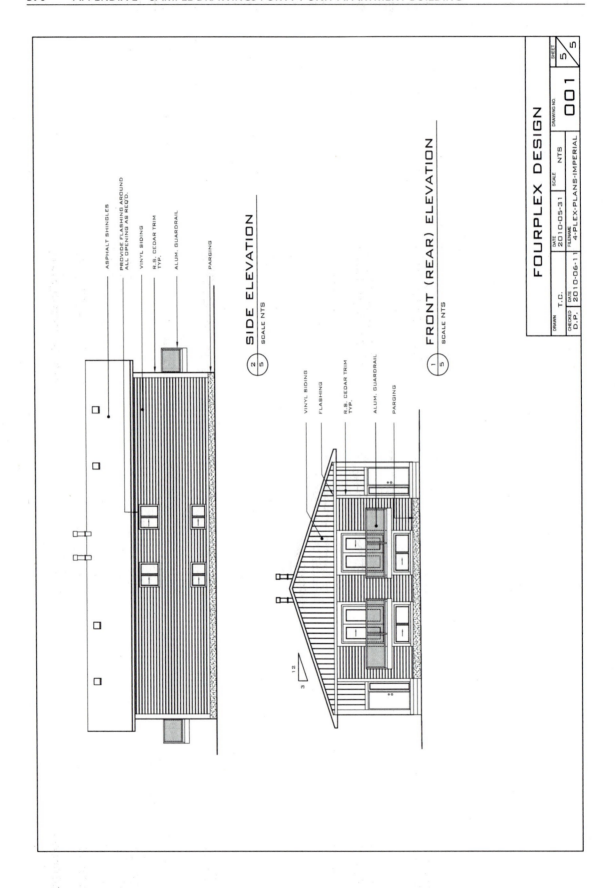

INDEX